Underwater Archaeology

The Nautical Archaeology Society also publishes *The International Journal of Nautical A* Publishing, a new series of monograph reports and a members' newsletter, *Nautical Ar* public participation projects.

Second Edition

Underwater Archaeology

The NAS Guide to Principles and Practice

Editor: Amanda Bowens, BA MA

THE NAUTICAL ARCHAEOLOGY SOCIETY
FORT CUMBERLAND
FORT CUMBERLAND ROAD
PORTSMOUTH
PO4 9LD, UK

This edition first published 2009
© 2009 by Nautical Archaeological Society

Blackwell Publishing was acquired by John Wiley & Sons in February 2007. Blackwell's publishing program has been merged with Wiley's global Scientific, Technical, and Medical business to form Wiley-Blackwell.

Registered Office
John Wiley & Sons Ltd, The Atrium, Southern Gate, Chichester, West Sussex, PO19 8SQ, United Kingdom

Editorial Offices
350 Main Street, Malden, MA 02148-5020, USA
9600 Garsington Road, Oxford, OX4 2DQ, UK
The Atrium, Southern Gate, Chichester, West Sussex, PO19 8SQ, UK

For details of our global editorial offices, for customer services, and for information about how to apply for permission to reuse the copyright material in this book please see our website at www.wiley.com/wiley-blackwell.

The right of the Nautical Archaeological Society to be identified as the author of the editorial material in this work has been asserted in accordance with the Copyright, Designs and Patents Act 1988.

Wiley also publishes its books in a variety of electronic formats. Some content that appears in print may not be available in electronic books.

Designations used by companies to distinguish their products are often claimed as trademarks. All brand names and product names used in this book are trade names, service marks, trademarks or registered trademarks of their respective owners. The publisher is not associated with any product or vendor mentioned in this book. This publication is designed to provide accurate and authoritative information in regard to the subject matter covered. It is sold on the understanding that the publisher is not engaged in rendering professional services. If professional advice or other expert assistance is required, the services of a competent professional should be sought.

Library of Congress Cataloging-in-Publication Data

Nautical Archaeology Society.
   Underwater archaeology : the NAS guide to principles and practice / author, the Nautical Archaeology Society; editor, Amanda Bowens. – 2nd ed.
      p. cm.
   Includes bibliographical references and index.
   ISBN 978-1-4051-7592-0 (hardcover : alk. paper)—ISBN 978-1-4051-7591-3 (pbk. : alk. paper)   1. Underwater archaeology.
2. Shipwrecks.   3. Underwater archaeology—Handbooks, manuals, etc.   I. Bowens, Amanda.   II. Title.

CC77.U5N39 2009
930.1′028′04—dc22

                                                                                               2007048319

A catalogue record for this book is available from the British Library.

Set in 10.5/12.5 Minion
by Graphicraft Limited, Hong Kong
Printed and bound in Singapore by C.O.S. Printers Pte Ltd

1   2009

Contents

List of Figures

List of Plates

Foreword

The Nautical Archaeological Society has for many years been a champion of best practice in archaeology under water through its internationally acclaimed training scheme, the *International Journal of Nautical Archaeology*, and a wide range of practical outreach projects. The original edition of the NAS Handbook was another invaluable and much-appreciated part of how the Society has supported practical involvement in nautical archaeology. Ever since it went out of print some years ago, there have been continual enquiries about when a reprint or new edition would be available.

It is therefore with great pleasure that I introduce this new edition of an invaluable publication. The whole text and most of the illustrations have been comprehensively revised and updated, and, as explained in the introductory first chapter, there are several entirely new sections on vitally important aspects of the subject. In many respects, as the new title indicates, this is as much a new publication as a revision, and it will be all the more valuable as a result.

To users of the first NAS Handbook, this second edition may have seemed a long time coming, but it has involved a huge amount of work by a large number of contributors who have very generously provided their expertise on a voluntary basis. I am sure that every reader will wish to thank them all for sharing their wisdom and experience. Our Vice-President, Colin Martin, has been very generous in providing a very large number of the images and we are also very grateful to Graham Scott for the majority of the line illustrations, which have intentionally copied or been inspired by Ben Ferrari's drawings in the original book. We are also grateful to all the other photographers and illustrators whose work is acknowledged.

This publication would not have come to fruition without the very patient and dedicated work of our editor, Amanda Bowens, who has shown great tact, persistence and skill in marshalling all this material. We are also grateful to Paula Martin for editorial assistance and compiling the index. Finally I would like to pay tribute to Lucy Blue and the rest of the NAS Publications Sub-committee for pushing the project ahead, and to Wiley-Blackwell, our very supportive publisher, who have been responsible for the design and production.

I am sure everyone who uses this book, whether in an academic, professional or a vocational context, will benefit from developing their interest and skills in nautical archaeology, thereby enhancing the quality of the work they do. If it is as successful as the original NAS Handbook, it may not be so many years before a further reprint or new edition is needed, so any comments and suggestions will always be gratefully received. In the meantime, may your explorations in nautical archaeology be a constructive and rewarding experience.

George Lambrick
Chair NAS

Acknowledgements

Except where otherwise stated, the drawings in this second edition are by Graham Scott, copied or inspired by Ben Ferrari's drawings in the first edition.

Contributors to the second edition

Jonathan Adams, BA DPhil FSA MIFA
Marc-André Bernier, BA MA
Lucy Blue, BA DPhil
Amanda Bowens, BA MA
Martin Dean, BSc MIFA (Hon) FSA
Justin Dix, BSc PhD
Joe Flatman, MA PhD PIFA
Craig Forrest, B.Comm LLB LLM PGCE PhD
Damian Goodburn, BA PhD AIFA
Alexzandra Hildred, BA MIFA FSA
Peter Holt
Kester Keighley, MSc
Stuart Leather, MSc
Gordon Le Pard, BSc
Colin Martin, PhD FRHistS FSA Scot MIFA MAAIS
Paula Martin, BA Dip Class Arch PhD FSA Scot MIFA
Douglas McElvogue, PhD MIFA FSA Scot
Seán McGrail, FSA MA PhD DSc MIFA Master Mariner
Gustav Milne, MPhil FSA
David Parham, BA MSc MIFA RPA
Leslie Perkins McKewan, MSc PIFA
Rory Quinn, PhD
Julie Satchell, BA MA MIFA
Graham Scott, AIFA
Amanda Sutherland, BSc ACR
Christopher Underwood, BEd MA
Steve Waring, BSc
Michael Williams, LLB

The editor is extremely grateful to the following for their help and support:

Mark Beattie-Edwards
Lucy Blue
Alan Bowens
Alistair Carty
Rebecca Causer
Alison Hamer
Mary Harvey
Stuart Heath
Jill Hooper
Kester Keighley
Colin Martin
Edward Martin
Paula Martin
Nick Rule
Julie Satchell

Contributors to the first edition

Jonathan Adams, Barrie Andrian, Adrian Barak, Martin Dean, Robin Denson, Chris Dobbs, Sarah Draper, Ben Ferrari, Robert Finegold, Antony Firth, Ian Friel, Debby Fox, Alison Gale, Cathy Giangrande, David Gibbins, Damian Goodburn, Alexzandra Hildred, Richard Larn, Thijs Maarleveld, Ian Oxley, Mark Redknap, Nick Rule, Paul Simpson, Kit Watson.

Note

Every effort has been made to trace copyright holders and to obtain their permission for the use of copyright material. The publisher apologises for any errors or omissions in the above list and would be grateful if notified of any corrections that should be incorporated in future reprints or editions of this book.

The NAS Handbook – Why It Was Written

The original *Archaeology Underwater: The NAS Guide to Principles and Practice* was first published in 1992. It was commissioned to help address a scarcity of information about 'how to undertake archaeological work under water while maintaining acceptable standards' (Dean et al., 1992:2).

As well as explaining fundamental archaeological principles, this book provides a general introduction to archaeology under water, detailing techniques and practices as they are applied in an underwater context. It provides the tools appropriate to tackle a variety of sites in different environments and emphasizes that archaeology is not just a set of techniques – it is shaped by fundamental principles and theoretical parameters. While this book is a comprehensive source of practical information, it is not a complete reference book and will not transform the reader into an underwater archaeologist. Its aim is to provide an awareness of the responsibilities that go with any form of fieldwork while outlining what is involved in achieving an acceptable standard of archaeological work in what can often be a challenging physical environment.

In the intervening years since the publication of the first edition, while the basic principles have remained the same, technological developments have resulted in new and improved archaeological techniques. Meanwhile, the World Wide Web and satellite television have helped take underwater archaeology into peoples' homes, feeding what appears to be an insatiable public appetite for all things associated with the past. In addition, links between countries with different approaches to archaeological investigation have strengthened and, as a result, the toolbox of techniques for archaeological work under water has grown.

In the light of such developments, an update to the text and graphics of the original book seems timely. The result is this long-awaited second edition of what is popularly referred to as 'The NAS Handbook'.

The body that eventually became the Nautical Archaeology Society (NAS) was originally incorporated and registered as a charity in 1972 under the name (The) Nautical Archaeology Trust Limited. The Trust was reconstituted in 1986 as the Nautical Archaeology Society, mainly to oversee the production of the *International Journal of Nautical Archaeology* (*IJNA*), the first volume of which had been published in 1971, and more generally to further research. The Society is committed to the research, conservation and preservation of maritime cultural heritage. The NAS is based in the United Kingdom but has a significant international profile. Membership is made up of a wide range of people who wish to promote and be involved in the preservation of their coastal and underwater heritage, in its broadest sense.

This second edition of *Underwater Archaeology* includes several new chapters covering such topics as photography, legislation and conservation. Additional chapters reflect significant developments or new approaches, particularly with respect to project planning, safety on archaeological sites, historical research, monitoring and maintenance and geophysics.

Each individual component of this book was written by someone who is an expert in his/her field. The production of this second edition has been a long iterative process involving many people, most of them members of the Nautical Archaeology Society. Text from the original book has been modified, supplemented and, where appropriate, replaced. This book therefore owes its existence to everybody involved in the production of this and all previous versions (please see the list of contributors in the acknowledgements). The Nautical Archaeology Society would like to acknowledge all contributors with grateful thanks.

The Nautical Archaeology Society would also like to introduce the reader to the real underwater treasure – a rich cultural heritage that has helped shape the world in which we live today. By outlining the principles and practices of maritime archaeology, this book will enable people to make informed and responsible decisions about how to get the most from their involvement with maritime archaeology above or under water.

2 Underwater Archaeology

Contents
- What is archaeology?
- What is archaeology under water?
- What is *not* archaeology under water?
- Closely related and complementary approaches (ethnography and experimental archaeology)

This chapter provides a short definition of what constitutes archaeology and an archaeological approach. It will briefly summarize the development of underwater archaeology as a distinct sub-discipline and consider some significant relationships between archaeology and other approaches and activities.

WHAT IS ARCHAEOLOGY?

Archaeology is concerned with the identification and interpretation of physical traces left by past ways of life. Archaeology is not just description, however; its primary aim is explanation. The process of archaeological investigation is similar to the detective work of police and forensic scientists. All traces, however unexciting or irrelevant they may at first appear, have the potential for providing a vital clue to understanding what happened before the detective or archaeologist arrived.

Evidence for the past survives both on land and under water, but the demarcation of 'wet' and 'dry' sites is complicated by the fact that boundaries change. Some areas that used to be sea-bed are now land while some areas that were once land are now under water. Maritime finds can therefore be discovered in quite unexpected places (see figures 2.1 and 2.2). As the title suggests, this book is concerned with the study of archaeological evidence that is under water although, apart from the use of specialized equipment to deal with the environment, the archaeological techniques are essentially the same under water as on land.

Anyone can call him/herself an archaeologist. What is of concern is whether that person does archaeology well or badly. Even the best-trained and most-experienced archaeologists will have limits to their knowledge and range of skills. Good archaeologists will be aware of their own limitations. This applies equally to professionals and unpaid members of the community doing archaeology for fun. Trained professionals have a better chance of achieving acceptable standards in their work because of the education and experience they have accumulated. Hobby archaeologists, however, can achieve equally high archaeological standards if they accrue the appropriate skills and experience. Apart from archaeological skills, one of the attributes of a good archaeologist is the ability to recognize the limitations of available resources. An archaeologist may turn down a project that involves the destruction of evidence (e.g. excavation) if adequate resources and support are not available. As will become clear throughout this book, resources are necessary to recover, record, interpret and look after finds and other evidence. There is also an obligation to arrange for the long-term care of recovered material and records in a museum or other suitable repository. In addition, resources will be required for publication and dissemination so that evidence from the investigation is available to others (see chapter 20).

Archaeology, as it exists today, has its roots in a curiosity about old things – the stories and legends about past events passed down over the generations, whether fact or fiction, and surviving objects which were associated with past events. This curiosity is common to many cultures

Figure 2.1 The bronze-age boat discovered 6 m (20 ft) below ground in Dover, UK during the building of a major new road. (Photo courtesy of the Dover Museum and The Bronze Age Boat Gallery)

and such interest is not a recent phenomenon. Medieval peasants are known to have collected stone hand-axes thinking they were of supernatural origin. Gradually, some of those interested in 'relics' attempted to explain what they were collecting and began to see that some of the material might have relevance to wider issues. For example, some tried to prove that early man was barbaric, whilst others tried to bend the evidence in an attempt to prove that some races were innately superior to others.

Fortunately, others were more enlightened and attempted to be objective about what the material might suggest. This really marks the beginning of archaeology as a discipline, separate from the 'gentlemanly pursuit' of curio collection (antiquarianism) or the study of individual objects against a historical background (art history). Workers began to borrow techniques from other, longer established disciplines, such as geology, and to look beyond the objects to their surroundings for more evidence.

This was the beginning of the realization that archaeological contexts are important in interpreting the past. Indeed, beginning with analytical techniques borrowed directly from geology, a great deal of attention was focused on the study of contexts and archaeological sequences. This led to an awareness of the factors that differentiate archaeological from geological deposits and has thus allowed more refined study of the subject (Harris, 1989).

Initially, the focus of attention was on individual sites but, as the discipline developed, archaeological research began to address questions such as the migration of populations, the development of agriculture and the structure of past societies. Over the past 200 years the discipline has accumulated increasingly sophisticated methods and a more refined theoretical base; each generation improving on the amount of evidence that could be collected from the physical remains of societies and cultures no longer in existence. Following an initial concern with the classification and description of objects, archaeology developed into a discipline concerned with using material evidence to make inferences about people and behaviour.

The past 30 or so years have seen a great deal of attention focused on the theoretical side of the subject. This has meant that as the body of scientifically collected evidence grows, fundamental questions about the past can now be addressed more effectively, and conclusions tested more rigorously.

Figure 2.2 On the banks of the River Usk in central Newport, Wales, the well-preserved remains of a Tudor ship were discovered. (Photo: Hampshire and Wight Trust for Maritime Archaeology)

Work conducted in the early years of the discipline recovered far less evidence about the past than can be recovered today. This is because early archaeologists unwittingly destroyed information that could have been retrieved with modern techniques. While it is too late to do very much about that loss of evidence, it serves as an important reminder that archaeologists of the future may look back on the work of today's archaeologists in the same way. Both professional and amateur archaeologists should feel a responsibility to hand on as much of the evidence as possible, so that future generations can make sense of the clues that cannot be understood today (plate 2.1).

Understanding the complexity and potential of archaeological sites (rather than just the objects) is a process that has taken a long time to develop, and it is not yet complete. A great deal of experience has been painfully accumulated over the centuries, and there is no excuse for someone curious about the past starting out today to make the same mistakes as those made 100 or 200 years

ago. Sadly, this does still happen. Some practitioners of underwater and foreshore archaeology become involved through the accidental discovery of archaeological remains, and may begin with little or no archaeological experience. Underwater archaeology is a comparatively new area of study and still has to prove its value to some traditional archaeologists. However, as it matures and learns from the experience of archaeology in general, priorities and principles can be developed and the overall quality of archaeological work under water will improve.

Archaeologists treat a site like the scene of a crime and carefully collect all the available evidence. The murder weapon, evidence of the break-in, the position of the body, traces of poison, the ballistics report, the systematic search, fingerprints and the fibres matched to the criminal's clothes, all have their parallels in archaeology. Indeed the methods and aims are so similar that the two disciplines borrow techniques from each other and sometimes work together.

If archaeology is the collection of evidence at the scene of a crime, its sister discipline, history (the study of documents), is the reviewing of witness statements. The two disciplines use different sources of information and different techniques but together they make up the evidence for the case. It is important to be aware of the potential of historical research and to use it where appropriate (see chapter 9). It is equally important not to be confused when the physical evidence appears to contradict the recorded views of witnesses. Each type of evidence has its own problems and limitations and the good detective will understand this and reach conclusions based on the merits of all the evidence.

An examination of our surroundings will soon reveal how little physical evidence of the past has survived. Activities such as building development, road construction and mineral extraction continue to eat away at the store of evidence that is left. In order to drive cars, have warm homes and new buildings, this is the price that has to be paid. With careful planning, however, the loss of information can be reduced. This can be achieved either by avoiding damage to the remains of the past where they exist or, if destruction is unavoidable, recording the sites archaeologically so that at least the evidence contained within them can be rescued and passed on to future generations.

Planned construction work is sometimes modified to avoid damage to archaeological material. If a site is to be destroyed by development then the rescue and recording of information may be done voluntarily by the developers, although occasionally a little encouragement from legislation is required. Although archaeological fieldwork on land is often related to anticipated site disturbance through development or changes in land-use, most sites are not recorded before they are destroyed in this way.

The reason for this is that there is a lot of archaeological work to be done but little money to pay for it. In these circumstances every archaeologist must think hard before undertaking any excavation (itself a destructive process) that is not rescuing information ahead of inevitable destruction.

As stated earlier, future generations will be able to infer more from sites than present-day archaeologists. At some point in the future, for example, it may not be necessary to excavate at all as methods of 'seeing' into the ground are becoming more and more sophisticated (see chapter 13). Fieldwork has not always been shaped by such considerations, and excavations have taken place in the past which might be difficult to justify now. That does not imply criticism of past workers – it simply means that archaeologists have learnt to ensure that every penny spent on archaeology today is money well spent, and that it is part of a co-ordinated and directed effort to understand our heritage.

There is more than enough non-destructive archaeological work available now to keep all those interested in the past busy for years. One of the most pressing is searching for and recording new sites. Whichever strategy for the conservation and management of the remains of the past is applied, one thing is vital – forewarning of potential problems. Sea-bed users, legislators and archaeologists need to know what significant remains exist/are known of in any one area before commercial development, or any other potentially destructive process, begins.

One of the areas of expansion within archaeology over recent years has been the compilation of inventories of sites by both regional and national governments. In the UK, these inventories are called sites and monuments records (SMRs) or historic environment records (HERs) and the information held in them is essential for the proper management of historic and archaeological remains. It enables the effective identification of sites and the appropriate allocation of limited resources for their protection.

Systematic 'stock-taking' of underwater sites is slowly advancing but it has a long way to go, and this is where members of the public, archaeologists and non-archaeologists alike, can help. Millions of sport-dives are made annually around the world so clearly divers have a vital role to play in finding out just what is on the sea-bed.

Registers of sites serve two main functions:

1　They provide information in a form that is convenient for researchers to consult and easy to manipulate. For example, with a computerized database a researcher should be able to find basic information on all the known sites on a particular date in a specific area or, in a more refined use of the system, be able to obtain information on only those from that period which contained specific types of material. Such a register can be a powerful tool for research as well as for the management of archaeological resources.

2　They provide the background information which allows an assessment of whether particular sites are in immediate danger, or likely to be damaged by new developments. If a company wishes to take sand and gravel from an area of sea-bed, a comprehensive register of sites will allow a very rapid and informed judgement about whether the extraction should go ahead in the intended location.

Many important discoveries have been made accidentally by divers, whereas deliberate searches for specific sites by underwater archaeologists have resulted in relatively few new finds. This underlines just how important recreational divers are in developing knowledge of the nature and distribution of archaeological remains on the sea-bed. The amount of time divers spend on the sea-bed can never be equalled by professional archaeologists. Consequently, the amount of information divers collect is crucial to the development of a representative database, but it can become even more valuable if certain basic observations are made.

For site inventories to fulfil their potential, there is clearly a need for a minimum level of information about each site, this should include:

- an accurate position (see chapter 11);
- an assessment of the age of the site;
- an assessment of the state of preservation of the site;
- factors that seem likely to threaten the site in the short or long term;
- any known historical associations or aspects of the site which make it particularly significant (but be wary of making a firm identification based on wishful thinking rather than hard evidence).

This information, together with any other relevant data, is obviously extremely useful. It is also often already known locally. Such knowledge held at a local level can be difficult to consult if it has not been passed to a historic environment record. This is particularly true in cases where information is not written down anywhere but held in divers' heads.

Methods have been developed, and are in common use, which allow information on sites to be recorded and consulted while still respecting the local sense of ownership and preserving appropriate confidentiality.

WHAT IS ARCHAEOLOGY UNDER WATER?

The study of the past is an extensive subject. Archaeologists often specialize in one or more aspect, such as the study of cultures found in a geographic location, or a specific period. Some archaeologists develop expertise in a class of archaeological material such as pottery or even ships. Less often do they develop skills for working in a particular environment, such as under water, and those who do would normally have specialist skills in another aspect of archaeology. The archaeology of ships and boats is a natural area of expertise for the archaeologist who dives, but some diving archaeologists will be more interested in submerged settlement sites or some other area of study appropriate to the underwater environment.

Archaeologists who work under water should have the same attitude to the available evidence as those who work on land and should have a familiarity with other areas of archaeological research. Since archaeology under water is not fundamentally different from archaeology on land, the standards applied should be no less stringent.

WHAT IS *NOT* ARCHAEOLOGY UNDER WATER?

Salvage: (This is not to be confused with the term 'salvage archaeology', a North American term which equates to the British expression 'rescue archaeology'.) Whereas archaeology is the collection of information, salvage is the collection of material for its monetary value. The salvor's role of returning lost material to trade is a valid activity but it can conflict with archaeology when that material represents surviving clues about the past. Archaeological material is only occasionally of sufficient economic value for commercial operations, and the conflict of interests between archaeology and reputable commerce is less common than might be thought. Unfortunately, there have been occasions when sites have been damaged just to keep salvage crews busy during slack periods.

Treasure-hunting and souvenir-collecting: On the fringe of salvage is treasure-hunting. While financial gain is normally the ultimate motive, the allure of the romance and glory can also play a significant part. It is surprising how many people invest in promises of easy pickings of treasure-fleet bullion or can be persuaded to support 'antique mining' expeditions on the flimsiest of evidence (Throckmorton, 1990). Compared with legitimate salvage, the activities of treasure-hunters tend to be less well directed, less financially stable and less accountable, although there are occasional exceptions. This means that such activity is often much more threatening to archaeological remains than salvage. Frequently, such projects are accompanied by exaggerated claims to entice potential investors, who help to keep many treasure-hunting organizations afloat. Few treasure-hunts are financially self-sustaining and so need the help of investors; in this way treasure-hunters usually risk other people's money in their schemes and not their own. The treasure-hunting community is always keen to promote its rare successes and play down the much larger number of failures so as to maintain potential investors' interest in future projects. Although some ventures make an attempt to reach acceptable archaeological standards (or claim to do so) during the recovery of objects, the majority do not. The outcome of most treasure-hunting expeditions is damage or destruction of irreplaceable parts of the heritage. The costs of such expeditions are high and the returns low, but the treasure-hunters simply move on to spend other people's money on the next project.

Another activity on the fringes of salvage is the collecting of artefacts as souvenirs. Many sites have been disturbed and partly or wholly destroyed simply because the finder has a 'general interest' in old things and wants a few souvenirs to display at home or in a small private 'museum'. The motive is often undirected curiosity rather than any destructive intent, but the activity is inevitably unscientific and evidence is lost for ever. To make matters worse, these individuals sometimes disperse material by selling it to offset the cost of collecting.

Although it would be wrong to equate cynical commercial greed with what is often a genuine and deep interest in the past, from an archaeological point of view there are few significant differences in the end results of treasure-hunting and souvenir-collecting. Projects which set out to make a financial profit, those which concentrate on the collection of souvenirs or personal trophies and those which subsidise a basically recreational operation by selling material, destroy important archaeological evidence. To some people the notion of a commercial recovery operation conducted to 'archaeological standards' appears achievable. The two approaches are, however, largely irreconcilable for three basic reasons.

Firstly, the major difference between archaeological investigation and salvage or treasure-hunting is that the principal aim of archaeology is the acquisition of new information that can be used now and is available for the benefit of others in the future. Although an increasing number of commercial projects claim to be attempting to reach this goal, very few ever achieve it. Archaeological work on a site is directed to this end and the final result is a complete site archive and academic publication rather than just a saleroom catalogue. Any unnecessary activity (treasure-hunting/antique mining/curio-hunting/incompetent archaeology) that results in the accidental or deliberate destruction of some of the few surviving clues about the past has to be viewed with profound dismay. Without preservation in the form of adequate, detailed records, that information about the past, which had survived for so long, is destroyed for ever.

Secondly, as will become clear later in this book, clues about the past can come from a wide variety of sources apart from recognizable objects. Archaeology involves far more than artefact retrieval. When a project is being funded by the sale of artefacts, attention is usually focused on the material perceived to have a commercial value. Other sources of evidence that archaeologists would consider vital to the study of the site, such as organic remains and even hull structure, are normally ignored and very often destroyed. Once the material reaches the surface, the commercial artefact-filter continues to operate. Conservation (see chapter 16) can be expensive and objects unlikely to reach a good price at auction are not worth the investment to the artefact hunter. They are often discarded. The end result is a group of isolated objects selected on the basis of commercial value, rather than a carefully recorded sample of the contents of a site, which can be studied as an assemblage of interrelated clues.

Thirdly, the result of the archaeologists' work, which is handed on to future generations (the site archive), is expected to include the finds as well as the records from the site (see chapter 19). Forensic science teams do not sell off the evidence from unsolved cases; rather, it is retained for reassessment. Something like Jack the Ripper's knife could fetch a high price on eBay but, apart from ethical considerations, the implement could still provide fresh evidence as new forensic techniques are developed. Archaeological sites are enigmatic, and the files on them have to remain open. No one interpretation of a site can be considered definitive and new methods and ideas must be tested against a complete set of the original clues if fresh, valid conclusions are to be drawn (Bass, 1990).

Dispersal of material makes re-evaluation virtually impossible. Sites cannot be studied in isolation, but must be compared with and linked to others (see chapter 4), and when the archive of evidence is incomplete, the usefulness of the site for comparison with new ones as they are discovered is greatly reduced. The damage caused by the selling of finds goes further than compromising the record of a single site. The self-sustaining system of promotion that brings in the investment required to fuel most treasure-hunting operations has already been mentioned. The glossy sales catalogues and publicity surrounding the sale of artefacts distorts the notion that the past is valuable. It is valuable, not as cash, but as a source of knowledge about 'what went before', an understanding of which is fundamental to all human cultures.

The NAS has drawn up a Statement of Principles (see the NAS website) that it would wish its members and others to adhere to in an effort to help vulnerable underwater heritage receive the care it deserves. Many other concerned organizations, both independent and inter-governmental, have published documents with similar aims and aspirations.

As treasure-hunting continues, sometimes officially condoned, those interested in archaeology are faced with a difficult choice. They can choose not to get involved, and so allow sites to be destroyed, or they can try to improve the standards of the treasure-hunting project, and then risk being 'sucked in' and exploited. There is no easy answer. The treasure-hunter will want:

- archaeological recording to a standard that will help convince officials to let their work continue and, in doing so, will provide a veneer of respectability that may help impress potential investors and others;
- validated historical background and provenance – to increase the monetary value of objects;
- the archaeologist to be a potential target of criticism about the project rather than themselves.

In return for this, the archaeologist will often receive a good salary and the opportunity to rescue information before it is destroyed during the recovery process. Many archaeologists do not feel that the working practices and imperatives of treasure-hunters can be modified sufficiently to make it possible to work alongside them. It cannot be denied that some treasure-hunting companies do attempt good field archaeological practice but they often restrict this to sites where there is external scrutiny and have lower standards on other sites. This suggests that the extra effort involved in disciplined archaeological work is not undertaken voluntarily but simply for expediency.

Any archaeologist considering working on a commercially motivated artefact-recovery project should consider the following points.

- Does an archaeologist have to be recruited before the project is allowed to go ahead? The archaeological community may be able to save the site from destruction simply by refusing to become involved.
- The archaeologist will need to be well qualified and have sufficient experience to make informed judgements under pressure. S/he will also require a strong character to deal effectively with any forceful personalities encountered. Operators will often approach inexperienced, under-qualified or non-diving archaeologists who may be more easily persuaded or misled.
- The archaeologist should not work for any form of financial rewards based on the quantity or monetary value of materials or objects recovered from the site. The archaeologist should not work under the control of the manager of the recovery operation, and should have the ability to halt the whole operation if adequate standards are not maintained.
- The archaeologist should not describe the recovery operation as 'archaeological' unless it is entirely

under his/her control and s/he is directly responsible for the standard of the investigation. Archaeologists should also retain the right to publish an objective and full report on the standards and results achieved and not contribute to the sanitizing of a treasure-hunting expedition by producing a glossy, popular volume masquerading as an academic publication.

- An archaeologist should not give up the right to campaign against treasure-hunting or actively oppose the dispersal of material.
- An archaeologist should always remember that while the funding for treasure-hunting usually comes from investors, the normal mechanism for topping up funds is for finds to be dispersed by sale. This is one of the key issues that separates proper archaeology from treasure-hunting and salvage.

Other archaeologists may find that as part of their work for government departments or heritage agencies they have to work alongside treasure-hunters and salvors. In such a situation honest and intelligent dialogue with all parties is advised.

Governments are often criticized for their relationship with treasure-hunters. Poorer countries have, on occasion, entered into financial agreements over potentially valuable wrecks in their waters. Sometimes it is because the country has no prospect of revenue from conventional sources and can see real short-term benefit in such deals. Unfortunately, sometimes it is simply because a senior government official is a diver and thinks it is a romantic notion. Even wealthy countries have entered into agreements with treasure-hunters, generally for pragmatic reasons rather than financial reward or romance. Rarely is a situation as straightforward or as simple as it might at first seem, so it is important for archaeologists to retain an open mind and engage in such debates calmly, taking care not to exaggerate claims or ignore evidence that does not support their case.

If the archaeologist faces a series of difficult choices in living with treasure-hunting, so must conscientious museum curators. They face a similar choice between saving a small part of the information for the general population, and so perhaps encouraging the treasure-hunter, or losing the little they could have saved in an attempt to reduce further destructive activity. By buying objects or even accepting them as gifts, the museum can give both respectability and, in the case of purchase, money, which will help the treasure-hunter to continue destroying sites.

Less well-informed or less scrupulous museums can sometimes become involved more directly. A narrow-minded view is to stock the walls and cabinets of an establishment without worrying about the effect on archaeological sites. Fortunately, this attitude has no place

in a modern museum and many institutions and international organizations have worked hard to develop codes of conduct to govern the acquisition of new material.

Further information on some of these issues and links to further resources can be found in chapter 7.

CLOSELY RELATED AND COMPLEMENTARY APPROACHES (ETHNOGRAPHY AND EXPERIMENTAL ARCHAEOLOGY)

Maritime ethnography is the study of contemporary cultures, their tools, techniques and materials. Maritime archaeologists employ ethnographic techniques by studying the material remains of contemporary seafaring and other waterside communities that use similar tools, techniques and materials to those found in archaeological contexts.

Maritime ethnography has three main applications:

1. as a record of a culture, its materials and tools;
2. as an artefact that is part of society, that ultimately reflects on aspects of that society; and
3. as a means of increasing an archaeologist's knowledge by visualizing past societies, their cultural practices and their use of materials and solutions to technological problems.

The applications of maritime ethnography cited above lead to a better understanding of the archaeological record. The study of contemporary fishing communities and boatbuilding traditions, for example, can provide a valuable insight into past practices and is particularly relevant as boatbuilding traditions are rapidly changing and wooden boats are increasingly replaced by metal and glass-reinforced plastic hulls fitted with engines. What McGrail expressed some years ago still holds true: 'Ethnographic studies can make the archaeologist aware of a range of solutions to general problems . . . Using such ethnographic analogies, the archaeologist can propose hypothetical reconstructions of incomplete objects and structures, suggest possible functions of enigmatic structural elements and describe in some detail how an object or structure was made' (McGrail, 1984:149–50).

Of course such an approach requires a certain degree of caution. The study of contemporary fishing communities does not necessarily directly determine the activities and use of materials in comparable archaeological contexts. People do not always use objects in similar ways and there may be numerous solutions to the same problem. The limitations and difficulties of using such evidence must be appreciated. However, in terms of investigating aspects of function and the manufacture of complex artefacts (such as boats and ships) the ethnographic record is invaluable.

When applied cautiously it can provide a baseline or launch-pad for retrospective enquiry (plate 2.2). Ethnographic evidence can also be very closely linked to experimental archaeology.

Experimental archaeology: Material on archaeological sites under water, as on land, can be studied and understood at a number of levels: as a part of the site, as a part of a functional assemblage within the site and as an object in its own right, which can provide information about the technology used by the society that made it. However, archaeological evidence is rarely complete. Objects can be broken and distorted, and they may be found in association with other objects and materials that have no relevance to the way they were actually used (see chapter 4). The evidence for the technology used in the object's construction may be hidden by other features or may simply be too complex to be understood through a visual inspection alone. It is necessary therefore to find ways of investigating these aspects of the evidence.

The phrase 'experimental archaeology' is often used in a very loose way to describe a wide variety of activities. Projects on land have ranged from cutting down trees using flint or bronze axes to the creation of earthworks that are surveyed and sectioned at regular intervals to examine erosion and site-formation processes. Projects beginning on land and ending up in the water have included the construction of water-craft varying in size from small one-person canoes (figures 2.3 and 2.4) to large sailing ships. The NAS regularly organizes experimental archaeology courses for members to learn how to cut out wooden frames for a ship or make things such as replica medieval arrows.

This field of study is not without significant problems, not least of which is the fact that it is possible to spend very large amounts of money building, for example, a replica ship, and actually gain very little useful information. Why is this so?

Figure 2.3 Experimental archaeology: building a replica of a logboat found in Loch Glashan, Argyll, Scotland. (Photo: Colin Martin)

Figure 2.4 Trials of the Loch Glashan replica logboat. (Photo: Colin Martin)

If a group plans to build a full-size model of a boat or ship and to investigate its construction and its performance, they are immediately faced with a problem: how accurate and how complete is the evidence on which they are basing their reconstruction? If the primary source of information is the excavated remains of a vessel, this evidence may be fragmentary and distorted, especially for the upper parts of the craft. If the evidence is mainly iconographic then other problems arise. Did the artists understand what they were drawing? Is the constructional information and scale distorted by perspective? Clearly it is important to consult as widely as possible and to collect information from as many sources as possible.

When a design has been decided upon and construction starts, what tools and materials will be used? It may be that using modern materials and tools will affect the way the vessel performs. Also, some constructional features may only make sense when the tools and technology used in the original are applied to resolving the problem in hand. It is important to record all aspects of the work, including the reasons for taking specific decisions, such as using a chainsaw to cut timber because of a lack of manpower.

The vessel is now complete and ready to take to the water. Who will sail or row her? Do the necessary skills exist? Sailors get the most out of their vessels by applying experience built up over many generations; to what extent can that human element be re-created? Once the sailing starts, how will the performance be recorded? After the effort of construction it is important to use measured criteria rather than casual observation, as these can be compared with other measurements taken elsewhere. Finally, how reliable are the results in the light of all the problems highlighted above?

This is not to suggest that experimental archaeology is a waste of time. It certainly is not. But it is very important to define aims and be honest about what is being attempted and what has been achieved. Due to the enormous expense involved in some larger projects, it is crucial to have transparent management and financial decisions so that others can judge whether money is being well spent. Some projects are mainly concerned with the appearance of the vessel or object – the main aim is display and communication. Some projects aim to study construction and therefore pay great attention to the tools and materials used. Such studies may involve only partial or scaled-down reconstructions. The most ambitious way to investigate the performance and function of a vessel is through a full-scale reconstruction, although tank-testing models and computer modelling are also very significant in this area. All are valid aims and have a contribution to make, even if on occasion it is more in terms of evoking the spirit of past endeavours rather than gathering useful data.

Experimentation is enormously valuable in encouraging people to look more closely at the material used to make inferences about the past. Often it is the only way to study the complex functional relationship between objects and to approach an understanding of the human element involved in their construction and use. Without an honest assessment of the aims and methods of a project and a detailed, objective record of the results, the usefulness of a reconstruction, whether a single object or an entire vessel, will be very limited.

This chapter has attempted to clarify exactly what archaeology is, where it has come from, what motivates archaeologists and how they approach their subject. For information about the basic principles behind archaeological practice, please see chapter 4.

FURTHER INFORMATION

Adams, J., 2002, Maritime Archaeology, in C. Orser (ed.), *Encyclopaedia of Historical Archaeology*, 328–30. Oxford.

Bass, G. F., 1990, After the diving is over, in T. L. Carrell (ed.), *Underwater Archaeology: Proceedings of the Society for Historical Archaeology Conference 1990*. Tucson, Arizona.

Delgado, J. P. (ed.), 2001 (new edn), *Encyclopaedia of Underwater and Maritime Archaeology*. London.

Gamble, C., 2006 (new edn), *Archaeology: The Basics*. Oxford.

Green, J., 2004 (2nd edn), *Maritime Archaeology: A Technical Handbook*. London.

Harris, E. C., 1989 (2nd edn), *Principles of Archaeological Stratigraphy*. London.

McGrail, S. (ed.), 1984, *Aspects of Maritime Archaeology and Ethnology*. London.

Muckelroy, K., 1978, *Maritime Archaeology*. Cambridge.

Renfrew, C. and Bahn, P., 2004 (4th edn), *Archaeology: The Key Concepts*. Oxford.

Throckmorton, P., 1990, The world's worst investment: the economics of treasure hunting with real life comparisons, in T. L. Carrell, (ed.), *Underwater Archaeology: Proceedings of the Society for Historical Archaeology Conference 1990*. Tucson, Arizona.

Getting Involved in Underwater and Foreshore Archaeology

3

There are many ways in which an individual can get involved in maritime archaeology (plate 3.1). The list below is by no means exhaustive but does give an indication of the range of opportunities that exist. Participation includes:

- visiting archaeological sites and exhibitions;
- undertaking the NAS Training Programme and other courses;
- historical research in libraries, records offices or on-line;
- searches for sites in seas, rivers, lakes and on the foreshore;
- surveys of harbour works, sites eroding from beaches and wrecks both on shore and under water;
- excavations on land, on the foreshore and under water;
- post-excavation work, finds-cataloguing, recording;
- researching, building and using reconstructed water-craft;
- publishing research and results;
- lobbying for better protection of maritime archaeological sites;
- attending conferences, talks and seminars.

An individual will decide on his/her own level of commitment, ranging from the occasional evening or weekend to those who fill every spare moment participating in activities such as those listed above. Some people even choose maritime archaeology as a career. Whatever the level of commitment, in the first instance groups or individuals might consider joining the Nautical Archaeology Society (NAS). The origins and ethos of the NAS are summarized in chapter 1. NAS members receive a quarterly *Newsletter* containing information about projects, courses, conferences and relevant issues related to maritime archaeology. The NAS is also responsible for the production of the *International Journal of Nautical Archaeology* (*IJNA*), one of the foremost academic publications in the field. The *IJNA* contains articles about academic research and fieldwork from all over the world. It is a bi-annual journal published by Wiley-Blackwell for the NAS and available at a preferential rate to NAS members.

The NAS Training Programme was instigated in the UK in the 1980s and has subsequently been adopted by many countries throughout the world. It is structured in progressive levels, beginning with a one-day 'Introduction to Foreshore and Underwater Archaeology' and ending with a 'Part IV Advanced Certificate in Foreshore and Underwater Archaeology' (see appendix 3).

The archaeological process is long and complex and includes documentary research, initial site-assessment, survey, recording, publication and dissemination of results. It may also include excavation, which will lead to finds-processing, recording, cataloguing, conservation, storage and, ideally, display of excavated material. Archaeological experience could therefore involve just about anything from heavy manual labour, to pot-washing, drawing, database management or museum work.

One avenue for part-time involvement in archaeology is to volunteer with an appropriate organization. Local museums are often grateful for volunteer help and this can provide the opportunity of working with archaeological material that is not usually on public display. Some people will be lucky enough to have a specialist maritime museum locally. Alternatively, archaeological units and charitable trusts may have opportunities for enthusiastic individuals to gain experience in many aspects of the archaeological process. For those intending to pursue a career in archaeology, volunteering is an excellent way to gain valuable experience that will enhance employment prospects.

Independent research is a flexible and readily achievable way for anyone to become involved in maritime archaeology. It enables an individual to pursue an aspect of the subject that particularly interests them, in their own time and in a way that most suits them. Such research can be carried out at home, in libraries, on the internet and via local and national libraries and archives (see chapter 9).

There are increasing numbers of conferences, lectures, talks and seminars on a maritime archaeological theme taking place around the world on a regular basis. These can be an excellent opportunity to broaden horizons, keep up to date with the latest research, projects and perspectives, and meet people who share an interest in maritime archaeology (see chapter 20).

Though an individual's primary interest may be maritime archaeology, an excellent way to gain archaeological experience is on a land (terrestrial) site. The archaeological process is exactly the same whether on land or under water. However, on a land site the learning process can be quicker as it is easier to communicate and ask questions. For those looking for hands-on involvement, it may be possible to join a training excavation. These are run by a variety of organizations worldwide and usually require at least a one-week commitment and a financial contribution.

The NAS organizes a range of maritime archaeological projects each year. These can be based in the UK or overseas, in lakes, rivers, the sea or on the foreshore and may include a training component. For more information, see the NAS website (www.nauticalarchaeologysociety.org).

With a good grounding in the archaeological process and survey techniques (perhaps through NAS training), it is possible to organize an independent survey project on a local site. Once permission has been obtained from the site owners or managers and the relevant authorities, the survey project can be planned and carried out. The NAS can provide advice about how to go about this and results can be submitted as a NAS Part II project (see appendix 3). Alternatively (or indeed additionally), a site could be officially 'adopted' under the 'Diving with a Purpose, Adopt a Wreck/Site' initiative (see NAS website). Information obtained can then be fed into local and national databases for archaeology and marine conservation, so that the information is available to all.

To participate in an underwater archaeology project as an unpaid diver, a diving qualification equivalent to at least CMAS 2 Star (BSAC Sport Diver, PADI Rescue Diver, SAA Club Diver) is required. Due to conditions on site, many projects will ask for more experience than the

minimum requirement, so it is wise to aim for a higher qualification and as many appropriate dives as possible while training. It is also worth remembering that few projects can supply diving equipment (with the possible exception of cylinders).

It is not necessary to be a diver to get involved in maritime archaeology. Many sites of a maritime nature are not under water at all (see plate 3.2); they may not even be very close to water any more. Even on an underwater project, for every minute spent under water carrying out archaeological work, there are many hours spent on the surface or ashore, studying and processing material and producing reports (figure 3.1).

People often become involved in maritime archaeology as a direct result of discovering an artefact or site, perhaps during a walk along the foreshore or during regular sport-diving activities. There are numerous organizations that can provide information about how to proceed in such a situation and these should be consulted at the earliest opportunity. Contact details are provided below.

Archaeology is a wide-ranging and varied subject closely related to other disciplines and, as such, can be an

Figure 3.1 Post-fieldwork activity at the Hampshire & Wight Trust for Maritime Archaeology. (Photo: Hampshire & Wight Trust for Maritime Archaeology)

appealing career choice. Archaeological jobs can include aspects such as survey, excavation, illustration, conservation, research, photography, database-management, education and display, geophysics, scientific examination of material, scientific dating techniques and, of course, maritime archaeology. However, in many countries, there are more people qualified to undertake these jobs than there are vacancies. Archaeological posts are often based on short-term contracts with modest salaries, and career prospects can be limited. A career in archaeology does therefore require a degree of determination, commitment and desire for intellectual rather than financial fulfilment.

Employing organizations would generally expect applicants to hold academic qualifications in archaeology. Archaeologists are employed in the UK by a number of organizations including:

- local government (county, district, city, regional or unitary authorities);
- non-governmental organizations (e.g. Council for British Archaeology, Institute of Field Archaeologists, National Trust, the Nautical Archaeology Society);
- statutory bodies (e.g. English Heritage, Historic Scotland, Cadw (Wales), Environment & Heritage Service (N. Ireland));
- archaeological units;
- contract archaeologists;

- Royal Commission on Ancient and Historical Monuments (for Wales and Scotland);
- universities/tertiary education bodies.

On site, specialist personnel on archaeological projects include photographers, geophysicists, illustrators, conservators and experts in particular materials such as pottery, glass, plants, seeds and bones. In the advanced stages of an archaeological project, specialist artists and editors may be employed to help with the preparation and publication of the final report. Specialists (see chapter 19) on a maritime archaeological site might also include experts in ship structure or marine ordnance.

Each country has its own regulations regarding diving qualifications for professional archaeologists. In the UK, anyone being paid to dive must comply with the Diving at Work Regulations 1997 (and subsequent amendments). This means that a diver must either hold a recognized Commercial Diving Certificate or an equivalent qualification as specified under the appropriate Approved Code of Practice (ACOP). In addition, the diver must hold a current Health and Safety Executive (HSE) recognized diving medical certificate (renewed annually). For further information, see chapter 6.

This chapter has summarized the range of opportunities that exist for involvement in maritime archaeology, from the interested amateur to the professional archaeologist. For further information on anything mentioned above, please see the sources cited below and/or contact the NAS office.

FURIDHER INFORMATION

MARITIME MUSEUMS
A list of naval and maritime museums in the UK: people.pwf.cam.ac.uk/mhe1000/marmus.htm
A list of naval and maritime museums world-wide (not USA): www.bb62museum.org/wrldnmus.html
A list of naval and maritime museums in USA: www.bb62museum.org/usnavmus.html

VOLUNTEERING
Archaeology Abroad, produced by the Institute of Archaeology, University College London, twice a year (April and November): www.britarch.ac.uk/archabroad/
British Archaeological News published every two months by the Council for British Archaeology: www.britarch.ac.uk/briefing/field.asp
British Archaeological Jobs Resource (BAJR): www.bajr.org/
Council for British Archaeology: www.britarch.ac.uk/
Current Archaeology – information centre: www.archaeology.co.uk/directory/
Earthwatch is an organization that promotes sustainable conservation of cultural heritage by creating partnerships between scientists, educators and the general public. Earthwatch puts people in the field: www.earthwatch.org/

ELECTRONIC DISCUSSION LISTS
Britarch (www.jiscmail.ac.uk/lists/britarch.html) is a discussion list to support the circulation of relevant information concerning archaeology and education (at all levels) in the UK.
MARHST-L (http://lists.queensu.ca/cgi-bin/listserv/wa?A0=MARHST-L). The purpose of MARHST-L is to promote communication between people with a serious interest in maritime history and maritime museums.
Sea-site (www.jiscmail.ac.uk/lists/SEA-SITE.html) aims to encourage multidisciplinary marine environmental research and fieldwork associated with submerged archaeological sites.
Sub-arch (http://lists.asu.edu/archives/sub-arch.html) is an electronic discussion list about underwater and marine archaeology. This list is used by both professional archaeologists and salvors which can lead to interesting and sometimes heated discussions.

DIVING
For UK information:
British Sub-Aqua Club (BSAC), Telford's Quay, Ellesmere Port, South Wirral, Cheshire, L65 4FL (www.bsac.com/).

Professional Association of Diving Instructors (PADI), PADI International Limited, Unit 7, St Philips Central, Albert Road, St Philips, Bristol, BS2 OPD (www.padi.com/).

Scuba Schools International (SSI) in the UK (www.ssiuk.com).

Sub-Aqua Association (SAA), 26 Breckfield Road North, Liverpool, L5 4NH (www.saa.org.uk/).

For international information:

Confédération Mondiale des Activités Subaquatique (CMAS) (www.cmas2000.org).

National Association of Underwater Diving Instructors (USA) (www.naui.org).

Professional Association of Diving Instructors (PADI) (www.padi.com).

Scuba Schools International (SSI) USA (www.ssiusa.com).

Basic Principles – Making the Most of the Clues

4

Contents

The term 'archaeological site' is a familiar one, but what is meant by the word 'site' and how are archaeological sites studied?

An archaeological site might take the form of a medieval castle, a neolithic trackway or, indeed, a shipwreck. Whatever its form, an archaeological site comprises material left behind by past societies. From the walls of a castle to the button from a shirt, the material remains form the archaeological record and associated material can be thought of as an archaeological site. An archaeological site could therefore be described as a concentration of material remains indicating the way people lived in the past.

It must be appreciated, however, that an archaeological site cannot be studied in isolation because it did not exist in isolation (figure 4.1). A castle formed part of a much wider social and economic community (materials and products would probably have been imported and exported locally, regionally or further afield). Taxes were paid to repair its walls, it was staffed using labour from the surrounding countryside and it had a defensive and protective role. Similarly, in the case of the neolithic trackway, the archaeologist would seek to answer questions such as: who built it, why and how was it built, where did the materials come from, who used it and how?

It is important to recognize that although archaeological sites are concentrations of evidence about past ways of life in one specific place, they have a relationship with other archaeological sites of similar date. If maritime trade is involved in the distribution of artefactual material, these other sites may be half-way round the world.

A wreck-site on the sea-bed contains a concentration of evidence about past activities. Even though the site represents the remains of a ship that was once a self-contained mobile 'settlement' (a warship is a bit like a floating castle), it is still linked to other archaeological sites. These can be both on land and under water, providing evidence about such things as its ports of call, the homes of the crew, the origins of the objects on board, the forests where its timbers grew, and the shipyard where it was made. When studying an archaeological site it is vital to explore its relationships and interdependences with other sites.

THE IMPORTANCE OF UNDERWATER SITES

Although a wealth of archaeological sites exist on land, a vast resource of information about past peoples and environments also survives under water. These sites have the potential to provide new and exciting information about the human past. Sites under water are important for two basic reasons: they are often unique in their nature and available nowhere else (e.g. shipwrecks) and certain materials are often much better preserved on underwater sites.

Some types of site are very rarely available on land. For example:

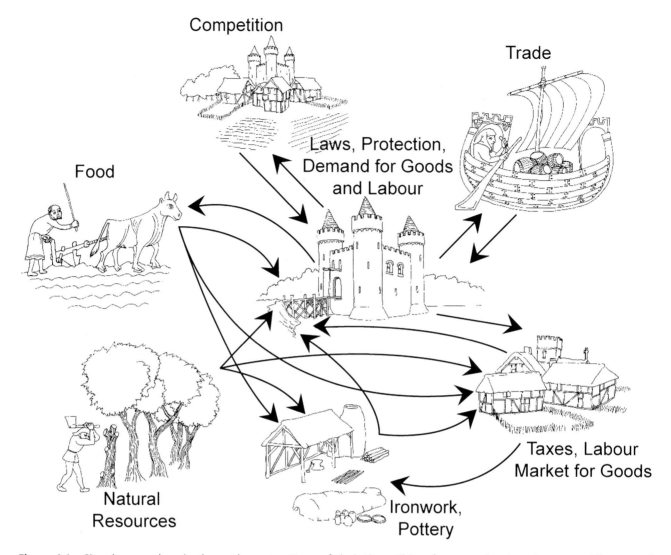

Figure 4.1 Sites have a place in the settlement pattern of their time. Ships, for example, though very mobile, are still just part of a worldwide system. (Drawing by Graham Scott)

- Sites of evidence lost or deposited while using the water. Shipwrecks are perhaps the most obvious example.
- Sites established on or at the edge of water, which are partly or wholly submerged. These often relate to maritime infrastructure such as quays, wharves or docks.
- Sites built in or over water are rarely completely accessible to investigations based solely on dry-land methods (e.g. crannogs and pile dwellings).
- Sites that were established on land but are now submerged (e.g. the prehistoric sites in the eastern Gulf of Mexico or the prehistoric sites which were submerged when the English Channel flooded).
- Sites which have continued to develop during a rise in water-level. Since the site will progressively retreat away from its original location, earlier ele-

ments of its development will now only be available under water.

The second reason for underwater sites being important is that clues about the past are often so much better preserved than on land (figure 4.2). However, if artefacts are left exposed to seawater they will suffer from natural processes of decay (see chapter 16). Nevertheless individual objects that do survive are, to some extent, better protected from recovery or disturbance by the barrier of water above them (plate 4.1).

Perhaps the most exciting example of potential preservation on underwater sites is a feature sometimes referred to as the 'time-capsule effect'. The clues usually available on land sites, which are often inhabited for long periods, do not necessarily give an accurate picture of what was happening at any specific moment; instead

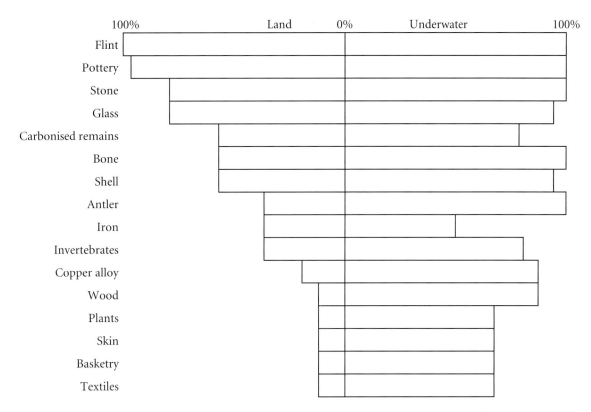

Figure 4.2 Survival of clues on underwater sites relative to dry sites. Information is often better preserved and protected under water. (After Coles, 1988, fig. 5)

they reflect changes and processes over time. It can therefore be difficult to see how a site functioned at any particular stage. An ideal scenario for archaeologists would involve a site being frozen at the height of its success. Not only would individual objects be preserved from decay or reuse but they would also be trapped in positions and associations reflecting the way they were used.

In reality, such 'time-capsules' are very rare, especially on the scale of a whole site. Very rapid burial, however, does create some of the characteristics of a time-capsule. While changes to the evidence do take place during and after burial, the number of clues trapped in a relatively undistorted way can still be significant. Only a very few land sites have been buried quickly enough for the 'time-capsule' effect to be a major factor, although more sites will have small-scale pockets of rapid burial (an event such as a fire can lead to the loss of a lot of material very quickly).

Rapid burial by water (sinking), however, has been a virtually daily occurrence for a very long time (figure 4.3). While this happens most frequently to ships, on occasions even towns have been trapped in this way. It would be simplistic to assume that every site under water contains nothing but groups of closely associated material. The nature of each site must be demonstrated by careful investigation. However, the possibilities are very exciting, especially in terms of the information such groups of material can provide about similar objects found in highly disturbed sites elsewhere.

In summary, archaeological sites under water are important because the water hides, preserves, protects and traps clues that are often not available elsewhere.

SITE TYPES

It is important to be aware of the great diversity and range of archaeological clues to be found under water. When sites on the foreshore are included, the list grows even longer and more varied; everything from wrecks and harbour works to prehistoric footprints preserved in inter-tidal mud. Some sites (e.g. shipwrecks) represent high levels of technical achievement; others, such as middens or simple fish-traps (figure 4.4), although apparently unexciting, provide important information about daily life. Indeed, the range of submerged material is such that there are few aspects of archaeological research on land that cannot be complemented or supported by information from underwater contexts (plate 4.2).

An account of all the classes of material to be found submerged by inland or coastal waters is beyond the scope of this book. However, for the fieldworker, the difference between the site types lies in the scale and

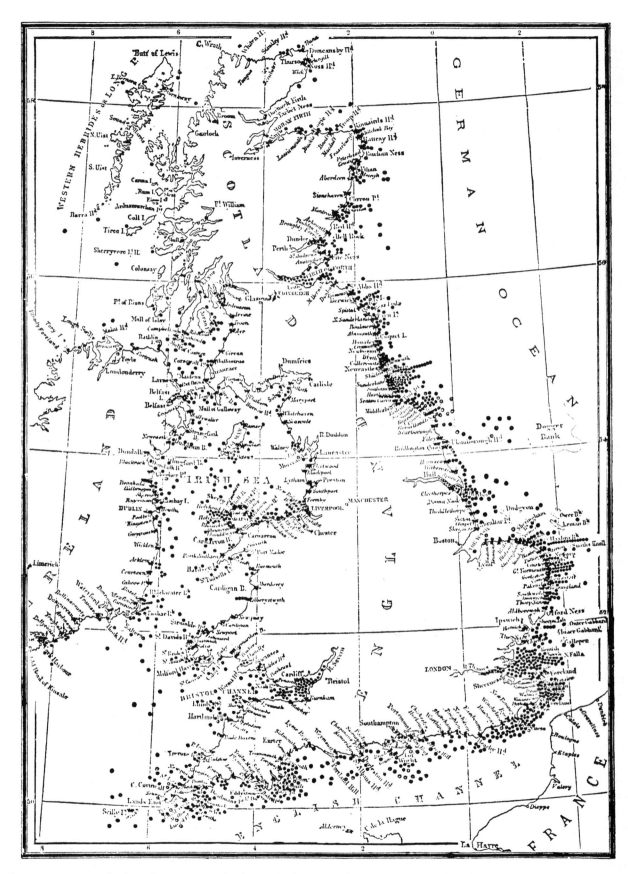

Figure 4.3 A wreck chart for 1876–7. The losses in that period were not unusual and give an indication of the large numbers of wrecks that may have occurred in British waters. (Reproduced by permission of the RNLI from the *Lifeboat* journal, vol. X, no.110)

Figure 4.4 Site types: aerial photograph of a stone-built fish-trap at Airds Bay, Scotland. (Photo: Colin Martin)

complexity of the subject, not the thoroughness of the investigation and recording that should be applied. Some classes of site, such as submerged landscapes, may require higher levels of specialist knowledge for successful recognition and analysis but all will benefit from a careful and systematic approach.

If wrecks are taken as an example of one type of archaeological site under water, it quickly becomes clear that even within this one category there is at least as much variety as there are boat and ship types. In discussing such material and comparing one example with another, it is therefore helpful to have some method of clarifying the situation through a system of classification. Such sites can be sub-divided according to age, constructional details or state of preservation, as well as simply in terms of what is known about them (e.g. exact location and full survey, or estimated date and area of loss). A number of online glossaries and thesauruses exist to provide structured word-lists to enable a standardized use of terminology (see 'Further information' at the end of the chapter).

The factors affecting the formation and preservation of sites are varied, complex and differ from site to site. As more work is done in this area it becomes even harder to generalize. Sites that appear to be in a similar condition on the sea-bed, and therefore fit easily into the same category in terms of state of preservation, may have arrived at that condition through very different processes. No site can be fitted neatly into a precisely defined category. To try to do this is to over-simplify the nature of archaeological material. But, as long as they are used sensibly, classification systems have much to offer in terms of formalizing vague ideas and theories within a framework (Gibbins, 1990).

It has also been shown that careful search and systematic survey can produce results that allow an interpretation of sites that apparently lack any pattern and are heavily contaminated by modern material (Parker, 1981). This means that although sites may be classified in terms of degree of survival, it does not necessarily imply that scattered sites deserve less attention or can be treated less sensitively. Information may be more difficult to extract from such sites but their potential has been amply demonstrated (Muckelroy, 1978; Tomalin et al., 2000). It might be suggested that the more scattered a site, the more careful the collection of the clues needs to be, because understanding the processes that scattered the site (and continue to modify it) becomes vital to its eventual interpretation.

THE RANGE OF EVIDENCE ON AN ARCHAEOLOGICAL SITE

As stated above, an archaeological site is a concentration of clues left by the hustle and bustle of life in the past. These clues exist in, and have been modified by, their surroundings and environment. The following section is a brief review of what makes up a site and the sort of information that can be extracted from the various clues. The evidence, in simple terms, comes in three groups: structure, sediments and contents. Sites come in many different forms but these basic components are the same.

Structures: What is the single most striking feature of a castle, a workshop or a merchant's ship? Very probably it is the structure – whether made from stone, brick or wood. Careful study of a structure can provide clues about levels of technology and methods of construction. What types of raw materials were used and where did they come from? What does this imply about the supply routes and transport systems available to carry these materials? The reasons for a particular construction method can be determined by studying the design: was it for defence or prestige and what does this suggest about the political situation at the time of its construction – a time of war or a period of peace and prosperity?

When examining a ship, the structural elements have much to reveal about the functional characteristics and performance of the vessel, vital to an understanding of its significance within the culture that produced it (e.g. speed, carrying capacity, manoeuvrability, and whether it could it be beached easily or even carried overland). Looking for faults or repairs in structures can also reveal much about the age, status and life-history of a building or vessel. In the case of a ship, perhaps even the reason for sinking can be determined. Just as buildings collapse through poor design today, not every design was successful in the past, so it pays to look critically at the evidence, though with due respect for past skills.

When examining prehistoric submerged landscape sites, the questions raised above in relation to technology, materials, resources and cross-cultural communication are still highly relevant. Although the 'structure' or 'feature' of a mesolithic hearth or a bronze-age trackway might not be as large or immediately obvious as the shipwreck example, such sites still have great potential to inform on periods of the human past.

Sediments: The arrival of a wrecked ship on the sea-bed is only one in a series of steps that turn the lost vessel into an archaeological site. Fortunately, clues that can help gain an understanding of changes on a site can be found in the sediments that cover it. For example, the nature and layering of sediments can provide information on the break-up of a vessel, movement of material within a site, later disturbance of the remains, and the stability of a site today. It is even possible to use the nature of sediments to help predict likely levels of preservation in different areas of a site.

On submerged landscape sites, the study of the nature and extent of sediments can reveal important archaeological and environmental evidence. As these sites were once dry land that has since been submerged, the sediments present have the potential to yield information on the nature and scale of inundation of the site and surrounding landscape, the possible effects on the people using that landscape, and implications for the preservation of the archaeological remains.

Contents: Within the structures and sediments are objects such as timbers, coins, pottery, flint-tools and bones.

They are important for the study of the past because they reflect how people were living their lives. They are the tangible remains of the views and knowledge of the people who made and used them to solve their problems.

Artefacts: Objects can shed light on the people of the time working, playing, worshipping, keeping warm, entertaining and decorating themselves. Characteristics such as shape, composition, method of manufacture, evidence of use of such items are all important. In addition, because objects operate with other objects and with their surroundings, their position on site and in relation to other objects (context) also provides important clues (see below).

The range of human-made objects is very large and readers will be familiar with the idea of dividing them into major groupings such as jugs made of clay, guns made of iron, shoes made of leather. It is beyond the scope of this book to explore these categories further but a wealth of information exists in current archaeological literature.

Ecofacts: A less obvious source of information among the contents of the structure and sediments are the non-artefactual remains, which are often referred to by archaeologists as ecofacts. Animal and plant remains associated with archaeological sites have become an enormous source of clues about the past. Insects, seeds, pollen, microscopic plants and animals, along with animal and human bones all provide evidence about the environment in which people lived (table 4.1). After all the quality of food, cleanliness, sanitation, pests, parasites, accidents and diseases in contemporary society, affect the way people live as much as the things they own. People in the past were no different.

Table 4.1   Types of ecofactual material that may be expected from archaeological sites and what can be learned from them. (After Spence, 1994, table 1 in section 3.2)

Material type	Information available
Human bone	Diet, disease, injuries, height, sex, lifestyles
Large mammal bone	Diet, husbandry, butchery, provisioning, disease
Small mammal bone	Natural fauna, ecology
Bird bone	Diet, natural fauna
Fish bone, scale	Species inhabiting the site or the remains of fishing activities, diet
Large molluscs (shellfish)	Diet, subsistence, trade, development of the site, shellfish, farming
Small molluscs (shellfish)	Past vegetation, local environmental conditions
Parasite eggs	Intestinal parasitic diseases, sanitation, identification of cesspits
Wood (charcoal)	Date (dendrochronology), climate, building materials and technology, fuel
Other plant remains, charred and uncharred (seeds, mosses, leaves, grain)	Vegetation, diet, plant materials used in building, crafts, technology, fuel, processing of crops
Pollen	Vegetation, land use, chronologies, container contents identification
Phytoliths	As above
Diatoms	Salinity and levels of water pollution
Sediment/soil	Information on how deposits were formed, development of the site

It is important to be aware that there may be evidence that is not always immediately visible. For example, insect fragments found on archaeological sites are usually in the size range of between 0.5 and 1.0 mm (0.02–0.04 in) and so are unlikely to be recognized during the excavation itself. Test samples of likely deposits should be taken and assessed without delay. It is important that the possible presence of environmental and scientific evidence is not overlooked, and that samples are taken of potentially valuable deposits (e.g. container and bilge contents).

Non-artefactual remains have received less attention in general and archaeological literature and readers may not be familiar with the range of material involved. For that reason the main groups are outlined below.

Animal remains: Animal remains appear on sites in a wide variety of forms. Ecofacts like bones are common and can provide much information on diet and, if examined for marks, on butchery practice and even organized supply systems. Animal hair is also a common component of weather- and waterproofing materials found on shipwreck sites (e.g. caulking). Fish bones found on an underwater site may be the remains of species that inhabited the site and as such they could be useful indicators of the characteristics of past environments. It is much more likely, however, that they are the remains of stored food, refuse or relate to fishing activities, particularly if they are found in any quantity on a shipwreck site. Both animal and fish bones can yield a great deal of information about diet and provisioning.

Human remains: If human remains are found on site, it is a legal requirement in many countries that the relevant government department is informed. The study of human bones by specialist palaeo-pathologists can yield information such as physique, sex, height and diet together with the identification of occupational diseases and injuries. Human bones may occur as burial groups on a flooded land site or as the remains of the crew on a shipwreck site. On well-preserved sites, material other than bone may survive (e.g. hair, tissue remains). Biological material that may be associated with human remains include stomach contents and coprolites (containing seeds, cereal fragments and parasite eggs).

Invertebrate remains: The study of insects, molluscs and parasites falls within the realm of 'invertebrate zooarchaeology'. The analysis of molluscs can provide information such as past climates and environments, diet and use as artefacts or tools. Molluscs have specific habitat requirements that reflect the contemporary environment and can be from land, rivers or the sea. Molluscs of

economic importance are usually found as food waste (e.g. oyster, whelk and mussel) although some may be collected for use in building materials or pottery, or to extract dyes. 'Single event' dumps can be analysed to determine the season of collection and information on the population being exploited (or even farmed).

The analysis of internal (endo-) and external (ecto-) parasites found within archaeological deposits can yield information on:

- the range and antiquity of various pests and diseases in both animals and human;
- the conditions under which people were living;
- the effect of these conditions and the parasites on peoples' health;
- determining the function of certain features (e.g. cesspits, bilges);
- examining methods of sewage disposal.

Internal parasites normally survive in anaerobic (without oxygen) deposits (e.g. cesspits) or are preserved in fossilized faeces (coprolites) in the form of ova (eggs). They consist of species that infect both humans and animals (e.g. tapeworm, whipworm). Examples of external parasites (e.g. fleas) have been recovered from wreck-sites. Other insect species can provide information on changes in local and regional climate, palaeo-environments, the infestation of food stores and an indication of the contemporary conditions (e.g. wet or dry).

Botanical material: Plant remains can be found on archaeological sites in a wide variety of contexts. Locational information and individual measurements together with the species identification can provide evidence of agricultural practices, pests/blights, provisioning, stowage and diet, nature and origin of cargo. A wide range of different plant components can be preserved including wood and bark, seeds (including fruit stones and grain), fungi and mosses. Ship's timbers can potentially reveal a great deal about past environments, timber resources and woodworking practices. Pollen analysis (palynology) is the study of pollen grains and spores, which have particularly resilient walls. Palynology can provide information about past environments and ecology, the dating of deposits, assessing the impact of humans on the environment and in certain cases the identification of residues within containers.

Phytoliths are microscopic particles of silica that occur within the cells of certain species of plants (especially the grasses) and as such they are useful aids to identification. They are particularly useful to the archaeologist because they survive when all other traces of the plant have disappeared and are also instrumental in imposing wear patterns on the cutting edges of tools such as scythes.

Information about botanical material has survived in some surprising ways. Imprints of grain and leaves have survived on ceramic vessels. Some of these impressions are so clear that the type of plant can be easily identified.

Micro-organisms: Micro-organisms, or microbes, include the bacteria and algae of the plant kingdom, the protozoa and viruses of the animal kingdom, and others which have some characteristics of both kingdoms (e.g. fungi). With the exception of viruses and a few other examples, micro-organisms, like plants and animals, consist of cells (unicellular and multicellular). Micro-organisms can survive in the archaeological record in a number of ways dependent on the nature of the organism's construction (some produce a resilient hard shell) and the nature of the burial deposit.

Diatoms are microscopic unicellular or colonial algae with a siliceous cell wall. They occur profusely in all moist and aquatic habitats in freshwater, brackish and marine environments. The study of diatoms in archaeology can yield information such as the nature of the environment, formation of different deposits and differing levels of salinity through time (Battarbee, 1988). Foraminifera are unicellular animals which secrete a test or skeleton. They are mainly marine benthic or planktonic forms, in which there is a considerable morphological variation, from a single, flask shape to complex chambered examples. Foraminifera are important zone fossils that can survive in a range of sediment types, providing information about changes of environment over time (e.g. variations in salinity in rivers and estuaries).

LINKS BETWEEN CATEGORIES OF EVIDENCE

Although a convenient way of thinking about the elements of a site, the categories of evidence do in fact merge with one another. A ship's hull is an object that combines artefactual and ecofactual information. Sediments can form part of the contents of the hull (e.g. ballast or bilge deposits). Sediments can also provide the evidence of structure that has long since decayed or been dug out for re-use (Adams, 1985). The contents and structure of the site can, like the sediments, show changes in the formation of the site over time. For example, the evidence of differential erosion of timber can often reveal past sequences of exposure and burial.

The types of evidence mentioned above will not all be present in every case. What should be remembered is that any investigation should involve the study and recording (see chapter 8) of all the surviving strands of evidence on an archaeological site. In the past, too much attention has been paid to the easily recognizable human-made objects. This is generally at the expense of the sometimes less glamorous, but equally important clues that often need a greater level of expertise to collect.

USING THE EVIDENCE

Once all the different clues have been collected and recorded, the next stage is to attempt to make sense of it all. This can be achieved if the clues are studied in a systematic and disciplined manner. Often methods of extracting information from archaeological material are adapted and adopted from other disciplines. This book cannot list all the techniques used in archaeology but by introducing some of the main techniques of 'getting answers' it can at least demonstrate what a broad-based discipline archaeology is. The methods conveniently split into: where (position and association), what (recognition, description and typology), how (context) and when (dating).

Position and association: Archaeologists are generally studying complicated elements that may have been used together. They therefore need to know where they were (their *position*) and what they were with (their *associations*). It would be extremely difficult to make sense of complex structures without an accurate plan and a description of the position and association of the various elements (see chapter 14).

In looking for clues about the past, the archaeologist has to make do with where things ended up; where they slid, fell, were carried or washed. It is vital, however, to record the position and associations for each clue so that archaeologists can attempt to determine where they originated and how they ended up in their final location.

Recognition, description and typology: How does an archaeologist identify what he/she has found? Some evidence will be immediately understood because it is within the archaeologist's own experience (e.g. 'I recognize that object as a sword'). Some clues cannot easily be identified because they are not immediately visible or because the particular analytical technique being employed is not suited to revealing them. In other situations, clues are not exploited simply because they are not recognized as being clues. *Recognition* has to be a co-operative process in which good communication, by publication as well as personal contact, is vital (see chapter 20).

The physical remains of the past are so complex that no one person has sufficient knowledge and experience to deal with every type of clue that is available. In fact the necessary specialist knowledge and techniques may not even be present in a team of researchers, but they are available somewhere. It is important to be aware of where and when relevant research is being conducted and to seek appropriate assistance when required. A simple maxim is: 'Everything that happened on this site has left a trace – it just needs recognition'.

A considerable number of clues can be found on every archaeological site, all over the world. How do archaeologists tell others about their evidence? Indeed, how do they remind themselves about their findings in ten years' time? Of course, it is vital to record a *description* of the clues, from the structure, sediments and contents, so that everyone can understand and use the evidence. If they are described reliably and consistently, clues can be divided not just into structure, sediments and contents but also into types of structure or sediment or contents. A sword is called a sword because it has certain characteristics that it shares with a group of edged metal weapons. That group can be called the sword group, providing a convenient and informative way of referring to all weapons sharing those characteristics. The sword group can be split into smaller groups in the same way (e.g. rapiers, sabres), based on common features shared by a particular group of swords within the general group.

This process of classification according to general type is called *typology*. The value of typology is that an unknown piece of structure or object with characteristics that coincide with a previously described type can contribute to all the research that has ever been done on that type of clue. This can include use, development, construction, date, origin, etc. A mystery find can be transformed from a headache to a source of information if the find is recorded and publicized appropriately.

Of course, it would be optimistic to assume that every typological series is totally correct. Such groups are usually built up using evidence from a wide range of sites. The more sites that produce evidence that supports the suggested typology, the more secure it becomes. Some typologies are based on very few finds and faulty assumptions. If a particular piece of evidence does not fit the accepted scheme, it should not be ignored or altered to make it fit. The information may prove important in improving the typology.

Context:　On any archaeological site, it is important to consider how the clues arrived there and how the site was formed. The dictionary defines 'context' as 'the circumstances in which an event occurs'. In archaeology, 'context' has taken on a particular and specific meaning that is central to the study of archaeological sites. It has come to mean the individual, recognizable steps in the build-up of a site.

The following is an example of how an archaeologist might detect a grave. When the grave is initially created, the original soil or rock is dug away, a body is placed in the hole and the soil or rock placed back in a more jumbled state. From the surface a slight difference in colour and texture may be apparent, showing the location of the filled-in grave. Excavation will reveal the fill of the grave within the hole, which was cut out of the original soil or rock, whose cut sides will be clear. Under the fill of the grave, but lying over the bottom of the hole, are the remains of the body. The original process of burial (*cut–body–fill*) are reflected in the archaeological remains.

This simple site has three steps; three sets of relevant circumstances or contexts:

1　*Cut*　Whatever was on the site before the grave (perhaps other graves) has been disturbed and dug away. Evidence has been lost (or moved and mixed up) as well as created. Different-shaped graves were dug at different times in the past. Bodies can also be placed in holes dug for other purposes. These holes can be left open for a long time before being filled in.
2　*Body*　As well as the remains of the body, other things such as 'gifts' for the afterlife, wrapping or coffins can be placed within the grave. These often reflect a particular set of beliefs. Sometimes two or more bodies are placed together at the same time. These activities are taking place well below the ground surface – possibly at the level of much older layers.
3　*Fill*　The excavated material is returned to the hole, possibly including any material from earlier graves or remains which were disturbed. Although material from earlier periods may be found in the fill of the grave, material in use after the grave was filled in should not appear. Some holes are filled in in stages or left to fill in naturally over time. Therefore they may have several different fills, reflecting the different methods and speeds of infilling. Air spaces in coffins or burial chambers may collapse and only be visible in the subsidence of the layers of fill above. Some objects, even some bodies, only survive as stains or fragments at the bottom of the fill.

The whole process is then repeated when the next grave dug. Some graveyards have been used for hundreds of years. Through the process of archaeological excavation, which examines each context in order, the complex series of inter-cutting graves or contexts can be broken down into the sequence of *cut–body–fill*.

The example of a simple grave indicates the importance of understanding contexts. By looking not just at the objects (the bones, the coffin or the grave goods), much more may be learnt. For example:

- why information is missing (dug away);
- why information appears (appears at bottom of a hole dug into surrounding layers);
- why information cannot be used directly (dug away and re-deposited later);
- unusual information (odd shaped hole, unusual orientation, not refilled quickly);
- unusual information arrived at same time (two bodies, mother/child, male/female – i.e. *cut–bodies–fill*);
- dating information (shape, orientation and depth of hole);
- social information (shape, orientation and depth of hole, location of body and artefacts in hole);
- artefact information (stains, fragments);
- structural information (stains, fragments, subsidence).

The human and natural processes that have the potential to disturb, remove or cover archaeological remains do not only occur at sites such as graveyards. Cultural and natural processes are constantly occurring across the landscape, foreshore and sea-bed. Through maintaining detailed archaeological records of 'contexts', it is possible to recognize processes that have affected the remains. This type of information is important for the full interpretation of archaeological sites. The detailed examination of contexts is just as relevant under water as it is on land. For example, throughout the life of a vessel, activities add and remove material (e.g. loading and unloading of crew, cargo and equipment, repairs, changes in design). Even before a vessel sinks, it is a patchwork of different events or occurrences giving evidence of relevant circumstances or contexts. On sinking, all this evidence is taken to the sea-bed where a whole new series of processes affect the ship and turn it into the archaeological site encountered today. Scouring, silting up, collapse, salvage, disturbance by trawling, burrowing organisms, looting, etc. are all processes that can affect the archaeological evidence.

DATING

Since archaeologists are studying the past and the passage of time, one of the main things they want to extract from clues is the point in time to which they relate. Chronologies or timescales provide the ability to relate events or features throughout antiquity and across the world. Widely separated cultures such as the South American civilizations and those in the Old World can be compared if dates exist for each. In the same way, through the use of techniques such as tree-ring dating (dendrochronology) and radiocarbon dating, sites such as submerged prehistoric settlements can be compared across the world. Dating techniques can be grouped into two main categories, absolute and relative dating, which reflect the ways in which the particular methods can be related to the present day.

Absolute dating

Methods of absolute dating can be related to calendar years and therefore the results of these techniques can be directly related to the present day. To say that an event happened 900 years ago is to give it an absolute date. Absolute dating techniques often require specialist scientific analysis. The range of dating techniques available for different evidence types is growing as new methods and approaches are developed. This section introduces three commonly used methods of absolute dating to demonstrate their potential and to show some of the associated problems.

Dendrochronology (tree-ring dating): Many wet sites will produce large quantities of wood, which fortunately can sometimes help in providing dating evidence. As trees grow they produce annual rings whose width varies according to the local conditions (figure 4.5). This pattern is similar amongst trees of the same type in the same area. This means that the same years can be recognized in individual trees. Overlapping the tree-rings from trees of slightly different dates can extend the sequence of years. This has been done for oak until the sequence extends from the present to, in some areas, 9000 years ago. Tree-ring dating is based on matching the pattern of growth rings found on a wood sample from a site with its place in the established local sequence of variation in growth ring size (figure 4.6).

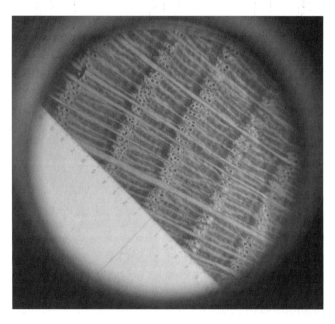

Figure 4.5 Tree-rings viewed through a microscope. (Photo: Mark Beattie-Edwards)

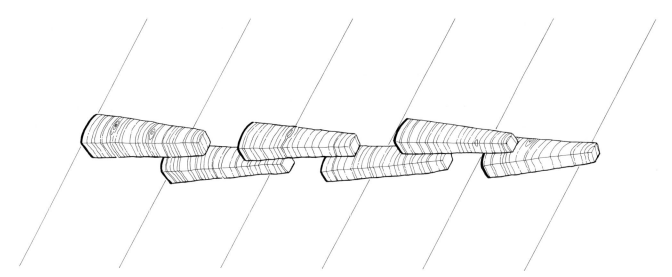

Figure 4.6 Tree-ring sequence. By using years of growth shown in the rings of individual trees from the same area, a year-by-year sequence can be constructed back into the past. (After Eckstein et al., 1984:10; drawing by Graham Scott)

Tree-ring dating can give very precise results in optimum circumstances, perhaps even to a season in a single year. However, once the sample has been matched with a point in the sequence it is important to understand what the resulting date actually means. Does the sample come from a context where a date would be useful to the understanding of the site? The wood could have been introduced on site any time between the site being formed and its discovery. Similarly, wood is often re-used. Ships' timbers are frequently found forming part of waterfront structures and other buildings, so care must be taken in how dating information gained from such elements is used. Ships and wooden structures in general require frequent repairs. A wood sample taken from a repaired area might be significantly later in date than the rest of the structure. The sample may have been found in a ship in UK waters, but the wood may have come from elsewhere. Also, it is very important to remember that the date provided by dendrochronology relates to the growing period of the tree (the period in which the rings were formed) and not its arrival on site. Timber can be stored for long periods to allow for seasoning before it is used. One also has to bear in mind that not all wood samples will be datable, even when a suitable number of growth rings exist. Sometimes a match cannot be found and dating is therefore unsuccessful.

Despite these problems, dendrochronology is sufficiently accurate to be used to check, or calibrate, other dating methods such as radiocarbon dating. Recommended procedures for dendrochronological sampling are given in chapter 15.

Radiocarbon dating: Radiocarbon dating is based on the known rate of decay of a radioactive carbon-14 isotope, which occurs in very small quantities in all living things. For the dating process, all that is needed is a sample of carbon that can be directly related to the evidence that requires a date. During their lives all living things absorb the naturally occurring carbon-14 isotope. The amount of isotope absorbed depends on its level of concentration in the organism's surroundings. When the organism dies, the carbon-14 level starts to reduce as the isotope decays. Since the rate at which this particular isotope decays is known, specialists can make an accurate assessment of how long ago the organism died by measuring the remaining amount, assuming that the levels of the isotope in the organism's surroundings have remained constant.

The method is generally less precise than dendrochronology but can be used to date much earlier material. Radiocarbon measurement is normally effective back to 30,000–40,000 BP (before present; i.e. before 1950) and up to 60,000 BP is possible. However, as with dendrochronology, radiocarbon dating relates to the living period of the organism, not to its arrival or use on site, so the same degree of care has to be used when the dates are interpreted.

One of the basic assumptions with radiocarbon dating has been that the amount of radiocarbon in the atmosphere has remained constant throughout time. Calibration work using techniques such as dendrochronology suggests that the level has in fact fluctuated and that calibration of all radiocarbon dates is now necessary to give an approximate date in calendar years.

Radiocarbon dating results include the following information that identifies uncertainties in the date given:

- Radiocarbon dates are normally quoted giving an error margin. For example, a date may be quoted as 1764 ± 100 years. This is a statement of statistical confidence of one standard deviation, meaning that there is a 68 per cent chance of the actual date lying within the range 1664–1864 years ago. Higher levels, such as two standard deviations, meaning a 95 per cent chance, or three standard deviations, meaning a 99 per cent chance, mean doubling and tripling the limits respectively, so the example above would be ±200 or ±300 years. Therefore, you can have a better chance of being within the limits, but the limits become wider.

- Radiocarbon dates can be quoted as either before present (BP, present being taken as 1950) or in calendar years (i.e. BC or AD).
- If the radiocarbon date has been calibrated (reassessed in relation to another dating system such as dendrochronology) then this is often indicated by adding 'cal' (i.e. 'cal BP').
- Methods of measuring and calibrating dates also vary between different laboratories. The name of the laboratory and the methods used are therefore likely to appear with published dates.

It is important that factors such as the meaning of the ± statistical confidence, the method of calibration and the actual radiocarbon technique used are all taken into account and fully understood when using radiocarbon dates. Recommended procedures for radiocarbon sampling are given in chapter 15.

Historical association: While the above techniques are relatively complicated methods of gaining an absolute date, the comparatively simple technique of using written records should not be overlooked. Written records allow historical association to provide archaeological clues with absolute dates.

While historical dates seem very attractive they do have their problems. Reliable written documents only go back a relatively short time into our past. Where they do exist, it is sometimes possible to misuse them or look to them for easy solutions. For example, written records provide information about the Spanish Armada but the fact that the event occurred and is recorded does not mean that every sixteenth-century vessel found in UK waters was associated with it. In addition, historical documents are the record of witnesses, some of whom may be biased or simply ignorant. Their accuracy should not be taken for granted. In the present day, reading a selection of daily newspapers can reveal conflicting interpretations of the same event, and the same can be true of other historical documents. Archaeological dates should not be ignored simply because they conflict with documentary evidence.

In addition, as with all dates, direct historical associations relate to one point in the development of the site – in the case of a shipwreck, the arrival of the material on the sea-bed. Subsequent events on the site must be dated independently. A problem can also arise with coins or cannon, which appear to give a clear date and historical association because of the inscriptions on them. A coin found on a site may commemorate a particular ruler, the period of whose reign is known from historical sources. This may well provide a good absolute date for the minting of the coin, but it gives little secure information about the date of the coin's loss or burial. Clearly it cannot have been lost before it was minted (this date is called a *terminus post quem*), but it could have been lost a long time afterwards.

Sources of further information on these and other absolute dating methods are given at the end of this chapter.

Relative dating

Relative dating can only indicate whether one process occurred before or after another one. It cannot reveal the length of time between the two events and neither can it provide a date in years that places the event in a conventional timescale. However, it is very useful for determining whether the information was deposited early or late in the development of the site, and for providing a framework into which absolute dates can be placed.

Typological dating: The value of typology as an aid to research has been noted above, but it also has a role to play as a form of relative dating. The form of objects designed to perform the same function often changes over time. If earlier and later characteristics can be recognized it is possible to reconstruct the sequence of development and give each object a relative position within it (figure 4.7). There is a real danger of such sequences being unsound because they are based on assumptions of early and late characteristics. Therefore, as much evidence as possible should be introduced to support any conclusions drawn.

Stratigraphic dating: It has already been shown how contexts can be used to identify the events in the history of a site. It is a simple process to start to recognize and study the order in which they occurred. This gives us a sequence for the events and a relative dating technique.

The ordering of contexts is known as a site's stratigraphic record. The study of this record or sequence is known as stratigraphy. The most basic principle of stratigraphy,

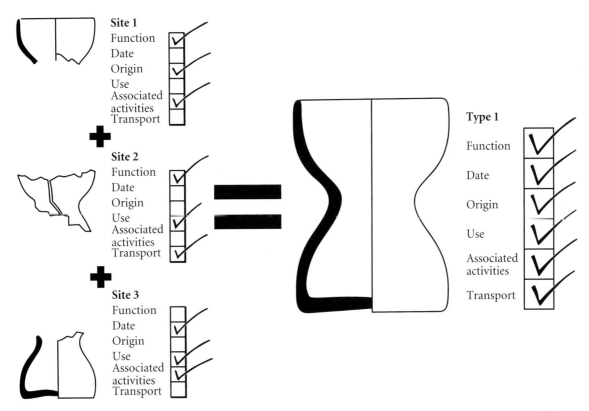

Figure 4.7 Typology: knowledge about the appearance and character of a type of object is accumulated from a number of sites, which each contribute different elements. (Based on original artwork by Kit Watson)

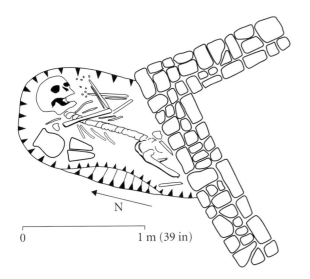

Figure 4.8 Stratigraphy from above: from the overlap it can be concluded that the grave is earlier than the wall. (Based on original artwork by Ben Ferrari)

which was adapted from the study of geological strata, can be summarized by the concept that a context which physically overlays another context is the later (figure 4.8 and 4.9).

On a particular site, the principles of stratigraphy are used to establish a sequence of above–below relationships, thereby placing all the contexts (and, therefore, events) into the order in which they occurred. Stratification can be studied at different scales using the same basic theory. Examining the layering of contexts in a scour pit may reveal many clues about large-scale changes to a deposit. Applying the same approach to the sediments between individual timbers can be just as revealing about equally fundamental processes in the formation of the site.

The principles of stratigraphy provide a framework within which archaeological investigations are conducted. They do not impose rigid boundaries on the way in which stratification is investigated. Applying them effectively requires a willingness to combine such principles with a good understanding of the nature of the contexts under investigation. For example, the nature of stratification in mobile sediments is likely to be very different to that in stable contexts. The application of stratigraphy to a site on a rocky sea-bed will not be exactly the same as its application to a deep urban deposit on land. But the value of the exercise, its aims and fundamental principles, will be precisely the same (figure 4.10). See chapter 8 for information about how to record stratigraphy (Harris matrix).

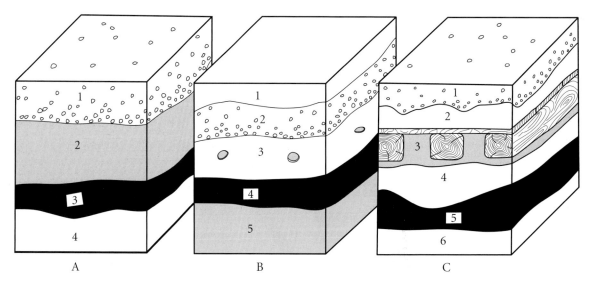

Figure 4.9 Stratigraphy: A) Context 1 is later than context 2. Absolute dates can be placed in the relative dating sequence. B) A coin dated AD 79 in context 3 indicates that 3, 2 and 1 arrived after AD 79. C) A floor constructed in AD 1322 indicates that the contexts below it must have accumulated before then. (Based on original artwork by Ben Ferrari)

Figure 4.10 The undisciplined recovery of material destroys evidence. This diver is only interested in the two metal objects. Many other clues have been destroyed along with much of the archaeological value of the finds themselves. (Drawing by Graham Scott)

ENVIRONMENT AND SITE-FORMATION PROCESSES

This book has so far introduced both the sources of evidence and some of the methods used to extract informa-

tion from those clues. In this section the site is viewed from a different angle. Before using evidence to build up a picture of the past, it is vital to develop an understanding of the processes that shaped the clues.

The following questions should be considered:

- What are the fundamental factors which shaped the past that is now being studied?
- What affects the way in which an object becomes part of the site?
- How does the evidence survive until the site is investigated?
- What are the biases and imbalances that such processes introduce into the evidence?
- Are these processes detectable and so understandable?

All these questions have to be addressed in order to use archaeological evidence to investigate complex aspects of past societies such as behaviour and social organization. These processes are not muddling factors to be filtered out in the final report. Their study is fundamental to archaeological research, not merely an interesting diversion from the main lines of inquiry (figure 4.11).

Environment is important because it affects the way people live and the survival of the clues they leave behind. Knowledge of the environment is more than just an aesthetic backdrop to events. Many human activities are centred on solving problems set by their surroundings. Therefore, much of the past being studied is a reaction to the environment. Climate, vegetation, wild animals, crops, water are all vital components that have to be studied before human activities can be explained.

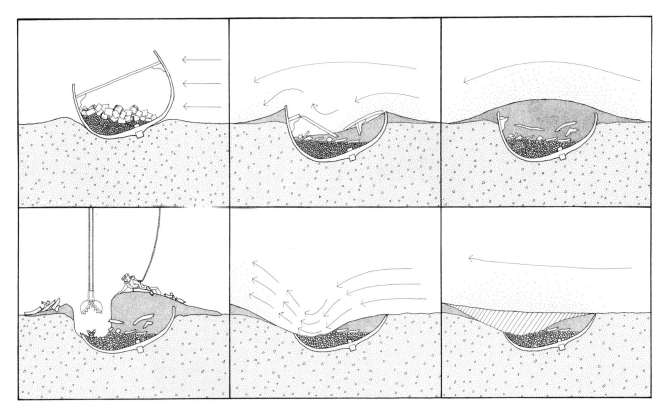

Figure 4.11 Simplified version of the site-formation process (top left to bottom right). The vessel lies across the prevailing current, which results in mechanical and biological degradation, leading to the breakdown of the superstructure and localized scouring. At the same time, the vessel sinks deeper into the sea-bed. In time the site stabilizes until human interference, which results in new scour patterns and infill that should be easily identifiable in the stratigraphic record. (Drawing by Graham Scott)

While a site was active or occupied, its environment would have influenced many aspects of life (e.g. the form of structures, the clothes that were worn, available sources of food). The environment of an area can change dramatically over time and reconstructing past landscapes is fundamental to interpreting the remains of past societies. For example, temporary campsites linked to ice-age migrations have been found in what are now temperate regions. Clearly they are not understandable by reference to their current environment. Boat-finds have been made in what appear to be landlocked areas. Studies of changes in the local environment have often shown that the area was previously closer to the sea or an inland waterway and so the find can be put in its proper context.

Archaeology under water, in the same way as archaeology on land, is likely to be concerned with the study of reactions to environmental factors. The sea, inland waterways and lakes can be seen to have had considerable influence on human populations. Seas and rivers offered a plentiful supply of food, but to obtain this food reliably, certain problems had to be solved: the construction of boats, fishing equipment, fish processing and storage techniques. Water offers the opportunity for travel and communication or trade, but again to take advantage of the potential opportunities, humans had to develop methods of surviving in that environment.

Having shaped activities in the past, the environment then shapes how evidence of the past survives. This can vary in scale from chemical changes in soil to erosion by sea or river action.

The environment can often be a major initial influence in terms of deposition. Material may move into the archaeological record from sites occupied in areas prone to flooding, earthquake or volcanic eruption in a very different way from those in a stable, temperate zone. The aquatic environment's influence on deposition can be very significant. The most obvious and dramatic example is that of a shipwreck. The sinking of a vessel results in a group of associated material arriving on the sea-bed in one event, although of course it may be scattered to varying degrees. This factor can be very useful for archaeologists, as has been discussed above. Water can also cause deposition by abandonment: rising sea-levels can force occupation sites to be abandoned. However, this process of deposition will take place over a period of time as opposed to the short-lived but intense process of a ship-wreck. This may mean that less material is eventually deposited, as the occupants will have the opportunity to remove what they want. However, the process of inundation is likely to be far gentler than the process of ship-wreck, so more material may survive very near to where it was actually used and in association with related objects rather than being scattered.

As soon as objects or structures have fallen out of use or are lost to become part of the archaeological record, the environment remains important in helping to determine what evidence survives, in what form and in what position. Certain specific conditions will promote the survival of particular material types. In general, the more robust materials (such as stone and pottery) survive better than others (such as wood, textile or leather). The more aggressive the environment, the less well will delicate materials survive. Studying the nature and impact of the environment of a site is vital to understanding the evidence that may eventually be recovered.

The deterioration of organic and inorganic materials on an underwater site is dependent on physical, chemical and biological factors. The initial formation of a wreck-site will of course be dependent on physical processes such as the nature of the wrecking, the area the ship is wrecked in (e.g. geographical and topographical) and type of sea-bed the wreck is finally deposited on or in. Thereafter, physical processes still play an important part in the formation of the wreck. The marine environment is dynamic; wave or tidal action and currents will affect the sedimentary regimes around a wreck-site and may cause scouring or silting around a site. Once the wreck has to some extent stabilized on or in the sea-bed, chemical and biological processes come into play and affect the long-term preservation of the wreck-site. These processes will of course differ depending on the material in question and there is certainly a need for further research to investigate the deterioration of all the different types of materials encountered on underwater sites. However, one material that is commonly encountered on underwater sites is wood and the processes of its deterioration serves as a good example to highlight the complex interactions of chemical and biological processes in the underwater and marine environment.

THE DETERIORATION OF WOOD

Wood exposed to seawater is rapidly colonized by a variety of biological agents (rather than chemical agents), including seaweed, barnacles, wood-boring molluscs (shipworm) and crustacea (gribble), fungi and bacteria.

Organisms such as seaweeds and barnacles will not cause extensive deterioration because they are merely using the wood as a substrate to attach to and not as a source of nutrients. However, even these can decrease the archaeological value of an artefact, as they will degrade the object's surface. The major problem under water is the activity of the wood-borers, fungi and bacteria. Under the right conditions, they will all rapidly colonize and utilize various components of the wood as part of their respiration. This colonization, if left unchecked, will lead to the complete deterioration of wood in a matter of years or decades rather than centuries.

This gives rise to the interesting question: why is archaeological wood preserved at all? The answer lies in the fact that the organisms that cause deterioration have particular environmental requirements for their survival. Shipworm (*Teredo* spp.) and gribble (*Limnoria* spp.) are among the wood-borers that cause the most deterioration (and concern). All wood-borers require specific salinity, temperature, depth and dissolved oxygen in order to respire and grow. As shown by the many well-preserved wrecks of the Baltic, salinity and temperature are important factors. In the Baltic, both of these parameters are too low to sustain wood-borer activity and as a result there are some fine examples of organic preservation (plate 4.3). However, of paramount importance to wood-borers is the dissolved oxygen within seawater. Even with optimal temperature and salinity, without dissolved oxygen the various wood-boring organisms cannot respire. Thus, if a wooden artefact is rapidly buried in sediments, or lies in anoxic waters, attack by these organisms will be limited.

This leaves us with the fungi and bacteria. These micro-organisms will utilize the various celluloses and lignin within the cell wall as part of their respiration. Fungi are the most destructive of these micro-organisms and it is the so-called Brown and White rot fungi (*Basidiomycetes*) which are more than likely the cause for deterioration of wooden finds in non-waterlogged terrestrial contexts such as the Viking-age ship-grave at Ladby in Denmark, the remains of which were merely an imprint of the former ship in the ground. This is because certain species are capable of completely destroying wood by utilizing not only the cellulose within the cell wall but also the lignin, which is the backbone of wood. These fungi are similarly influenced by environmental factors. In their case, adequate moisture and oxygen are essential for decomposition to occur. Most Brown and White rot fungi will tolerate low oxygen concentrations but they will not grow under anaerobic conditions and, importantly, they will not tolerate water-logged conditions. In addition to the *Basidiomycetes*, what is termed 'soft rot' can be caused by *Ascomycetes* and *Fungi imperfecti*. In most cases, deterioration by soft rot fungi is confined to the cellulose of the wood cell wall while the lignin is often not degraded. Observations of soft rot in waterlogged wood suggest that the responsible fungi are able to attack wood at levels of oxygen lower even than those required by wood-degrading *Basidiomycetes*.

Bacteria are ubiquitous in nature and can survive in environments with or without oxygen; some are even capable of living in both types of environment. Although many different bacteria can be found on wood, not all species possess the necessary enzymes to degrade the wood. Those that can cause deterioration may degrade only certain pit membranes within the wood, while others have the capacity to degrade lignin and cellulose within the cell wall. Bacteria that can degrade woody cell walls produce specific degradation patterns and have been separated into three forms: cavitation, tunnelling and erosion. Cavitation bacteria form cavities that often develop perpendicular to the long axis of the cell wall. Tunnelling bacteria penetrate the secondary wall and form minute tunnels. Erosion bacteria are characterized by erosion of the cell wall from the lumen toward the middle lamellae. Very little is known about the influence of environmental factors on bacterial decay and the occurrence of bacterial degradation of wood under completely anaerobic conditions has not been unequivocally demonstrated. The occurrence of bacterial attack on wood in waterlogged conditions that could be characterized as near anaerobic (e.g. buried within sediments) suggests that erosion bacteria are the main bacterial cause of deterioration and are capable of degrading wood in situations where fungi are completely excluded because of the limited oxygen content. Fortunately, these bacteria are only able to attack the cellulose within the cell wall of the wood and leave the lignin in the middle lamella behind.

From a conservator's perspective, waterlogged archaeological wood is generally considered to be poorly preserved. It is often only the water that has replaced the degraded cellulose that provides the shape of the surviving wood and the wood cell is only held together by the skeleton of lignin remaining in the middle lamella. However, from an archaeological perspective, the fact that the wood has been deposited within an environment that limits biological attack leaves us with the chance of finding well-preserved archaeological clues about our past.

Although this section has only dealt with wood, it demonstrates how important it is to understand the type of archaeological material being studied and what factors will affect its deterioration. Only by understanding this will it be possible to understand the effects the under-water environment will have on preserving archaeological clues about the past.

CULTURE AND SITE-FORMATION PROCESSES

As well as the physical environment, it is important to consider how cultural aspects affect site formation. It would be far too simplistic to suggest that an understanding of the environment leads to an understanding of humankind. One is not the only factor determining the behaviour of the other. It could be said that people do not react to the environment but rather they react to their view of the environment. Levels of technology will shape the response to environmental challenges. Factors such as religion will shape attitudes (e.g. to issues like the eating of meat or birth control). A consideration of all the motives and actions that go together to make up 'human nature' clarifies just how complex the problem of understanding cultural influences on the archaeological record can be.

How does an object go from being used to being recorded as part of an archaeological site? The idea of throwing out rubbish or trash is a familiar one. Archaeologists are often experienced in painstakingly investigating the equivalent of ancient dustbins. A great deal of archaeological material was buried through being discarded, thrown into pits or left in abandoned buildings that collapsed around them. Items are regularly lost by accident – coins, wallets, car keys. Sometimes they are found while on other occasions they may lie where they fell for many years. The same has been true since before people first had holes in their pockets. Isolated, accidental losses – sometimes traumatic, such as during a fire or battle – are therefore responsible for much material entering the archaeological record.

Deliberate burial of groups of associated material is much less common. There are examples of material buried for posterity like the 'time capsules' buried by schoolchildren. Graves may seem a more common example of burial for posterity, but in many circumstances this could be seen as another example of rubbish disposal. Hoards of coins or other valuables are sometimes deliberately buried for safekeeping. However, it could also be said that those that are found are examples of 'accident' since the owner was not able to reclaim them once the trouble had passed.

There is considerable evidence of ritual activity around lakes and rivers (Bradley, 1990). The superstitious nature of fishermen and sailors is also well attested and is often considered to be a cultural response to the uncertainties of the environment in which they operate. The carrying of talismans for good luck, or the deliberate

Figure 4.12 Re-used ship's timbers in an open barn on the Turks and Caicos Islands. The heavy transverse beam is part of a keel, showing a hook scarph. Roman numerals (XIX) identify the mating pieces. Below is a small knee. (Photo: Colin Martin)

deposition of an ornament or weapon into a river or the sea to appease the spirits, may both be represented in the archaeological record, but will these objects be understood when they are found? Moreover, there is some debate as to which objects were deliberately placed in the water and which have eroded out of riverbanks or lakesides. The evidence is unlikely to allow any simple or easy interpretation.

The effect of the environment on humans has been considered above but another major source of evidence is the traces left by humankind's effect on the environment. The changes that can be studied range in scale from the disturbance of ground, building of houses or digging of pits, to large-scale deforestation. Much of this evidence is again created 'accidentally' by human activity and so is one step removed from the activities of real interest. This evidence may not, therefore, seem immediately relevant until its causes are traced back. Of course, to appreciate the impact of humans on the environment it is also necessary to understand the character of the environment before it was changed.

Occupation sites are often inhabited for long periods and activities carried out on the site will vary over time, leaving behind often very complex sets of clues. These clues provide a record of the changes. It can be extremely valuable to study long periods of occupation in this way, precisely because changes and continuities in the society concerned should be detectable in the evidence. Yet in this situation the surviving clues do not represent a total picture of the site at any one moment. Clues left by earlier occupation might be altered or destroyed by later activities on the same site, such as pit-digging

or the preparation of deep foundations for modern buildings.

The recycling of material is also an important factor in modifying the evidence archaeologists eventually study (figure 4.12). There is often a conscious selection of what is taken away and what is left behind. This will depend on many factors. The occupants might only remove what they consider valuable (which will not necessarily be the same things that are considered valuable today). The material removed may depend on what can be carried with the available means of transport. Perhaps objects will be selected on the basis of sentimental or ritual value.

It is very difficult to define all the processes that might result in material being removed from a site, but it is important to consider as wide a range of potential factors as possible. Underwater sites are no different from land sites. On submerged settlement sites, after inundation it may be more difficult for material to be recycled or disturbed by later activity but they will have been modified by long periods of habitation before they are inundated. Shipwrecks may result in a group of closely associated materials being deposited together. However, as surviving documentary records make clear, many vessels have been partially or wholly salvaged, involving selective removal of material from the site. Use of the seabed for fishing, anchorage or dredging will remove and add material. Finally, sites are known where a shipwreck lies on top of prehistoric remains (and sites comprising multiple shipwrecks), producing the effect of later activity on a site blurring the clues left by earlier occupation (Murphy, 1990).

The influence of environment and the influence of cultural factors are interlinked. It would, for example, be difficult to interpret an umbrella successfully without reference to the environment (in this case shelter from the rain or sun). However, a detailed study of the environment will not reveal why the umbrella was coloured red, green and yellow or had a certain design on the handle – these factors may simply be dictated by personal or community preference.

The fact remains that most of the material found buried in archaeological sites represents the rubbish and chance losses which have survived retrieval by later occupants and the various natural processes that cause objects to deteriorate. Any investigation that tries to use archaeological evidence by assuming that a site, even a submerged one, has been free of the processes and mechanisms that modify the way that clues appear is adopting a very simplistic approach. Complex questions must be asked of the evidence and any biases within the evidence must be rigorously evaluated. To achieve this, it is essential to document carefully the nature of the processes that interact to form the archaeological record.

FURTHER INFORMATION

For a list of standardized terminology relating to maritime craft type, maritime cargo and maritime place-names, see thesaurus.english-heritage.org.uk/newuser.htm

Adams, J., 1985, *Sea Venture*, a second interim report, part 1, *International Journal of Nautical Archaeology* **14**.4, 275–99.

Battarbee, R. W., 1988, The use of diatom analysis in archaeology: a review, *Journal of Archaeological Science* **15**, 621–44.

Bradley, R., 1990, *The Passage of Arms: An Archaeological Analysis of Prehistoric Hoards and Votive Deposits.* Cambridge.

Gibbins, D., 1990, Analytical approaches in maritime archaeology: a Mediterranean perspective, *Antiquity* **64**, 376–89.

Greene, K., 2002 (4th edn), *Archaeology: An Introduction.* Oxford (www.staff.ncl.ac.uk/kevin.greene/wintro3/).

Muckelroy, K., 1978, *Maritime Archaeology.* Cambridge.

Murphy, L. E., 1990, *8SL17: Natural Site Formation Processes of a Multiple-Component Underwater Site in Florida*, Submerged Resources Center Professional Report No. 12, National Park Service, Santa Fe, New Mexico.

Parker, A. J., 1981, Stratification and contamination in ancient Mediterranean shipwrecks, *International Journal of Nautical Archaology* **10**, 4, 309–35.

Tomalin, D. J., Simpson, P. and Bingeman J. M., 2000, Excavation versus sustainability *in situ*: a conclusion on 25 years of archaeological investigations at Goose Rock, a designated historic wreck-site at the Needles, Isle of Wight, England, *International Journal of Nautical Archaeology* **29**.1, 3–42.

5 Project Planning

The process of archaeological investigation is a collection of relatively simple tasks that can only be efficiently and safely undertaken if they are adequately resourced and carried out in the correct order, with appropriate use of specialist knowledge. With the ultimate aim of publicizing results (see chapter 20), any project that involves archaeology must ensure that staff, resources and specialist knowledge are in place at the right time. This requires efficient and effective project planning.

THE PROJECT DESIGN

Any project should start with a project design. This is a document that details how the major parts of the project will work together to try to address either the central research questions that the project aims to answer or the way in which the site is to be recorded to mitigate its destruction.

A project design should detail the aims of the project, the objectives required to achieve these aims and make an assessment of the human, material and financial resources required to support these operations, including any particular specialist facilities and expertise. It should also cover health and safety issues and contain a thorough risk assessment of the work to be undertaken. If any diving is to take place, a dedicated diving project plan must be prepared (see chapter 6).

As well as providing a key project-planning tool, a thorough project design is likely to be a prerequisite for any application for funding or permission to work on a site that is protected by heritage legislation. There are a number of formalized approaches to producing a project design. The most common in use in the United Kingdom is English Heritage's *Management of Research Projects in the Historic Environment* (MoRPHE) (English Heritage, 2006b).

The project design should start with a 'desk-based assessment'. This is a study aimed at establishing the current archaeological knowledge for a site or area that is to be investigated. It should establish, as far as possible, the character, extent, date, integrity and relative quality of the archaeological resource. It should also make an assessment of on-site environmental conditions. These will be important for the planning and safe management of field operations and will influence a site's preservation. The desk-based assessment should assemble, collate and synthesize all available data, consider its relative importance and comment on its reliability. Ultimately it will form the basis of a full project design.

In the case of major archaeological projects it may be necessary to undertake a number of surveys or limited excavation projects. These are termed evaluations and aim to increase knowledge and enable sufficient planning. Evaluations are archaeological projects in their own right and should be approached in the same way as a major project.

The project design should detail the following issues.

Background: This should clarify the area to be investigated and should consist of an area defined by geographic co-ordinates and shown as a polygon (usually a rectangle) on a map. The age of the site should be stated along with the nature of any archaeological deposits known to be present on the site.

Previous work: This should include a précis of previous work conducted on the site, including the location of existing site archives (if any) and the extent to which they have been consulted. This section should also detail the results of any previous work in such a way that they are easily accessible to the current project.

Project details: This should include information about the legal status of the site (if the site is protected or controlled by any form of legislation) and what permissions are in place, or are required, to allow the work to take place. It should detail the project timescales, dates, any arrangements for access agreed with landowners and site occupiers (if required) and the proposed reinstatement of the site (after any excavation).

Archive deposition: It is vital that consideration is given to the ultimate location and curation of the project archive (see chapter 19) at the project planning stage. Details of recipients of finds and records, curatorial arrangements and associated information should be included in the project design.

Justification for the project: This section will explain why the project should take place. It should detail what research questions the project aims to answer and balance the destruction of the archaeological record with the added knowledge the project will bring to our understanding of the past. Where work is prompted by an external threat to the archaeological record, such as coastal erosion, or development, this section should provide an assessment of the nature and scale of the threat (short- and long-term) and detail how the proposed archaeological work will mitigate this.

Publication and presentation: This section will summarize the likely format of the published report (e.g. monograph, article or note), the intended place of publication and any arrangements made for display and public access to the site archive.

Methods statement: This section should explain what work is to occur on site and how the data required to answer the research questions will be collected and recorded. It should justify why these are the best methods for the task in hand and, if the methods are experimental, why it is considered that they will work in this case.

The following should be considered in the formulation of a methods statement:

- the components of a site which will be investigated and those which will not;
- the different types of data-gathering methods to be used;
- the recovery and recording strategies to be used;
- any discard policies (these should be related to the different classes of data anticipated from the site – e.g. structural elements, artefacts or environmental material);
- the necessity for developmental work (for those cases where a suitable methodology does not appear to exist);
- estimated post-fieldwork activity (although this will almost inevitably alter once fieldwork has been completed).

Resources and programming: This section should detail the structure and size of the project team and the levels of expertise represented by its members. It should match the team's expertise to the needs detailed in the methods statement and identify whether any further training is required prior to work taking place. Whether professionals or volunteers, each team member should have a clear understanding of exactly what his/her role is in the project and be competent to carry it out. Prior to the start of the project, each team member must have read and understood the project design. They should know where their role fits within the entire process. In a discipline that is completely reliant on teamwork and where the rewards are often more intellectual than material, it is important that each individual gains as much from the process as can reasonably be expected. For each team member, due credit and a sense of ownership is essential.

Details of the materials and equipment needed to undertake the work are required. This includes materials and equipment necessary to ensure that the archaeological data collected is appropriately stored and curated. It may be appropriate for this section to be compiled in association with a conservator and relevant museum professional (see chapter 16).

If the project involves diving operations, these will have to be conducted from a platform of some description. Careful thought should be put into what kind of platform is used. The final choice will probably depend on a combination of what is required and what is available to the project. The ideal platform for any project is one that has the space and shelter on board to achieve all of the project's tasks in comfort and be sufficiently seaworthy to travel to the site and stay on station in all likely sea conditions. It should also meet any relevant local and national safety regulations. The exact needs of every project differ: some sites are easily reached from the shore, some may be approachable from small inflatable boats while others may require a large diving support-vessel or barge. It should be remembered that the bigger the platform, the larger and often more complex the mooring operations required to keep it on station. Experience has shown that a suitable mooring system that can be left on site is often a wise investment, saving many hours wasted in the re-laying and retrieval of moorings. Where a platform is designed to remain at anchor for the length of the diving operations, consideration should be given to the use of a small safety boat. Consideration should also be given to the use of a 'live aboard' platform, where the team stays on board for the duration of the project. This can save time and effort by avoiding a daily trip to and from the site. Staff can eat and sleep around their working schedule and the cost is often less expensive than accommodating personnel ashore. Of course, issues such as obtaining supplies, accessing conservation facilities, and allowing the team some time off must all be considered if a 'live aboard' platform is to be used.

Timetable: The project timetable should be planned through to completion. This should include fieldwork, assessment, analysis, dissemination and curation.

Budget: The costs of undertaking the work programme should be presented, making clear the basis of any calculations. Costs should be related directly to the methods statement and sufficient funds should be allowed for the competition of all post-fieldwork activity through to dissemination. It should not be assumed, unless confirmed by those involved, that equipment or specialist skilled work will be provided for free.

Logistics: It is vital that the tools, equipment and facilities required to undertake archaeological work, manage it safely, and provide living and subsistence for the project team in the field are available at the correct time throughout the project. This may mean ensuring that all equipment is available from day one, or arranging for equipment to be brought on site only for the period it is required. Each piece of equipment should be supplied with a suitable level of spares and consumables to ensure it is safe and in full working order. There should also be a suitable level of training and experience within the team to manage safely the operation of all equipment and the inevitable maintenance issues that all projects face. Large projects may consider including specialists within the team who take on specific roles, such as project engineer or cook, to allow the archaeologists in the team to spend their time more efficiently as archaeologists.

Post-fieldwork: It is important that while fieldwork is under way all records and processing of finds should be kept up to date and not left until the final stages of the project (see chapter 8). Experience shows that neglecting such work during a project will result in many wasted hours trying to sort out problems at a later date. Such issues should be considered in the planning stages of any

project and appropriate resources allocated. Once operations in the field have been completed, all records and finds should be collated and archived in a suitable manner (see chapter 19). It is not unusual for there to be a delay of many months, or even years, before these records are accessed again and this work may well be undertaken by an entirely different team.

Once the archive has been completed, an assessment of the work carried out should be undertaken to see if the results matched the aims of the original project design. This work should include an assessment of the data in terms of its potential, given further analysis, to satisfy the original aims of the project. This assessment should detail the volume, nature, context, method of recovery (where appropriate) and the possibility of contamination. The latter is particularly likely for sites located in areas with high volumes of historic shipping traffic. In such cases it is not unknown for archaeological layers to become contaminated by material subsequently deposited onto the sea-bed from passing ships.

Following this assessment it may be necessary to modify the project design based on what has been learnt about the site from the work carried out. At this stage it will also be appropriate to summarize the potential for any recovered material to answer other research question that have come to light as a result of the work undertaken. It should also set out the potential value of the site to local, regional, and national research priorities.

Storage and curation: From the outset, the project design should identify who the legal owners of the archaeological (if any) material recovered will be and where the project archive will eventually be deposited. It should be compiled following consultation with conservators (see chapter 16), the appropriate museum professionals, and the material's legal owner. It should also consider the immediate and long-term conservation and storage requirements for the data held in the site archive. Consideration should be given to how electronic data is to be stored and accessed in the future (see chapter 8). Recommendations should be made about selecting a representative sample of material from poor contexts, whose conservation and long-term curation costs far outweigh their further research potential. In such cases it is vital to ensure that all required recording and analysis has been completed prior to their disposal. In formulating a discard policy, due regard must be given to the views of the eventual recipient of the archive and the legal owners of the material. Care should be taken to ensure that discarded archaeological material does not contaminate other archaeological deposits.

Dissemination: The eventual aim of any archaeological work is dissemination of the results. Results can be disseminated through a variety of formats and media and these are discussed in detail in chapter 20. The project design should indicate when and how the results of the project will be disseminated.

Health and safety provision: The main consideration of any archaeology project should be the health and safety of its personnel. With this in mind, key sections in a project design are a risk assessment and a diving project plan. These and other important issues relating to health and safety on archaeological sites under water and in the inter tidal zone are covered in chapter 6.

This chapter has taken the reader through the process of planning for an archaeological project. Well thought-out and efficient project planning is essential for any archaeological project. This is equally true for a large underwater project involving multinational teams of professional divers and for a small foreshore investigation carried out by a team of enthusiasts over the course of a weekend. While the size of the document produced for each project will differ considerably, the NAS considers it best practice to produce a project design following English Heritage's guidance provided in the aforementioned MoRPHE (English Heritage, 2006b).

FURTHER INFORMATION

English Heritage, 1991, *Management of Archaeological Projects (MAP2)*. London (www.eng-h.gov.uk/guidance/map2/index.htm).

English Heritage, 2006[b], *Management of Research Projects in the Historic Environment (MoRPHE)*. London (www.english-heritage.org.uk).

Institute of Field Archaeologists 2001a (rev. edn), *Standards and Guidance for Archaeological Excavation*. Reading.

Institute of Field Archaeologists, 2001b, *Standards and Guidance for the Collection, Documentation, Conservation and Research of Archaeological Materials*. Reading.

Institute of Field Archaeologists, 2001c (rev. edn), *Standards and Guidance for Archaeological Desk Based Assessment*. Reading.

Institute of Field Archaeologists, 2001d (rev. edn), *Standards and Guidance for Archaeological Field Evaluation*. Reading.

Note: the IFA Standards and Guidance documents can be freely downloaded at www.archaeologists.net.

6 | Safety on Archaeological Sites Under Water and on the Foreshore

Contents
- Risk assessments
- Diving project plan
- Codes of practice
- Control of diving operations
- Working under water
- Potential diving problems and solutions
- Safety during excavation
- Inter-tidal site safety

The importance of safety during all archaeological work under water and on the foreshore is paramount. While it is not possible, nor an objective of this book, to provide an exhaustive guide to safety on such sites, it is relevant to summarize a few essential points related to archaeological work in a foreshore or underwater context. The importance of a thorough project design for any archaeological project was covered in chapter 5. Within each project design, it is essential that issues relating to the safety of those involved with the project are considered.

Divers are entering an alien environment and each will respond differently to the challenge. No matter how much preparation and assistance is provided by the organizers of a diving project, each participant will carry much of the responsibility for his/her own safety once they enter the water. This is particularly true of untethered scuba-diving, where once in the water, divers are largely independent from surface supervision, although through-water communication systems can provide a link with a surface supervisor. In addition, it should be remembered that the requirement to work under water means that the diver must be competent to complete the assigned task. Additional task loading must be planned for before the dive takes place.

Furthermore, each diver has a duty of care towards his/her fellow participants and because they are likely to be carrying out activities that have additional effects – other than those associated with normal sport diving – it is everyone's duty to ensure safe diving practices are followed.

Participants should:

- familiarize themselves with the site and the scope of work to be undertaken;
- be familiar with all safety information and code of diving practice;
- realistically assess their own training, experience and capabilities;
- be aware of responsibilities to other team members;
- ask for advice and communicate any reservations about safety issues, or their competency to undertake an assigned task;
- remain within the limits set by the project and by their own capabilities.

It is essential that project organizers do not place pressure on participants to undertake dives that are beyond their experience, or where they are not confident. Peer pressure can be dangerous and will also lead to poor archaeological standards.

RISK ASSESSMENTS

In everyday life, individuals regularly assess their own personal safety. Crossing the road, for example, involves identifying the possible dangers and taking appropriate precautions. Precautions might involve looking in both directions before crossing or, if the road is busy, a sensible precaution might be to move along the road to a designated crossing-place. If a safe means of crossing the road cannot be found, the decision may be taken not to cross the road at all. Normally this process is carried out in our heads. A hazard is identified, the level and type of risk assessed and measures put in place to control the likelihood of being hurt (up to and including simply avoiding the hazard).

As part of any project design it is essential that there is a formalized process of risk assessment that raises everyone's awareness of potential risks and hazards. While most people are familiar with crossing a road, not all project members will have the experience to understand all the risks associated with a project. The process of identifying potential risks and means of mitigation and committing them to paper enables everyone to benefit from the knowledge of more experienced project members. This has numerous advantages to the organizer of a complex operation like an archaeological project, including the following:

- Hazards and risks are identified in advance and assessed in a systematic manner.
- Safety information can be communicated to all project members quickly and concisely.
- Control measures can be included in the planning stage.
- In the event of an accident, it will help identify whether the problem could have been avoided in the planning phase.

Note: Project diving supervisors consider it best practice to prepare risk assessments and diving project plans (see below) for each part of the diving operation. Licensing authorities, funding bodies and government agencies, and even insurers, may require it. Once the project is under way, the risk assessment should not be forgotten. It should be referred to frequently; as conditions change, different control measures may be triggered. It is also considered good practice to carry out additional risk assessments as part of individual dive plans.

DIVING PROJECT PLAN

Where diving operations are involved, the project design should also include a diving project plan. This details the diving work, resources and equipment required to achieve the project design's objectives. To avoid confusion it is important that diving operations are conducted in a standardized and safe manner and the plan should clarify what recognized diving code of practice is to be used (see below). It should also contain a risk assessment that explains how the identified risks are to be controlled. A site-specific risk assessment should be completed prior to the start of any individual diving operation. Everybody involved in the project should be aware of the diving project plan, which should, as a minimum, consider:

- who is to be in control of diving operations;
- sea conditions, underwater visibility, pollution, depth and temperature;
- access to and from the shore/boat/platform (which must include the emergency recovery of an incapacitated diver, ideally within 3 minutes of the casualty reaching the surface);
- breathing-gas mixture and equipment needed;
- number of personnel (including those who are not part of the dive team) and their experience;
- emergency procedures, including the means of recovering an incapacitated diver, the location of and proximity to emergency facilities (e.g. decompression chambers) and medical expertise;
- the method chosen for the dive (i.e. surface supply or scuba), stating the safety reasons for the choice, which should be made with consideration of the task at hand and without slavishly following one particular method or another;
- a casualty evacuation plan and an identified means of summoning further emergency assistance without requiring essential personnel to leave the dive site.

CODES OF PRACTICE

One widely used method of benefiting from the experiences of other fieldworkers is to use a 'code of (best) practice'. These comprise a list of recommendations or standards, usually issued by a relevant organization, which give a guide to what has worked or been found acceptable in the past. It is recommended that a 'code of diving practice' is adopted for each project. The recording of diving operations must be given as much emphasis as any other technical part of the project. Personal dive-logs should be maintained as well as a project log of dive times and conditions.

Most codes of practice are either generalized or very site-specific, which makes the universal adoption of any existing code difficult. It is advisable that project organizers carefully consider all safety aspects of their project and write a project-specific code of practice. An existing code is unlikely to fit another site exactly and its adoption could lead to gaps in safety rules and procedures. As new techniques and procedures are adopted and developed, codes of practice will require updating. In addition, the regulations of governments with regard to the health and safety of those employed on diving operations will vary from country to country. It is important that project diving supervisors fully understand the regulations relating to the operation of archaeological teams, particularly where there is the possibility of a team of professionals and amateurs working together, when specific regulations might apply. Codes of practice for archaeological sites under water would normally be written with reference to numerous sources (e.g. commercial, scientific and sport-diving manuals). Some suggested sources for further information are provided at the end of this chapter.

Some codes will be more suitable than others and in unusual circumstances more than one code may be in use on the same site. This can happen when separate diving teams are involved in one project and responsible for different aspects of the work on site. It is far less complicated and preferable to have one specified code of practice but whatever code is in use, it is important that everyone involved is familiar with its application. As with the risk assessment, a code of practice must be regularly reviewed for appropriateness during the course of a project.

Both the risk assessment and the diving code of practice developed and adopted for a specific project are effectively statements of competence. In the (hopefully unlikely) event of an accident, these documents will be referred to by the authorities. The project manager will have to prove to the authorities that all reasonable care was taken and health and safety requirements were met.

CONTROL OF DIVING OPERATIONS

The appointment of a controller of diving operations, who has no direct responsibility for archaeological work, is a sensible arrangement on larger projects. This person, often referred to as the diving supervisor, diving officer or dive marshal, has to organize, regulate and record the diving operations in a way that creates as safe a working environment on site as possible. It is a job, however, that requires tact as well as experience; it takes considerable skill to ensure safe diving without seeming overbearing or patronising. A suitably qualified and experienced diving supervisor may control each individual diving operation.

The diving supervisor should be accessible to divers with queries or complaints about diving practice on site and it is often useful to have full and open discussions about any incidents that occur, no matter how minor. Incidents should be recorded in the project operations log and in some countries there is a requirement to report incidents to health and safety executives or an alternative safety organization. Some people need prompting to vocalize anxieties, especially if they feel overawed or intimidated by other, more confident, but not necessarily competent, individuals. Do not wait for a serious incident before having an honest appraisal of how safe people feel with working arrangements.

WORKING UNDER WATER

It has been mentioned above but it is worth re-emphasizing that archaeology is the objective of the diver rather than the diving – i.e. the diver is there to work and diving is simply a means to get to work. Therefore, any diving system should enable the diver to safely carry out archaeological work to an acceptable standard. A diving system that requires the diver's constant attention just to stay safe is not acceptable. In choosing an appropriate system, the list of factors to consider will include the following:

- Site environmental conditions – including accessibility, visibility, depth, tidal strength and temperature (e.g. a relatively deep-water site in an isolated location would require additional safety equipment).
- Type of work to be undertaken – excavation should normally be considered more strenuous work than survey and it is advisable to use a diver-to-diver or diver-to-surface communication system.
- Size of vessel – this will have an impact on the way the divers enter and leave the water as well as influencing the method of recovering equipment or finds. In addition, the ability to recover an incapacitated diver is essential.
- Divers' experience and qualifications – some sites are best investigated using a commercial diving system but specific diving qualifications or experience will usually be required.

Putting together the right diving system and diving team should lead to higher archaeological standards. To achieve this, the project organizer needs to have a thorough understanding of both archaeological and diving issues (plate 6.1).

POTENTIAL DIVING PROBLEMS AND SOLUTIONS

During the planning stage of a project there are numerous factors that need to be considered.

Physical problems: A wide range of factors can affect the physical well-being of a diver working on an archaeological project. Examples include: nitrogen narcosis, decompression, quality of breathing-gas, air and water temperature, exhaustion, equipment malfunction, poor communications, injury, bad weather, water movement and visibility.

Cold (hypothermia) is a common problem in archaeological work because the diver is often required to be stationary for long periods. Even in the tropics, long dives can lead to divers becoming cold. It is advisable to use more insulation than would be used for a normal dive; alternatively shorter dives can be considered. The extent to which efficiency deteriorates as the diver becomes chilled and the safety problems associated with cold are well known. Clearly, there is nothing at all to be gained from allowing divers to become colder than is absolutely necessary.

It should not be forgotten that there are equally serious problems associated with divers and even surface crew becoming over heated (hyperthermia), which can result in dehydration. Divers can be tempted to spend long periods in diving suits, maybe waiting for their next dive, filling in logs or washing finds. Divers should be aware that dehydration could also exacerbate decompression problems.

Diving supervisors should assess all the factors that might be associated with a particular site and take measures to avoid any of them becoming a problem (see risk assessment). It should also be stressed that the diving supervisor should have the authority to restrict any diver (even the director) from diving if, in his/her opinion, the diver is unfit to do so, for whatever reason. It is advisable during intense projects involving long, continuous diving periods that divers are encouraged to take scheduled non-diving days to recover and reduce the risks associated with residual nitrogen build-up. A non-diving day can enable divers to catch up with paperwork and help with surface tasks.

As the sport of diving has developed, new and more complex diving systems are becoming popular, particularly enriched air (nitrox), trimix and re-breathers. These systems can offer distinct advantages to the archaeological diver, such as extended no-stop times, safety buffers and clarity of thought associated with some breathing mixtures, which is an important factor in achieving high standards of archaeological work. However, there can be associated technical and logistical problems that need to be considered. It is important that those responsible for the diving-related aspects of a project consider the benefits of all available diving systems. Whichever one is chosen, it is equally relevant to go through the process of risk assessment and put in place appropriate safety measures and codes of practice.

Psychological problems: There are a number of factors that can have an impact on a diver's ability to make decisions effectively under water. The effects of nitrogen narcosis are well known but there are other things that can have similar effects, such as anxiety, stress, alcohol/ drugs and even over-enthusiasm. Some of these factors can result in a disregard for personal safety, while others (e.g. anxiety) can result in panic and an abandonment of proper diving practices. In extreme cases, it has been known for divers to suddenly rush for the surface, even though their training would tell them that this is potentially dangerous. Any impairment in the ability to make good decisions will affect safety and the standard of archaeological work. It is therefore important to consider taking measures that will alleviate these problems, such as communications, which make it unnecessary for the diver to remember detail or measurements. Using an alternative breathing-gas, as discussed above, is also an option. A combination of training, experience, common sense and an appropriate and well-maintained diving system will help prevent many of the consequences associated with psychological problems.

Diving techniques: Sport-diving training should be adequate for allowing a diver to operate safely within most projects, but there are some aspects of underwater work which are not covered in sport-diver training. Some of these may even run contrary to the normal diving practices taught to recreational divers.

The advantages of neutral buoyancy are rightly emphasised in terms of sport-diving and good buoyancy control is at a premium when engaged in photography or the investigation of a delicate deposit. However, to be effective, some tasks and environments require the diver to be negatively buoyant while actually working. On occasions, provided it is within the code of practice in use and prevailing environmental conditions allow it, divers may also find it easier to remove their fins when working. This can help to avoid accidental damage to archaeological material and often increases the comfort of a diver when supported by a grid.

Similarly, diving with a partner is a mainstay of safe sport-diving, but this is not always possible or practical in a working situation and alternative safety procedures will need to be implemented. It is possible to draw a distinction between diving alone and working alone with other divers in the vicinity, but this must take into account all local site conditions and the experience of the diving team.

Diving alone can have its advantages – for example, in very poor conditions where the presence of a buddy diver would impede the working diver, or where the task only requires one diver and the buddy would be ignored and effectively be diving alone. Lone divers must have an effective means of communication with their surface team and be competent to use the equipment (e.g. lifeline, hard-wire or through-water communications). There should also be other team members who will be responsible for tending the diver's lifeline, a stand-by diver ready to provide assistance and a dive supervisor. Many countries have specific regulations that govern archaeological diving practice, which can provide useful guidance for those divers not at work.

The most effective communication system is a hard-wire telephone via an umbilical between the diver and the surface (plate 6.2), but rope signals are reasonably efficient if the team is experienced in their use. Through-water communication systems are improving and are less restrictive than an umbilical, but they can be affected if the diver is in the shadow of rocks or similar upstanding features. In addition it is advisable to equip the lone diver using through-water communications with a surface marker-buoy or lifeline, so that s/he can be located if there is a problem. Alternatively, in some very benign diving situations it may be possible to allow divers to dive alone. However, this should be incorporated within the risk assessment and there must be adequate safeguards.

It is important that divers are not pressurized into adopting any practice they are not comfortable with. Equally, divers should not feel worried about refusing to adopt procedures that they consider unsatisfactory. Some very experienced diving archaeologists prefer not to work alone unless they are connected directly with the surface, while others are happy to consider it if there are other divers on site. It is also vital to remember that any diving procedure adopted conforms to the code of practice in use on the project, and that the procedure also complies with any local or national legislation that might apply. The following basic points are suggested as being the minimum necessary for safe diving operations involving lone divers:

- Divers operating alone must have an effective method of alerting the diving supervisor if they get into trouble.
- There must be a standby diver in full equipment on the surface ready to give assistance.
- There must be an effective way of locating the lone diver in distress.
- Divers operating alone must have an adequate and separate reserve of breathing-gas for use only in emergencies.

SAFETY DURING EXCAVATION

If the work of the project involves excavation, there are additional safety factors that must be considered. An excavation site can be a daunting and unfamiliar place for the inexperienced worker. The site is likely to have air, water or other power supplies for the airlift, water-dredge or other tools. In poor visibility these can be potentially treacherous and divers must have a clear understanding of where these hazards are in relation to the route to their place of work. Project diving supervisors or those responsible for safety should rationalize

the way that equipment lines and hoses are placed on the site to reduce the potential dangers associated with snags.

Likewise, the process of excavation can concentrate a diver's mind to the point where the contents gauge and dive timer are not given due attention. Complacency is just as problematic as anxiety and neither contributes to a high standard of work. It therefore makes sense to introduce less-experienced team members in a progressive way rather than to simply hope that they will cope. As with many working situations, it is very useful to establish a shot- or down-line, which is used for all ascents and descents. By establishing a fixed route to the site divers can get to work with minimum fuss.

During an excavation, care must be taken to ensure that no items of equipment or large objects capable of causing blockages are sucked into the mouth of an airlift or dredge. When the lower end of an airlift becomes blocked it rapidly becomes buoyant and will suddenly rush to the surface if not tethered. Less-experienced divers should be trained to ensure familiarity with the safety procedures. A means of shutting off the supply to the excavation equipment quickly must be within easy reach of the diver operating the equipment. It is worth considering having a wire-cross over the suction end to help prevent blockages. For safety, many divers have an alternative air supply as a secondary breathing source, but these can be dangerous, particularly when used with an airlift. If the secondary source gets sucked into the airlift, the breathing supply will be rapidly emptied. Incidents of this nature have led to fatalities. To reduce this danger, it is important to consider the position of an alternative breathing source and other extraneous pieces of equipment.

There is no archaeological investigation in the world that is worth the health or lives of those involved in it. Those responsible for a project must avoid generating an atmosphere where people become willing to take risks and push their luck 'for the sake of the project'. Get the job done, but do it competently, professionally and safely, even if that means taking a little longer.

The following is a list of top tips for safety (best practice):

- Ensure that participants fully understand the aims and objectives of the project.
- Make sure participants are sufficiently competent to carry out their tasks.
- Complete a comprehensive written risk assessment that covers all aspects of the project.
- Provide additional training if required.
- Keep written dive-plans.
- Keep records of diving times.
- Conduct daily equipment checks.
- Keep a daily project-log that includes details of team members, diving supervisors and individuals

responsible at any given time, weather and diving conditions, equipment checks, any significant event during a project day (e.g. dives, change of supervisor, non-diving periods, equipment breakdown, other incidents). NB: the project director or dive supervisor should sign the project log on a daily basis as a record of events.

INTER-TIDAL SITE SAFETY

Potential risks and hazards associated with inter-tidal sites should not be underestimated. Before any work is conducted in the inter-tidal zone, a risk assessment should be completed to identify any potential hazard associated with working or accessing a site. If a site is situated in a tidal area, access may well be governed by the rise and fall of the water level. There are also various risks associated with working on structures that are likely to be in a very poor state of repair. Consequently, stranded or abandoned vessels should be considered a hazardous working environment, with potential risks including stranding, structural collapse, eroding iron or other fastenings and even entrapment by falling internal fixtures and fixings. Heavy items such as engines or boilers can move without much notice.

The dangers of working in the inter-tidal zone need stressing, and the following precautions should be taken:

- It is essential to assess all safety aspects of the work site before contemplating carrying out any work on the site.
- Consider having a team member who is responsible for keeping a look-out for changing conditions.
- Never undertake work on an individual basis. For a large group, ensure a list of team members is checked before leaving the site.
- Check the local tide-tables before venturing out onto the foreshore. Plan the work schedule carefully to ensure that there is enough time to return to the shore.
- Get local advice regarding tidal conditions near the site and make sure that all team members know where the safest access routes are.
- Where possible, each team member should have a mobile phone (with fully charged battery) and ensure that everybody has everyone else's number.
- Make sure that everyone has access to numbers for the emergency services as well as maps giving directions to the local accident and emergency hospital.
- Consider laying some form of mattress or walkway to the site, as it will make the work easier and less tiring for team members who otherwise might struggle in the mud.

- Consider the use of a safety-boat.
- Make a plan to evacuate a casualty, or someone who is taken ill.
- Inform a responsible person when to expect team members ashore or home and consider informing a relevant authority when work will begin and end.
- If working in a tidal area, be aware that tide-tables are not necessarily accurate for all locations or weather conditions. The tide may turn earlier or may not recede as the tide-tables suggest. Keep a watch at all times, especially if the team is engaged in absorbing work on the foreshore. Remember that the tide can also come in faster than it goes out.
- Check that watches are functioning: take a spare or ensure that someone else in the team has one. If it has an alarm facility set it for the time the tide is scheduled to turn.
- Always wear slip-on, hard-toecapped boots rather than lace-up shoes because it is much more difficult to free feet from lace-ups if they become stuck in the mud.
- Consider wearing hard hats, high-visibility jackets and any other safety equipment that will help prevent accidents.
- Be aware that the weather can change quickly and that team members could suffer from weather-induced problems such as sunstroke or hypothermia.
- Be cautious about wading through water-filled creeks, as they are often deeper than they look. It is advisable to go around or over such features, even though such action may take up valuable working time.
- If team members become trapped by mud that is particularly treacherous and the tide is coming in, it may be safer to crawl or 'swim' over the mud rather than trying to walk across it.
- In more remote areas it may be advisable to take distress flares or alterative signalling equipment.
- If in danger of being stranded by the incoming tide, abandon equipment and leave as fast as possible. Such action may be unpopular, but safety must be the first consideration. Most equipment will survive a soaking.
- In some regions, sewage is discharged around the coast so check the positions of outfalls near the survey area. It is advisable to check with the relevant authority in the planning stage of a project. Wear disposable gloves and take water to enable the washing of hands before eating or drinking. It is important to clean up cuts, however small, with disinfectantl. Check that everyone has had a recent tetanus injection.
- If team members feel ill during or after working in the inter-tidal zone, they should consult a doctor.

(Based on 'Inter-tidal Site Safety Guidelines' published in Milne, McKewan and Goodburn, 1998, updated by G. Milne in 2006.)

Other essential equipment includes a first aid kit, spare clothing and sufficient ropes and lines to work the site effectively and safely. Fresh drinking water and food should also be provided in ample quantities, with spare water for washing hands and tools.

Remember that an inter-tidal risk assessment should include details of how emergency situations will be dealt with. Particular attention should be given to identifying the shortest route to solid ground and to evacuating a casualty. All access routes and the time taken to walk between different points should be known before serious work begins.

FURTHER INFORMATION

Please note, a number of the following publications will only be relevant to those working in the UK. The reader is advised to seek out local regulations relevant to their area of work.

The UK Health and Safety Executive (HSE) sells copies of the Diving at Work Regulations and the Approved Code of Practices (ACOPs). The HSE also issues an annual list of approved diving qualifications suitable for diving under the various ACOPs. Health and safety enquiries: HSE Information Centre, Broad Lane, Sheffield, S3 7HQ; tel: 0541 545500; www.hse.gov.uk/diving/information.htm.

Bevan, J., 2005, *The Professional Diver's Handbook*. London.

Health and Safety Executive, 2004, *Guidelines for safe working in estuaries and tidal areas when harvesting produce such as cockles, mussels and shrimps* (www.hse.gov.uk/pubns/estuary.htm).

Joiner, J. T., 2001 (4th edn), *National Oceanic and Atmospheric Administration Diving Manual: Diving for Science and Technology*. Silver Spring, Maryland.

Larn, R. and Whistler, R., 1993 (3rd edn), *The Commercial Diving Manual*. Melksham.

Lonsdale, M. V., 2005, *United States Navy Diver*. Flagstaff, Arizona.

Milne, G., McKewan, C., and Goodburn, D., 1998, *Nautical Archaeology on the Foreshore: Hulk Recording on the Medway*. RCHM, Swindon.

Scientific Diving Supervisory Committee (SDSC), 1997, Advice notes for the Approved Code of Practice (www.uk-sdsc.com).

International and National Laws Relating to Archaeology Under Water

Contents

The legal issues relating to the discovery, survey and excavation of underwater cultural heritage were once described as 'a legal labyrinth' (Altes, 1976). This is certainly the perception of many amateur and professional underwater archaeologists. The purpose of this section is to chart a path through this maze of national and international laws and to establish an understanding of the legal issues that might be encountered when interacting with underwater cultural heritage.

JURISDICTION – WHERE DO THE LAWS APPLY?

When determining the legal regime that is to apply to any given situation, the first step is to identify where one is. In the world's oceans, this is no easy matter. To determine the legal regime that applies, one has to refer to international law and, in particular, to the United Nations Convention on the Law of the Sea (UNCLOS). Adopted in 1982, this Convention sets out to divide the world's oceans into different zones and specifies what legal regime is to apply in each zone. Five different zones were established:

- deep sea-bed and the high seas;
- continental shelf;
- exclusive economic zone;
- contiguous zone;
- territorial seas.

UNCLOS establishes the extent of each zone and specifies what states can and cannot do within each of them. When this Convention was negotiated, underwater cultural heritage was not particularly high in the order of priorities and therefore did not receive the consideration that was necessary to establish a strong protective regime. Nevertheless, the Convention is important in that it specifies what each state can regulate in each zone. It is therefore necessary to consider each zone in turn.

It is also important to be able to determine where each of these zones begins because UNCLOS starts where the land ends and the sea begins. This, however, is never constant given the ebb and flow of the tides and indentations caused by bays, estuaries etc. In international law, the sea begins at the low-water mark as marked on large-scale charts. This is called the 'baseline'. It is from this line that all the maritime zones are measured. The baseline does not, however, always follow the low-water mark around the coast, and may be drawn across river mouths, bays, harbours and along islands or sandbanks that appear at low tide. Each zone is then measured from this agreed baseline.

Deep sea-bed and the high seas: The deep sea-bed is defined as the area 'beyond the limits of national jurisdiction'. Therefore, it is the area that exists beyond the territorial sea or any other maritime zone claimed by a particular state, such as the contiguous zone or exclusive economic zone. In this area, no state can claim unilateral jurisdiction, and the age-old principle of the 'freedom of the high seas' applies. However, for some resources, for example the minerals of the deep sea-bed and living resources such as fish stocks, UNCLOS regulates what each state can do and establishes a system in which each state can share in those resources. During negotiations on the Convention, it was decided that underwater cultural heritage should not be included in the definition of resources, so that, in principle, the freedom of the high seas applies in regard to the search, survey and excavation of underwater cultural heritage found in this area. The Convention did, however, declare in Article 149 that 'all objects of an archaeological and historical nature found in the Area shall be preserved or disposed of for the benefit of mankind as a whole, particular regard being paid to the preferential rights of the state or country of origin, or the state of cultural origin, or the state of historical or archaeological origin'. Unfortunately, the Convention did not define any of the terms used in Article 149 and generally international lawyers have agreed that this Article is too vague and imprecise to act as a regulatory provision. Therefore, on the high seas and the deep sea-bed, all states have the right to search for, survey and excavate underwater cultural heritage and can authorize their nationals and vessels flying their flag to do so. The only possible exception to this is the right of other states to prohibit disturbance of state-owned non-commercial vessels that have sunk, under the internationally recognized principle of sovereign immunity.

The continental shelf: The continental shelf extends from the shallows of the territorial seas down to the deep sea-bed. The length of the continental shelf varies geographically. In the case of coastal states that have very long continental shelves, the state can make a claim for up to 350 nautical miles from the baseline used to calculate the territorial sea or 100 nautical miles from the 2500-metre isobath. Due to the complexity of these and associated rules, states must submit their proposed delineation of the continental shelf to a special commission established under UNCLOS. In cases where the continental shelf is short, the state can claim up to 200 nautical miles, even if the actual shelf is shorter than that. Each coastal state is given the exclusive right to explore and exploit the natural resources of this area. Underwater cultural heritage was not included in the definition of 'natural resources' so that the coastal state does not have the exclusive right to regulate the search, survey or excavation of sites on the continental shelf. Article 303 of the Convention, however, requires states to 'protect objects of an archaeological and historical nature found at sea and shall co-operate for this purpose'. In order to give effect to this requirement, a number of states have extended their jurisdiction so as to regulate underwater archaeology on their continental shelves. These include Australia, Ireland, Seychelles,

Cyprus, Spain, Portugal, Norway and China. While these may appear to be contentious claims, no other state has, to date, objected to these extensions of jurisdiction. It is therefore important to note that if a project is to take place within 350 nautical miles of the coast, one of the coastal states may impose a regulatory regime on the operation.

Exclusive economic zone: The exclusive economic zone was created to allow coastal states the exclusive right to exploit the natural resources of the seas up to 200 nautical miles from the baseline used to calculate the territorial sea. Other states will continue to have certain freedoms in this zone, such as freedom of navigation, over-flight, laying of submarine cables and marine scientific research. Underwater archaeology is not regarded as marine scientific research and is therefore not automatically regarded as a freedom of the high seas in this zone. The right to regulate underwater archaeology is therefore uncertain, and is the basis of continued debate between a number of states. Both Morocco and Jamaica, for example, claim jurisdiction to regulate underwater archaeology in their exclusive economic zones. Most other states, however, do not.

Contiguous zone: The contiguous zone is an area adjacent to the territorial sea in which the coastal state has limited control to prevent or punish infringements of its custom, fiscal, sanitary or immigration laws. The maximum breadth of this zone is 24 nautical miles from the coastal baseline. Article 303 of the Convention allows a coastal state to consider the recovery of any underwater cultural heritage from this zone as having taken place in the coastal state's territorial seas, thus giving the coastal state the exclusive right to regulate such activities. A number of states have extended their national legislation to take into account underwater archaeology in this zone. The United States is the most recent state to have done so, having declared a contiguous zone in 1999. This increasing tendency means that if a project is to take place within 24 nautical miles of the coast, it is likely that the coastal state will regulate these activities.

Territorial sea: The territorial sea extends up to 12 nautical miles from the coastal baseline. In this zone, the coastal state has the exclusive right to regulate all activities relating to underwater archaeology. The nature of these regulatory laws differs from state to state, and a number of these will be considered in a later section.

THE REGIME IN INTERNATIONAL WATERS

The Law of the Sea (UNCLOS) therefore stipulates that in the territorial seas and contiguous zone, the coastal state can regulate activities directed at underwater cultural heritage, while beyond that, no state has the exclusive right to regulate activities directed at the underwater cultural heritage. This does not mean that a state has no powers in these zones. Under international law, all states have jurisdiction over nationals of that state, and over vessels that are registered in that state and fly that state's national flag. A state can therefore adopt laws regulating the conduct of its vessels and nationals in international maritime zones. This could apply to underwater archaeology. A state could not, however, prevent other nationals and vessels flying foreign flags from interfering with the underwater cultural heritage in these zones.

It is therefore important to determine in which zone an underwater cultural heritage site is situated in order to determine which state has jurisdiction. Having done so, it is then possible to determine what legal regime the regulating state applies. While each state is unique and with unique legal systems, a number of states have agreed that in relation to certain activities in international waters, a common regime will be applied. Although there is no common regime in relation to underwater cultural heritage, there is a common regime in relation to salvage law, which may be applied to the underwater cultural heritage.

INTERNATIONAL SALVAGE LAW

Salvage law has a long history, which begins in the Rhodian Maritime Code of 900 BC. Since then, it has developed relatively uniformly in most maritime nations. To ensure that similar laws are applied to salvage operations that take place in international waters, a number of states entered into an international Salvage Convention in 1910, which was subsequently updated in 1989. These states have therefore agreed that certain uniform principles will apply. Under this regime, salvage is defined as 'the compensation allowed to persons by whose voluntary assistance a ship at sea or her cargo or both have been saved in whole or in part from impending sea peril, or in recovering such property from actual peril or loss, as in cases of shipwreck, derelict or recapture'. The policies that form the foundation of salvage law are to encourage individuals to voluntarily save lives and property at sea and to return such saved property to its owner. By so doing, the salvor ensures that valuable commercial goods are not lost and are able to re-enter the stream of commerce. Before salvage law may be applied, three criteria must be satisfied:

1 property in marine peril on navigable waters,
2 voluntary or contractual efforts to rescue the property; and
3 partial or total success.

Once these three criteria are satisfied, the court will grant a salvage award. In assessing the salvage award, the court will take into account a number of factors, such as:

- the salved value of the vessel and other property;
- the measure of success obtained by the salvor; and
- the skill and efforts of the salvors in salving the vessel and other property.

It does not, however, take into account the extent to which appropriate archaeological techniques have been used in the excavation and recovery of historic wreck.

UNDERWATER CULTURAL HERITAGE AND SALVAGE LAW

What should be obvious about this salvage regime is that it is designed to save vessels and other property that are in immediate marine peril. Many would argue that it is not appropriate to consider underwater cultural heritage, which has been submerged for a long time and reached a state of near equilibrium with the marine environment, as being in 'marine peril'. This argument was offered in a number of court cases in the US involving the salvage of historic wreck, only to be mostly rejected by the US Admiralty Courts. The US therefore considers salvage law to be appropriate to the recovery of underwater cultural heritage. Not all states, however, agree, and a number, such as Canada, the Republic of Ireland and France, will not apply salvage law to underwater cultural heritage.

The 1989 Salvage Convention makes no specific mention of sunken vessels or their cargo in the definition of 'vessel' or 'property'. During negotiations, the question of salvage of underwater cultural heritage was raised. France and Spain attempted to have underwater cultural heritage excluded from the Convention, but were only partially successful. Article 30(l)(d) of the 1989 Convention allows a state to enter a reservation which reserves the right not to apply the Convention 'when the property involved is maritime cultural property of pre-historic, archaeological or historic interest and is situated on the sea-bed'. The 1989 Salvage Convention therefore does apply to underwater cultural heritage unless a state specifically chooses not to apply it. Not every country that enters a reservation will refrain from applying the Convention to the salvage of underwater cultural heritage. The UK, for example, entered a reservation in accordance with article 30(l)(d) which gave it the right to enter a reservation in the future. As such, the reservation does nothing more than allow the UK to enter a reservation not applying the Convention at some future date. France, on the other hand, has entered such a reservation, and underwater cultural heritage that is recovered in international waters by French vessels, or by French nationals, and subsequently landed in France will not be subject to salvage law.

OWNERSHIP OF UNDERWATER CULTURAL HERITAGE

Salvage law does not affect ownership rights in any property. The salvor does not become owner of the salvaged property and is considered to have saved the property and to be holding the property for the true owner. However, problems arise when the owner either expressly abandons all property rights or the owner cannot be established. In the case of very old vessels, it may be that the owner is deemed to have abandoned all property rights in the vessel because the owner has done nothing during that period to assert continued rights of ownership. A further problem arises when the vessel is owned by a state rather than by a private individual. International law is not clear with regard to a number of these problems, and states have adopted different approaches to dealing with them.

ABANDONMENT OF OWNERSHIP

State-owned vessels: International lawyers generally agree that a state will only be regarded as having abandoned ownership of a vessel when the state clearly and expressly does so. Abandonment cannot be inferred from the circumstances. As such, any project that might be undertaken on a state-owned vessel should be regarded as still being owned by the state. States have continually claimed ownership of vessels sunk in international waters or in the territorial waters of other states. For example, the wreck of the CSS *Alabama*, sunk in 1864 off the coast of Cherbourg, France, was claimed by the government of the United States, while the French government has claimed ownership of *La Belle*, which sank in 1686 off the coast of Matagorda Bay, Texas. Not all states, however, readily accept this rule of international law.

Privately owned vessels: While abandonment of ownership of a privately owned vessel can be made through a clear and express declaration to that effect, it may also be lost if, considering all the factors, it would be reasonable to imply that the owner had abandoned ownership. Factors that might support an implication of such an abandonment of ownership could include the passage of time, inactivity on the part of the owner to undertake salvage operations, or destroying documentary proof of ownership.

Acquisition of ownership after abandonment: Where the owner is deemed to have abandoned ownership, either the state or the finder will acquire ownership.

State ownership: Most states claim ownership of vessels that have been abandoned. This occurs both when the owner has expressly abandoned ownership or when abandonment can be implied. The time allowed differs amongst states, as does the point from which this period is measured. In the UK, the owner has 1 year in which to claim ownership after a discovery has been reported. Failure to do so vests title in the Crown. In Spain, the state claims ownership after 3 years. Alternatively, some states will only acquire ownership of vessels that are deemed archaeologically or historically important, effectively applying the law of finds to those that are not.

Law of finds: The law of finds is based on the principle of 'finders, keepers'. If ownership has been abandoned, finders are entitled to become owner once they have taken the property into their possession. The state most often associated with the law of finds is the United States, where it has been used as an alternative to salvage law. This is particularly important in the case of historic wreck where ownership is difficult to determine.

NATIONAL LEGISLATION

While most states' legislation will only apply to the state's territorial waters, some states have extended their legislation to cover other maritime zones over which they have some competence, such as the contiguous zone, exclusive economic zone or continental shelf. Some legislation, such as the United Kingdom's (UK) *Protection of Military Remains Act 1986*, may also apply to a state's nationals and flagged vessels in international waters. Each state will have a unique legal system and regulate underwater archaeology differently. There are a number of features that one can expect to find in most state legislation.

Scope of protection: Some states' legislation applies only to historic wreck (US and UK), while others' applies to all underwater cultural heritage, including historic submerged landscapes, such as Palaeolithic sites. In the case of those which apply only to wrecks, some apply to all wrecks over a certain age, such as 50 years (South Africa), 75 years (Australia) or 100 years (Republic of Ireland), while others only apply to wrecks which are deemed to be of archaeological or historical significance (US and UK). Some legislation will also apply to all wrecks, irrespective of ownership, while others, such as the *US Abandoned Shipwreck Act 1987* applies, as the name suggests, only to abandoned shipwrecks.

Ownership rights: Different states have different provisions for the rights of owners in wreck. Most states will allow the owner a certain time in which to claim ownership. If no owner comes forward, then the ownership will vest either in the state (UK) or in the finder (US). States such as Turkey deem all cultural heritage to be owned by the government and no private trade in these items is allowed.

Rewards: Those states which apply salvage law to underwater cultural heritage will reward the recovery of an historic wreck with a salvage award (US and UK). In the UK, this award may be extremely high, often over 75 per cent of the value of the recovered artefacts. This high award is said to encourage the reporting of finds and recoveries, in what is termed a policy of 'incentive to honesty'. If the recovered artefacts are not of archaeological or historical significance, the Receiver of Wreck in the UK may award ownership of the artefacts to the salvor in lieu of a salvage award. This, in effect, applies the law of finds. In those states which do not apply salvage law or the law of finds to historic wreck, such as Australia and France, the finder will often be awarded a 'finder's reward'. This provides an incentive for divers to report finds without recovering anything.

Search licences: Most states will not necessarily require a licence to search for underwater cultural heritage (e.g. UK, Bahamas, South Africa and Canada). Some states, such as Greece and Turkey, restrict search and diving activities, and a licence is required to undertake underwater searches.

Survey and excavation licences: Most states will require a licence to survey and/or excavate an historic wreck. The stringency of the licensing requirements varies from state to state. Many Mediterranean states, such as Greece, Turkey and Italy, have extremely strict licensing requirements, while many developing states have few, if any, licensing requirements.

Penalties: The penalties for not abiding with regulations also vary between states. Some states, such as Turkey, will impose heavy fines and confiscate equipment.

Export of cultural heritage: Most states impose export or import restrictions on cultural heritage. This applies to cultural heritage found on land and within the territorial seas of the coastal state. Export without an export licence will normally result in a criminal conviction. Unfortunately, it is difficult to police the territorial sea and it is often possible for clandestine excavation to take place in a state's territorial waters with the artefacts being taken directly to another state.

INTERNATIONAL CONVENTIONS

UNESCO Convention for the Protection of the Underwater Cultural Heritage

It should be evident from the above discussion that there is little regulation of underwater archaeology beyond the limits of state territorial jurisdiction. Salvage law has tended to be applied in most cases, which is clearly not appropriate to underwater cultural heritage. Concerned that valuable archaeological information was being lost in cases where inappropriate techniques were being applied to the excavation of underwater cultural heritage in international waters, the International Law Association and the International Council of Monuments and Sites (ICOMOS) drafted a proposed Convention that would regulate underwater archaeology in international waters. This draft Convention was forwarded to UNESCO for consideration and, with some amendment, was adopted by UNESCO in 2001. It is currently awaiting ratification by a sufficient number of states to bring it into force. The salient features of the Convention include the following:

- The introduction, as an Annexe to the Convention, of an archaeological code of good practice and appropriate archaeological techniques, which are to be applied to the excavation of underwater cultural heritage.
- The controversial extension of coastal-state regulatory jurisdiction over aspects of underwater cultural heritage co-extensive with the exclusive economic zone or the continental shelf of the coastal state.
- Restricting the application of salvage law to underwater cultural heritage to those instances where prior authorization for disturbance has been given. This effectively introduces a system of state-controlled excavation.
- The introduction of a system of penalties and confiscatory powers for items recovered in a manner not consistent with the archaeological code and/or items illicitly exported or imported.
- A very wide definition of what comprises underwater cultural heritage.
- A duty upon states to co-operate to implement the provisions of the Convention.

European Convention on the Protection of the Archaeological Heritage (revised)

Commonly referred to as the 'Valletta Convention' because it was opened for signature by member states of the Council of Europe and other states party to the European Cultural Convention on 16 January 1992 at Valletta in Malta, the aim of the Convention is to protect the

European archaeological heritage as a collective memory and for historical and scientific study. The Convention was made under the aegis of the Council of Europe (not the European Union), and is a revision of the 1969 European Convention on the Protection of the Archaeological Heritage.

During the 1960s clandestine excavation was seen as the major threat to archaeological heritage, whereas during the 1980s large-scale construction projects were seen as the greater danger. At the same time the professional emphasis in archaeology shifted from the recovery and display of objects to their preservation *in situ*. Recovery was seen as a last resort and equal prominence was given to examination of the context as well as the object itself. As the Preamble to the Convention makes clear, the heritage is not inviolate but any disturbance must be conducted using appropriate archaeological methods, thereby securing the archaeological information arising from the context as well as the object itself.

Accordingly, the Valletta Convention:

- defines the archaeological heritage extremely widely;
- applies under water as well as on land, although its application under water may well be an afterthought – indeed, there are indications that this aspect was poorly drafted and took little or no account of the unique nature of the maritime context;
- seeks to remove or mitigate the threat posed by commercial developments and to reflect this change of emphasis and procedures in archaeology;
- contains provisions for, *inter alia*, the identification and protection of the archaeological heritage, its integrated conservation, the control of excavations and the use of metal-detectors;
- requires states to control illicit excavation and ensure that any intrusion into the heritage is conducted with appropriate and, preferably, non-destructive methodology;
- sets out the measures required of each state for the identification and protection of that heritage;
- requires each state to provide a legal system for protection of the archaeological heritage and make certain stipulated provisions;
- specifies additional measures that are to be adopted (i.e. the maintenance of an inventory, the designation of protected monuments and areas, the creation of archaeological reserves, the mandatory reporting of finds, which must be made available for examination).

The Convention states that the heritage is comprised of things within the jurisdiction of the state parties. If a state exercises jurisdiction beyond its territorial sea for any purpose then it is arguable that the Valletta Convention applies to the extent of that jurisdiction. For many states this will include any heritage located out to the edge of the Continental Shelf.

CASE STUDIES

Case study: *The Protection of Military Remains Act 1986* (UK)

This piece of national legislation illustrates the manner in which a state can regulate activities directly relating to underwater cultural heritage in international waters. The Act was adopted to protect the integrity of crashed military aircraft or vessels, both on land and under water, irrespective of whether they contained human remains. The genesis of the Act centred around concerns that a number of excavations of historically important wrecks containing human remains had not been conducted with the appropriate respect for those remains. These included the salvage of HMS *Edinburgh* in 1982, and the recovery of personal belongings of casualties of HMS *Hampshire* in 1983. The sinking of a number of vessels during the Falklands conflict in 1982 also raised concerns about the sanctity of 'war graves'. The vessels sunk included HMS *Ardent* and HMS *Antelope* in the Falkland Islands' territorial waters, and HMS *Coventry*, HMS *Sheffield*, MV *Atlantic Conveyor* and RFA *Sir Galahad* in international waters. In UK territorial waters, both UK and foreign vessels may be designated, as the UK has complete jurisdiction in this maritime zone. Because international law allows a state to regulate the activities of its nationals and flag vessels in international waters, the UK can also designate UK wrecks in international waters. However, in this case, the Act will only apply to British nationals or British-controlled vessels conducting excavations on a protected or designated site. The Act cannot allow the UK to prevent foreign nations from interfering with a designated site. Thus, the Act allows the UK to regulate the conduct of British nationals who undertake any activities on such military remains even if those remains lie in international waters.

Case study: Marine peril in US Admiralty Courts – the *Espiritu Santo*

On 9 April 1554, a fleet of Spanish vessels left Vera Cruz, Mexico, homeward bound for Spain. Twenty days later, the fleet was hit by a storm off the coast of Texas and a number of vessels were lost off Padre Island, including the *Espiritu Santo*. There the vessel lay undisturbed until it was discovered by salvors in 1967 and excavated without any archaeological recording. The state of Texas claimed ownership of the vessel and the artefacts, though the salvors sued for a salvage award. In determining whether the salvage regime was applicable, the court noted that 'the artefacts came to rest on the clay bottom of the Gulf of Mexico, thirty to forty feet under water. Eventually four to ten feet of sand covered them. Under these conditions, the items were effectively impervious to weather conditions above the surface of the sea, and the sand prevented deterioration under-water. The items remained in this state of equilibrium until 1967 when Platoro commenced recovery operations.' It would appear that the court recognized that the artefacts were in fact protected and in no immediate danger. However, the court ruled that as a matter of law, marine peril did exist. The court stated that 'the *Espiritu Santo* was still in marine peril after its position was discovered' as 'it is far from clear that the sand would remain sufficient protection from the various perils of the Gulf of Mexico'.

Case study: Marine peril in Canadian Admiralty Courts

In a Canadian case, Her Majesty *v.* Mar-Dive 1997 AMC 1000, the court held that a wreck embedded in the bottom of Lake Erie was not in marine peril, and therefore the salvors of a number of artefacts from the wreck were not entitled to a salvage award. In fact, the court held that the activities of the salvors had damaged the wreck and significantly damaged its archaeological integrity, and that proposed further action would not save the vessel but cause the wreck to be in even greater danger.

Case study: RMS *Titanic*

Probably the most famous maritime disaster in the history of western society, the *Titanic* sank in 1912 with the loss of over 1500 lives. She came to rest in international waters close to Newfoundland, Canada. The discoverers of the wreck believed that she should remain as she was found as a memorial to those who had perished in the disaster. However, there was no international mechanism to enforce this, and in 1987 a salvage company undertook an expedition to the wreck and began to recover items. These items were taken into the US, where the US Admiralty Courts applied the doctrine of constructive possession in order to establish jurisdiction over the wreck-site. The doctrine of constructive possession means that if part of an item is in the jurisdiction of the court, the court will consider the entire item to be subject to the court's jurisdiction. This was an unprecedented extension of national maritime jurisdiction, which not all international lawyers accept as valid in international law. The court found the salvors to have a possessory right to the wreck and were entitled to a salvage award, or if the wreck was not claimed, to

ownership of the items recovered on the basis of the law of finds. The US, UK, France and Canada were concerned that salvage would be undertaken in an inappropriate manner and entered into negotiations to adopt an international convention which would allow these states to prevent their nationals or their vessels from undertaking such inappropriate salvage activities. The salvors of the wreck, however, have agreed that they would only recover items from the debris field and would not sell any of the recovered artefacts. The negotiations between these four states have subsequently been successfully concluded. The four nations are now incorporating the resulting agreement into their national legislation. However, the agreement cannot be enforced against any citizens or flagged vessels whose state is not a party to the agreement. Consequently, the case of the *Titanic* can still be said to highlight the lack of any regulatory measures in international waters.

Case study: HMS *Birkenhead*

HMS *Birkenhead*, an iron-hulled paddle-wheel frigate, was carrying troops to the eastern frontier of South Africa when it was wrecked off the coast of the Cape Colony in 1852. The vessel was crowded with troops and passengers when it began to sink. The British troops stood fast while the passengers were taken off on the lifeboats. From this heroic stand, the naval traditional of placing the women and children first in the lifeboats was born. Of the men, 445 were lost in the disaster. In 1983 the South African Government, who considered the vessel to be the property of the Government as it lay in South African territorial waters, issued a salvage permit. The UK Government, however, also claimed ownership. The dispute was later resolved by means of an agreement between the two countries. This agreement allowed investigation and salvage of the vessel to continue but acknowledged that ownership remained vested in the British Crown.

Case study: The *Central America*

The *Central America*, a side-wheel steamship, was lost in international waters off the coast of South Carolina, USA, in 1857. Lost with the vessel were 423 lives and US$2,189,000 in gold from the California gold rush. The gold had been insured with a number of insurance companies in the US and UK, who subsequently paid out on all claims. In 1987 a salvage consortium, the Columbus-America Discovery Group, discovered the wreck. On recovery of a vast amount of gold, a number of insurance companies who had either paid out in 1857 or had at some point taken over older insurance companies, claimed ownership of the gold. The salvors argued that the insurance companies had abandoned ownership, particularly as the companies had destroyed all evidence such as bills of lading and insurance policies. The Admiralty Court initially ruled that the insurance companies had indeed abandoned ownership. This decision, however, was overturned by the Court of Appeal in 1992, which ruled that abandonment can only be proved by clear and convincing evidence, and that the salvors had not done so in this case. The insurance companies were therefore considered owners of the gold. As salvage law was applied in this case, the salvors were entitled to a liberal salvage award. The US court took the then-unique step of taking into account the extent to which the salvors had followed appropriate archaeological practice during the excavation. The court finally awarded the salvors over 80 per cent of the value of the gold as the salvage award.

Case study: The *Nuestra Señora de Atocha*

The *Nuestra Señora de Atocha*, the vice-flagship of a Spanish Fleet bound for Spain, was lost off the coast of Florida in 1622. The salvage firm Treasure Salvors Inc. searched for the wreck for a number of years before finding the 'mother lode' in 1985. Large numbers of artefacts, and large amounts of gold and silver were recovered. Because the site was in international waters, being approximately 50 km (27 nautical miles) off the Florida Keys, the state of Florida and the US Government were not able to exert ownership over the wreck. Spain made no claim of ownership and so the Admiralty Court ruled that the wreck had been abandoned and that the law of finds was to apply, rendering the salvors owner of the wreck and artefacts.

Case study: The *Doddington*

The *Doddington*, a British East Indiaman carrying a shipment of gold belonging to 'Clive of India', foundered off the coast of South Africa in 1755. The wreck was discovered in 1977. The South African National Monuments Act 1969 applies to all vessels which have been under water for more than 50 years, and prohibits, among other things, the export of artefacts without an export permit. In 1997, a London Auction House advertised 1200 gold coins for sale as having come from the *Doddington*. As the South African Government had not issued any export permits for items from the *Doddington*, it was concluded that the gold must have been exported illegally. The South African Government has subsequently applied for the restitution of this gold and it has now been taken off the auction listings. However, few states will enforce the public laws of another state, which includes export laws, so the South African Government may have to rely on other rights in order to obtain possession of the gold. Such rights might include ownership rights.

Case study: The *Geldermalsen*

The *Geldermalsen*, a Dutch East Indiaman, sank in the South China seas in 1751 while carrying a consignment of Chinese porcelain for the European market. In 1985, the site was discovered and excavated by salvors, paying little regard to appropriate archaeological standards. Once the porcelain was recovered, it is rumoured that the salvors destroyed the remains of the site in order to hide its position. Two possible positions have been suggested. The first is that the site lay on the continental shelf of China. As China regulates underwater cultural heritage on its continental shelf, the site would have been subject to Chinese jurisdiction. China, however, had no knowledge of this recovery operation – hence the need to hide the site. The second possibility is that the site lay in the territorial waters of Indonesia. The Indonesian Government brought a lawsuit against the salvors, which remains unresolved. The porcelain was sold at Christie's Auction House in Amsterdam.

FURTHER INFORMATION

Brice, G. and Reeder, J., 2002 (4th rev. edn), *Brice on Maritime Law of Salvage*. London.

Caminos, H., 2001, *Law of the Sea*. Aldershot.

Dromgoole, S., 1996, Military Remains on and around the Coast of the United Kingdom: Statutory Mechanisms of Protection, *International Journal of Marine and Coastal Law* 11.2, 23–45.

Fletcher-Tomenius, P. and Forrest, C., 2000, The Protection of the Underwater Cultural Heritage and the Challenge of UNCLOS, *Art Antiquity and Law* 5, 125.

Fletcher-Tomenius, P. and Williams, M., 2000, When is a Salvor Not a Salvor? Regulating Recovery of Historic Wreck in UK Waters, *Lloyd's Maritime and Commercial Law Quarterly* 2, 208–21.

Forrest, C. J., 2000, Salvage Law and the Wreck of the R.M.S. *Titanic*, *Lloyd's Maritime and Commercial Law Quarterly* 1, 1–12.

Kennedy, W. and Rose, R., 2002, *The Law of Salvage*. London.O'Keefe, P., 2002, *Shipwrecked Heritage: A Commentary on the UNESCO Convention on Cultural Heritage*. Leicester.

Williams, M., 2001, Protecting Maritime Military Remains: A New Regime for the United Kingdom, *International Maritime Law* 8.9, 288–98.

Zhao, H., 1992, Recent Developments in the Legal Protection of Historic Shipwrecks in China, *Ocean Development and International Law* 23, 305–33.

Archaeological Recording 8

Contents

Recording is the absolute dividing line between plundering and scientific work, between a dealer and a scholar . . . The unpardonable crime in archaeology is destroying evidence which can never be recovered; and every discovery does destroy evidence unless it is intelligently recorded. (Petrie, 1904:48)

THE NEED FOR RECORDING

The range and quantity of data collected during an archaeological project are vast. In order for current and future generations to learn from archaeology, this information must be made available to all in an organized and accessible form. This chapter looks at the different types of evidence likely to be encountered on an underwater archaeological site and summarizes what to record and how to do so. It also emphasizes the importance of choosing and maintaining an appropriate recording system to keep track of all evidence and records generated during the course of an archaeological project.

Ideally, archaeologists or their successors should be able to 'reconstruct' the site from the archive (see chapter 19) or records of a site. This is particularly important after excavation, which destroys the site and prevents future investigation. It is good discipline, even for non-destructive survey, to record all relevant data for future reference, as sites can change over time. To convince people of the value of your conclusions it is necessary to show the detailed results of the investigations on which they are based.

The aim of archaeological recording is to note what is there as accurately and completely as possible, giving each piece of information equal weight without allowing the interpretation of information to affect the method of recording. Recording should be an objective process. Of course, it is also important to know what the excavator thought about the site as s/he recorded it – after all, flashes of inspiration on the part of the excavator can often explain objects and their relationships as they are uncovered. However, such comments and ideas should be kept separate from the data record of the site. The way information is interpreted is likely to be affected by an individual's background and culture, which give everybody a set of ideas and experience against which things are judged. This may lead archaeologists to make assumptions about an object very different from the ideas of the society that used or created it in the past. It is vital, therefore, for everything to be recorded in a way that best avoids prejudices or influences that may affect archaeological interpretation.

RECORDING SYSTEMS

The choice of an appropriate recording system should be made in the early stages of project planning. A description of the system and the reasons behind its choice should be included in the project design (see chapter 5). The recording system must be capable of recording information relating to aspects as diverse as the location, identification and interpretation of all the evidence from a site. A recording system should be constructed in a way that stores and manages the recorded clues in a manner that is simple to understand. The system should make it easy for the user to cross-reference information that exists in a variety of forms (e.g. individual observations, photographs, drawings, etc.).

The following list gives an indication of the diverse and extensive range of evidence types that the recording system will need to incorporate:

- the results of desk-based research (see chapter 5) and historical information (see chapter 9);
- geographical information;
- environmental information;
- survey data;
- relationships between artefacts and the site (i.e. Harris Matrix – see below);
- drawings (artefacts and site);
- photographic records (artefacts and site);
- artefacts and samples (finds);
- conservation records (see chapter 16);
- interpretations (artefacts and site);
- resources for further investigation (a bibliography of reference material, specialists, museums, etc.).

A recording system must allow for:

- obtaining consistent, reliable and accurate information (usually from many different people);
- storing the information in an appropriate format (in terms of accessibility and long-term survival);
- cross-referencing between different categories of archaeological material;
- efficient and effective interrogation of all the information held, during the project and long after it has reached completion.

When deciding on an appropriate recording system for a project it is best to consult existing examples and texts to identify common pitfalls and best practice. It is important, however, to remember that each existing system has been designed for that organization and project's particular approach and recording task. It may also have been designed for use in conjunction with a specific recording manual (Spence, 1994). Do not faithfully adopt any existing system without understanding how and why it was designed.

PLANNING THE RECORDING: WHAT TO RECORD

Familiarization with all aspects and categories of information likely to be encountered on any given site is essential. Does the system only have to record boats, pilings and harbours, or other types of structure as well? Compile a list of information that it is necessary to record on each site. It is important to remember to record both observations (what is seen) and interpretations (possible meanings) as fully as possible, but not to confuse them.

It is very important when thinking about what to record to consult the specialists who require the information. For example, if the work involves recording cannons (see appendix 2), consult the current authorities and documentation to determine exactly what should be recorded in addition to the normal level of artefact recording.

Recording should begin as soon as any category of evidence is found. At the earliest possible stage, all archaeological material should be given some form of unique identifying number (e.g. artefact number – see below).

RECORDING INFORMATION ON SITE

The following section will summarize some of the important points to consider when recording and storing information from an underwater or foreshore archaeological site.

Site notebook: Traditionally, the director only entered details of a site into a single (or several) site notebook(s). This has the advantage of being easy to set up, portable and the format can be flexible. The disadvantage of this method is that it is difficult to record all aspects of archaeological material consistently and objectively. On longer projects, the range and quantity of information can become overwhelming and impossible to organize. It is also harder to extract the information when the time comes to analysing it.

Pro-formas: To help overcome the problem of organizing overwhelming quantities of data, it has become normal practice to use pre-printed forms (or 'pro-formas') for recording information on an archaeological site. The recording forms are generally completed by site-workers and each form asks the recorder questions, prompting the recording of a consistent level of information. Regardless of the number of pieces of information that have to be recorded, or how many different people are doing the recording, the same details should be noted. Entered in an ordered, standardized manner, the information should also be easier to consult and analyse. The disadvantages of forms are that they require more preparation in advance and their purpose and meaning must be clearly explained to all site-workers who will be using them. A previously completed form can be used to provide an example of appropriate use and a set of guidelines to follow if any confusion or ambiguity exists concerning the pro-forma itself. Badly filled-in forms produce information that is as incomplete and garbled as any single source such as a site notebook.

Not all details will need to be recorded about each artefact or feature. An ideal form would be limited to only those questions that are necessary. A clear and logical layout for the form is important. Ideally, relevant details will be grouped together into easily recognized categories of evidence (e.g. context details, object details, sample details). Specialized sheets can then be used for recording specific categories of evidence, such as timber (Milne et al., 1998), vessel remains on the foreshore (Milne et al., 1998) or guns (see NAS website for recording forms).

Pro-formas are also used for recording the existence, location and relevant information relating to plans, photographs, video footage and survey information from the site. As long as all the forms are cross-referenced, using pro-formas makes the task of recording on site much more manageable. However, in general, every project should also have a general site project-book that serves as a diary, records a variety of non-structured information, and can be used for notes during planning. Such a book is particularly useful for noting the reasons behind decisions that were taken and documenting non-archaeological but significant events during a project (e.g. compressor malfunction, personnel problems or simple flashes of inspiration) which might otherwise go unrecorded.

All recording systems should be as simple and straightforward as their purpose allows.

Archaeological dive-logs: The primary record of work under water will be the archaeological dive-log. Dive-logs are the primary source of first-hand observations and, as such, will be referred to frequently during post-fieldwork processing. They will also provide an important insight into the effectiveness of the diving operations and the effect of working conditions on the information recorded. It is important to enforce the completion of dive-logs as soon as possible after the dive. They should include information on:

- the diver (name, equipment);
- the dive (time, depth, temperature, decompression);
- the conditions (visibility, current, environment);
- the planned work (tasks, equipment);
- the results (measurements, observations, sketches, cross-references to other records);
- any thoughts on interpretation;
- any finds recovered, providing a description, a location in a sketch, and measurements from survey points to the object;
- artefact numbers (these should be recorded on the dive-log once the number has been assigned, whether it is assigned on the sea-bed or at the surface).

On some sites, dive-logs are restricted to personal diving-related information while archaeological information is recorded on a drawing board, which may be worked on by several people during the working day (just like the records kept within a trench on a land excavation). This means data does not have to be transcribed or remembered.

Recording objects/artefacts: It is important to keep an open mind and record all evidence with equal care. Animal bones, fish bones, clam shells, etc. should receive as much attention as gold coins. It is important not to discard or destroy materials/deposits simply because they do not appear to be of immediate value. The most unattractive or unlikely items could be ancient packing materials or the last traces of a delicate object. It is particularly important to record the associations of finds; such information may be crucial to determining whether the material was the contents of a container or the packing around it. Materials relevant to such questions can be very insubstantial, so all details observed must be recorded, even if their true significance is not yet understood. Even discoloration on an artefact can indicate the previous presence of something else that has since eroded away. For example, the presence of a black 'inky' staining or residue on an artefact might indicate the presence of gunpowder nearby. Or a white layer of sediment on wood might be the remains of a whitewash.

The ways in which objects can be recorded in detail once on the surface are described in chapter 18 and the appendices. However, it is often not necessary to raise objects to record them properly. Guns, structural features and even pottery fragments have been effectively recorded *in situ* without damage to the site and the subsequent risk of information loss that can often be caused by raising them to the surface. A good *in situ* record of an object can sometimes be enough for specialist analysis and even for initial publication. If the material is to be raised, and it appears to be particularly delicate or fragmentary, it is worth spending more time on a detailed record of it while still *in situ*.

In situ recording should include the unique identification number (artefact number), a description of the object, measurable dimensions and a sketch plan with details of location, orientation, associated material/finds, appropriate survey measurements and any important features visible (figure 8.1). A more detailed *in situ* record would include measured drawings and more intensive photographic recording. Sometimes a quick photo with a label will be helpful in clarifying the written information of a particular find, even before it goes to specialists

or conservation. The final photograph of the artefact, however, should have appropriate lighting, the artefact cleaned, and scale and labelling consistent with publication standards where possible.

Certain classes of object and material are particularly common in maritime archaeology, for example, guns (which are common on wrecked vessels), anchors and timber in the form of elements or complete structures such as ships, harbours, bridges and wharfs etc. For specifics concerning the recording of guns and anchors, please consult appendices 1 and 2. For timber recording see below.

There are general factors that need to be recorded for each object:

- Position: for example, site name/code, trench code, location measurements/position co-ordinates.
- Unique identification number (artefact number).

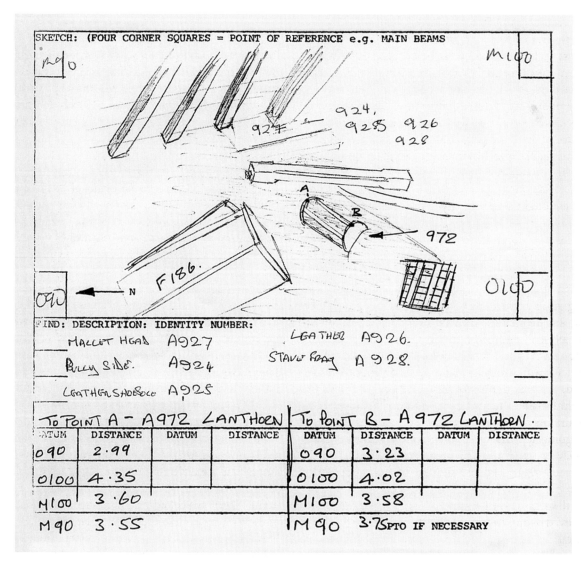

Figure 8.1 *In situ* recording: a copy of part of a diver's recording form completed during excavation of the *Mary Rose* in 1982. (Mary Rose Trust)

- Description: name of object (including hierarchal listing, such as metal, iron, nail), form/shape, sediment colour, sediment texture, sediment compactness, composition (material type), condition, dimensions, date found, relationship to other objects.
- Relationships: how the different layers and features of a site relate to one another (see Harris Matrix below), and how it is determined which layer or artefact is of earlier or later date than its neighbour.
- Associations: for example, associated with, types of object found in the context, timber jointed to, timber fastened to, or even lying next to another. Orientation of objects is also a very important detail to record.
- Interpretation and motives: for example, methodological comments, reasons for taking sample, method of excavation, notes on circumstances of recording, excavation, recovery of material, interpretive comments, etc.
- Annotated sketches, drawings, plans, scans.
- Co-ordination of records: for example, relevant plans and sections, photographs/video footage, dive-logs, conservation records, storage records, scientific analysis.
- A significant aspect in the interpretation and co-ordination of records is knowing who has been doing the recording and interpretation. For this reason, useful fields in the recording system include: who made a comment, who recorded a feature, who checked the recording, which expert opinion was sought.

RECORDING TIMBERS

Two main methods of drawing timbers are commonly used. Making scaled drawings and making full-sized tracings (which are then often scaled down by hand or photographically). Equipment required may include clear polythene and spirit-based pens (i.e. water-resistant) for tracing. Perspex sheet and chinagraph pencils can also be used. A variety of coloured pens are very useful to highlight features such as treenails, iron nails, concretions or repairs on full-sized drawings. However, if the drawing is going to be reduced photographically it may be necessary to use black pens and establish clear conventions to distinguish between the various features. Supports for timbers are useful. If the timber is at a comfortable height then excessive bending can be avoided. If timber is removed from wet storage for drawing it must be kept wet.

Scaled drawings: Wood should be drawn at a scale that is most useful for the level of detail required. This is often 1:10 but larger scale drawings are frequently made, in particular to record complex relationships between timbers. It is not uncommon for 1:1 tracings to be made. At this level of recording, an attempt should be made to show all the major structural features such as holes, notches, joints, fastenings and damage (such as that caused by wood borers or charring by fire). It is important to ensure that all the timbers on the drawing have been individually marked with their unique timber numbers.

There are several methods of drawing timbers in plan. The shape of individual components of a hull can be recorded by means of measurements added to sketches. These enable the main features of construction to be recorded and act as a control on shrinkage and distortion that may subsequently occur. The direct measurements can be used as a basis for the artist's reconstruction in conjunction with 1:10 drawings made from tracings. A drawing frame can also be used, though some control measurements and additional notes are needed to eliminate possible inaccuracies and to allow the incorporation of less-visible features.

On a scale drawing, many of these details can be shown using a standard set of conventions and symbols. In selecting conventions, care must be taken to ensure that they are consistently used and that symbolic representations are not confused with realistic representations of actual features on the timber.

Tracing timbers: An alternative to scale drawing is to trace the details of the timber onto transparent film at 1:1. This can be done directly onto polythene sheeting, or acetate film, which is actually laid onto the timber being drawn (plate 8.1). Waterproof pens are used to trace the features. Laying the sheet directly onto the timber reduces distortion caused by parallax.

Drawing onto clear PVC rigid sheeting, supported horizontally over the timber, with a chinagraph pencil works well under water and on land. Once on the surface the drawing can then be photographed and/or transferred onto polythene sheeting and the PVC sheet wiped clean ready for re-use.

Care is needed to reduce distortion when using any tracing-based method to record more complete three-dimensional shapes, and the drawing should be backed up with linear measurements. For example, a tracing made by laying a polythene sheet directly onto a very curved timber (such as a rib or frame) will not produce an accurate plan view, but an expanded view of the timber surface. To be useful, such a tracing must be accompanied by a side tracing of the timber or measurements to describe the curvature.

The results can be reduced photographically for redrawing and publication. Before reduction, a standard scale should be imposed onto the drawings to allow the accuracy and consistency of the reduction to be checked very carefully. It is very important to remember to label all drawings with identification numbers and to mark the position of any cross-sections that have been drawn.

Some points to remember:

- Care should be taken to ensure that distortion in the drafting film or movement during tracing is minimized. Light pinning or weighting can help but care must be taken not to damage the timber.
- Reference points should be marked in all directions and measurements should be made between these points to check for distortion. The marks are also useful for checking the accuracy of subsequent copies.
- The tracings are waterproof and useful for checking the originals for shrinkage. However, even waterproof ink on polythene will rub off and care should be taken in their use and storage.
- Tracings are useful for working out displays where full-size paper templates are often used, and for checks during conservation.
- Tracings must be accompanied by section or profile drawings, preferably on a level datum to record twist. Section lines should mark the position of section profiles.
- Tracings can be fast and cheap. Symbolic conventions may be used to designate particular features such as details of fastenings (e.g. iron nails may be shown as red circles; wedged treenails as hatched circles, etc.). Measurements of fastenings should be marked on the sheet.
- With plank-type timbers it may not be necessary to draw edge views – only faces and profiles.
- End views of planks showing the orientation of the rays of the grain are an important aid to reconstructing the method of timber utilization.
- A photographic record should accompany the drawings.

Further information about recording timbers, including a timber recording form can be found in *Nautical Archaeology on the Foreshore* (Milne et al., 1998).

RECORDING CONTEXTS

What do archaeological contexts look like? Contexts can be categorized as structures, cuts (e.g. scour-pits), fills and layers (see chapter 4). The easiest context to recognize on a shipwreck is the vessel's hull, which survives as a coherent structure and is an obvious indication that the vessel sank to the sea-bed. Collapsed parts of the hull may represent the events in the disintegration of the ship structure. Collections of objects (e.g. a pile of cannon balls or a galley oven made up of a large number of bricks) must also be recorded as contexts as this helps the overall future interpretation. The digging of any hole on an archaeological site, by nature or by humans, is obviously a very important event or process and, as such, should be recorded as a context. These holes or voids may fill up with sediments and other material and the hull can become buried, sometimes in recognizable stages. Each layer of the infilling material should be designated as a context, as each represents a specific episode in the history of the site (figure 8.2). In general, stratigraphy (see the section on dating in chapter 4) under water can be as complex, or more so, than on terrestrial sites, due to factors such as movement of tides, scour, human intervention and the attention of the local flora and fauna, to say nothing of the conditions of the site itself, such as low visibility and dynamic water movement.

Most stratigraphic layers are recognisable because the material in them is slightly different (in composition, texture, or colour) to the neighbouring areas. Renfrew and Bahn (2004) provide a good discussion of stratigraphy and its appropriate recording. The differences between types of sediment may be so subtle that they are almost undetectable, so great care is needed to recognize them (plate 8.2). This is where personal interpretation and experience comes to the fore. All the variables that make up the distinctive character of a context should be recorded for each context encountered on a site. For deposits, this might include parameters such as: colour, texture, consistency, particle size (for sediments), bonding agent, sedimentary structures, shape, dimensions and precise location. Dinacauze (2000) gives further details about the nature of these parameters and their recognition on land sites.

RECORDING STRATIGRAPHY

When recording contexts, it is very important that the position of each one in relation to those around it is recorded. This can be done by a written description (below, above, within, etc.) supplemented by a diagrammatic representation detailing the sequence of individual contexts. The method of presenting such a sequence is known as a Harris Matrix (Harris, 1989; also see www.harrismatrix.com). Such diagrams are constructed as an investigation progresses to clarify relationships between contexts within the site (figure 8.3). An alternative means of representing stratigraphy is shown in figure 8.4. Measured drawings (or site-plans) of the physical relationships between contexts are also fundamental to the site record and must be cross-referenced to any other documentation relating to stratigraphy.

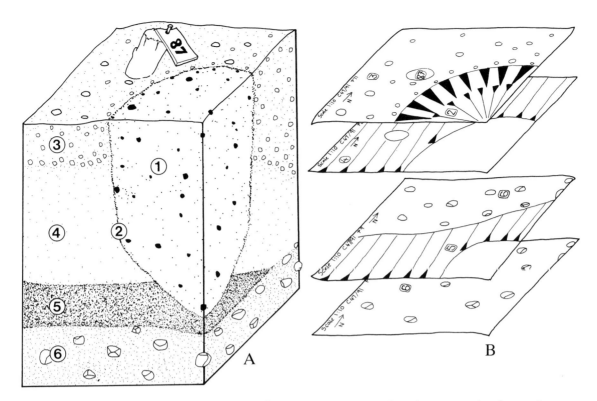

Figure 8.2 Planning contexts: A) a representation of some contexts on a site; B) an example of recording contexts in plan. After each context has been removed/excavated, a plan is made before the next one is removed. (Based on original artwork by Ben Ferrari)

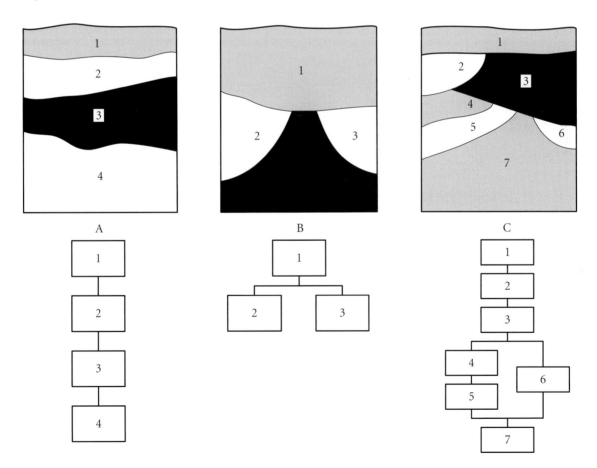

Figure 8.3 Stratigraphical diagrams (e.g. Harris matrix). A) a simple sequence of direct relationships between contexts; B) relationships can be demonstrated between contexts 2 and 1 but not between 2 and 3; C) using the basic principles shown in A and B, the direct relationships in sequence C can be clarified in diagrammatic form. (Based on original artwork by Ben Ferrari)

RECORDING ENVIRONMENTAL EVIDENCE

The way in which the environment of a site influences the survival of evidence has already been discussed (see chapter 4). Having identified the significance of this area of study, it is important to develop effective recording strategies for it. Consulting the specialists whose work might be most directly influenced by environmental factors is essential. The conservator will want to know the details of an object's burial environment so that the optimum conservation treatment can be decided on. Scientific dating methods may be influenced by factors in the site's environment and those responsible for dating the objects will want the relevant information in a usable form.

Space should be provided on dive-log sheets, or survey and excavation record sheets, to record details of localized environmental factors around individual objects or areas of structure. Studies of general environmental factors affecting the site may require a specifically designed form to accommodate all relevant information.

RECORDING SAMPLES

Investigating environmental characteristics of a site might involve taking samples for subsequent analysis in controlled conditions. Non-artefactual deposits might also require sampling for study. It is very important to

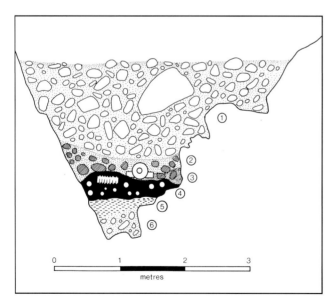

Figure 8.4　Section through a gully on the wreck of *El Gran Grifon* (1588). The stratigraphy is as follows: 1) mobile stones and shingle; 2) partially stable stones and shingle with organic staining; 3) a bronze gun in contact with level 4; 4) concretion level with abrading top surface and loose lead shot also present; 5) organic sludge, deriving from pre-1588 weed; 6) pre-1588 deposit of stones and shingle. (Colin Martin)

record what proportion the sample is of the whole and to record that proportion carefully on the sample record (e.g. 15 litres recovered of an estimated 50 litres). The sampling procedure used to recover the samples should be noted in detail. The sampling strategy adopted in the field together with some indication of the density of material collected over the site or the concentration within particular features should also be included. In addition, details relating to the length of time and condition of storage, together with the current location of the original samples (and all sub-samples if located elsewhere) should be recorded. Each sample should be assigned a unique identification number (see below).

RECORDING SURVEY RESULTS

The results of any survey need to be recorded and stored as carefully as any other evidence. See chapter 14 where the survey process is discussed in detail.

RECORDING PLANS AND SECTIONS

Drawings in general should be of a standard scale, usually 1:1 for small objects, 1:2 or 1:4 for larger finds, up to 1:10 for most site-plans (though more reduction may be necessary before publication). The following information should be recorded on each drawing:

- code/site name
- plan number
- subject (plan or section of what?)
- recording person
- draughtsperson
- date
- scale
- position (e.g. grid coordinates for plan square)
- orientation (e.g. north indicated on plans, and the direction sections are facing).

This information should also be recorded alongside the drawing number in a drawing register, a catalogue of all drawings relating to a site/project.

RECORDING PHOTOGRAPHIC RESULTS

Archaeological photography is discussed in detail in chapter 10. However, it is worth emphasising a few important points regarding how photography relates to the artefacts and the project itself.

A record of all the photographs taken on any dive should be made on a log sheet for the project. This is essential, as many details of pictures will often be known only by the photographer and may easily be forgotten in the future. Experience has shown that it is best to make brief notes when possible about the photographs while still under water, and then write up the log sheet soon after the dive. The documentation will then be finished when the results have been downloaded, or the film has been developed and the exact frame or image numbers are known. Each sequence of shots should be numbered and dated, with the name of the photographer, the type of camera, its number (if more than one is used), whether the shot was black and white or colour, and any other technical aspects, such as resolution (for digital cameras), f-stops, lens type, etc. Of course, the most important information is the subject matter for each shot, but the angle of view, the orientation and the area being photographed are also important for cross-referencing, and need to be recorded on the photographic log. Specific technical information regarding each camera can be recorded separately from the logs and act as the baseline dataset, which is particularly necessary for recording the test exposures of a site, and especially when using film.

Even when using a digital camera, it is important to record the technical details: type and make of camera, lens setting, flash or not, resolution (in terms of dots per inch or pixel-by-pixel file size), macro or 35 mm or telephoto lens. These details will allow tracking of the source if any distortions appear in the shots. These images can be noted in the database (see below) as a link to the image file. Artefact photographs for the database should include the site code, artefact number and a scale.

Video footage (digital or analogue) should be dealt with in a similar way but also requires a written description of the subject, especially if there is not a verbal commentary. It is easy to be disoriented in murky waters, and one close-up of a timber can look very much like any other. A log of the video footage should be compiled on a running-time basis, with each tape numbered and dated, with time-coding and subject annotation, as this will also be of great value when editing. When preparing the edited version of a video survey it is essential to note which tapes the footage was taken from and store this information with the edited compilation. If someone later sees an area of interest, then the video footage can be easily traced to the specific tape which might show more detail. Mark the video cassettes and the boxes clearly so that if they are separated the tape is still easily identifiable. Ensure the tapes are stored appropriately. In the case of digital footage, the following information should be included: the resolution, file size, compression type and media type (i.e. mpg, mpg2, avi files).

Be sure to download and check each video on a generic computer (whether Mac or PC) to ensure that the camera coding has not failed. This should be done before the end of the project, so that if something is not good, or has become corrupted, the recording process can be repeated. When storing each digital video sequence, include an appropriate 'reader' (i.e. the software that can play back the recording). Technology changes too fast to be complacent about this step.

CONSERVATION RECORDS

Once in the conservation laboratory, or on-site base, recording/registration of each find must take place before embarking on any treatments. Copies of records will be needed for reference during treatment, especially if conservation is being carried out by specialist laboratories. Details of how the object was stored while awaiting transport to the conservation facility must be noted, along with details of any on-site initial treatment.

Prior to undertaking conservation treatments, or before the long-term storage of un-conserved objects, it is essential that objects or assemblages of objects be photographed with a scale and label. A small thumbnail or contact-print of the artefact should be attached to the record cards or added to the database record to aid identification. A full record of all the treatments applied to an object should be kept. Almost all objects will require further treatment in the future and this will be more effective if the conservator knows the detailed history of the item including the specific solutions, adhesives, chemicals or solvents employed. Chapter 16 looks in detail at archaeological conservation.

IDENTIFYING ARCHAEOLOGICAL MATERIAL

All archaeological material identified on an archaeological site should be assigned a unique number. In certain circumstances, such as a group of identical objects (e.g. musket balls found together), a single group or collection number may be assigned. Finds should be allocated their unique number at the earliest opportunity.

The project-numbering system for finds should be as simple as possible, but include the site code, the year, the artefact number, and possibly the trench name. For example, [SHIP00 A001] would represent 'Scarborough Harbour International Project, 2000, Artefact 001'. All numbers should be assigned to an object because gaps in numbering can lead to confusion during post-excavation analysis. A 'number register' or master list should be kept for each project so as to record when a number is not used, lost, or voided. The last number used should also be clearly noted on the master list. A string with pre-numbered garden tags is an effective way of keeping track of which numbers have and have not been used. These combined techniques help to ensure that items are not lost at any stage of the project. Separate sequences have been used on large sites to define whether the recorded item is a large timber, environmental sample or artefact, etc. On smaller sites only one sequence may be necessary.

TAGS AND LABELLING

Each item will need to be labelled with its unique identification number/code. It is best practice to do this as soon as possible, to cut down the risk of loss or confusion. Numbers can be assigned and labels attached while the material is still on the sea-bed (plate 8.3). If this is not possible, however, then a system for doing this on the project boat or work-platform is necessary. A supply of relevant labelling materials will be required, along with the number register or master list. The sooner this information is transferred to the main site recording system the better.

Labelling should be attached to the material as securely as possible without causing any damage. Labels can be attached to archaeological material by means of nylon fishing line, or packed into Netlon plastic mesh along with fragmentary objects. In the case of large timbers, the labels can also be attached using dipped galvanized or copper nails, or stainless-steel pins. Never use nails that are barbed on the end; when these are removed (for photography, for instance) the wood will be damaged. Samples and other materials that are double bagged should have a label inside the bag and both bags labelled on the outside. Garden tags attached with poly-twine seem to be very successful when used under water and with waterlogged objects. These should be marked with a waterproof indelible marker, preferably one that is also light-proof so that it will not fade over time. Staedtler permanent Lumocolor markers have been recommended, but other similar products will be available.

It is vital to ensure that archaeological material and the associated number stay together, especially during recording and analysis when the object may be handled a number of times by different people. The final step is physically to mark the object with its number and site code to be sure that its identity cannot be lost and so that any researcher can refer back to the original evidence. If material is destined for a particular museum, consult the curator about how and where objects should be marked (plate 8.4).

STORING THE INFORMATION

As discussed above, information from an archaeological site will take many forms, from original dive-logs to the artefacts themselves. Each class of information must be

stored in an appropriate manner with due attention to its long-term survival and accessibility for future research.

A unified storage facility can be created for all information from a site/project in the form of an electronic database (figure 8.5). In archaeology, as in other disciplines, computerized record-keeping is now standard, and is often used in conjunction with paper (or waterproof-paper) records because it is difficult to enter data directly into a computer when working in a wet environment. The key advantage of entering information into a computer is that, provided the information is entered using an appropriately designed database program, it is easier to interrogate and analyse the records quickly and effectively. This is the case regardless of the volume of data and their complexity. Although some would think it a disadvantage to use both computer and paper records – as the data must be double-handled – the amount of effort required has two advantages. First, the data entry acts as a double-check and decreases the likelihood of errors creeping into the records, and, second, it provides an immediate alert should data or measurements need to be confirmed or further investigated while still on site.

A database is the fundamental collection of information, but the database program makes the manipulation

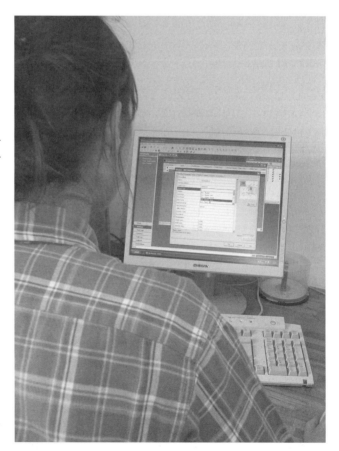

Figure 8.5 An archaeological database containing all project information. (Photo: Kester Keighley)

and presentation of these records much more efficient. Use of a database allows for comparison between evidence both from different parts of the site and between different sites, the formation of bibliographic files and the cataloguing of objects for further study. A database program further enhances the more mundane aspects of site management: site records, work rosters, artefacts, samples, surveys, site reports, budget/finance and payroll. Many good systems are currently available and custom-made designs are ubiquitous. The important consideration is that they fulfil the requirements for record-keeping, thereby speeding up the interpretation process and making the data more widely accessible.

There are, however, a number of important points to consider in relation to the use of computers in archaeology.

COMPUTING OPTIONS AND ISSUES

The field of computers is very dynamic and rapidly changing. The hardware and software considered powerful today is often outdated or obsolete in a very short time. Those responsible for setting up a computerized recording system must be familiar with the latest technology and software or consult those who are. The best approach is to think about systems that can also be 'future-proofed' (i.e. carried over into the future due to the versatility of the software or the simplicity and flexibility of its output). Not all of the most popular software will still remain prevalent in 20 years' time. The Archaeological Data Service (ADS – see below) will be able to offer advice in this area. Research on the internet is also advisable because there are various options available for even the smallest budget, both for the current project and to provide for data to migrate into future computer software.

These days many archaeological projects have pre-designed recording systems from the outset. Putting all the information in one place can be very convenient. Regardless of the system adopted, however, the information becomes vulnerable to total loss if a problem occurs with the machine being used. A carefully programmed system for backing-up must be established, making copies of the information on a separate storage device in order to minimize any loss of records resulting from a problem with the main device. The copies should then be stored in a safe place well away from the original so that a single disaster will not destroy both. Two separate hard drives, with one acting as a back-up, is far more convenient than copying onto smaller-capacity disks or memory sticks.

Technical issues, compatibility issues, and preventative measures must all be considered as well. If an archaeological site is away from standard power supplies, investment in an uninterruptible power supply (UPS) can be critical. In addition to a UPS, guarding against electronic disasters involves surge-protectors, copies of all relevant

software necessary to run the project as well as the computer and, of course, carrying out daily back-ups.

GEOGRAPHICAL INFORMATION SYSTEMS

Geographical information system (GIS) software allows display and analysis of archaeological and other data in a graphical form and in four dimensions. Once the data is in the system, a GIS package allows interrogation of datasets in a wide variety of ways. Layers of information can be placed one on another so that spatial, chronological, and other trends can be observed in an area, through time, on one site, between sites or even in one small part of a site. Statistical analyses, viewshed (a map of the line of sight from any given location) analysis, as well as historical comparisons are possible using a GIS-based archive with a database. Currently, there are many examples of GIS software, but the most common are ArcView, and MapInfo.

EXPLAINING, DOCUMENTING, AND SUPERVISING THE SYSTEM

There are a few guidelines to consider when deciding on a recording system:

- The more sophisticated the recording system, the more explanation it will require for people to be able to use it and the more room there is for error.
- It is very important when designing a system for recording a site that future researchers will be able to understand how it works. It must therefore be effectively documented, including the database, which ideally should have a schematic layout of how the database is designed (i.e. what the relationships are, what the identification names are, etc.).
- Ambiguous terms should be avoided. With any system it is advisable to use (or if necessary create) a specific glossary or reference book that will provide a list of terminology to be used. This will ensure consistency and common understanding among project workers. Online glossaries and thesauruses are now available (see Further Information below).
- It is important that information is recorded fully and reliably for each part of the site. Mistakes and ambiguities do occur but data-entry can provide a check system and is therefore a good idea. On a small project, this is likely to be the job of the project leader but on larger projects it is more effective if a single person is appointed. This role is usually combined with administering the records and the allocation of unique identification numbers, such as artefact numbers, sample numbers, etc.

- Pull-down menus in a database help to standardize descriptions and avoid spelling mistakes – a database search will only produce a list of, for example, all the pocket sundials, if they are all described in an identical way, including correct spelling.

While the recording side of any archaeological project can at times seem tedious and daunting, it is important to remember how both the accuracy and thoroughness of such records can provide an enormous payback when the project reaches publication stage. Only with meticulous attention to detail and a well-designed, well-managed and well-maintained recording system will it be possible for research to be recognized, validated, and presented to the academic world to a high professional standard.

FURTHER INFORMATION

Archaeology Data Service, Department of Archaeology, University of York, King's Manor, York, YO1 7EP. Telephone: +44 (0)1904 433954; fax: +44 (0)1904 433939; e-mail: help@ads.ahds.ac.uk; website: http://ads.ahds.ac.uk/.

Dinacauze, D. F., 2000, *Environmental Archaeology, Principles and Practice.* Cambridge.

Harris, E. C., 1989 (2nd edn), *Principles of Archaeological Stratigraphy.* London.

Institute of Field Archaeologists, 2007, *Draft Standard and Guidance for Nautical Archaeological Recording and Reconstruction.* Reading (www.archaeologists.net).

Petrie, W. M. F., 1904, *Methods and Aims in Archaeology,* London.

Renfrew, C. and Bahn, P., 2004 (4th edn), *Archaeology: Theories, Methods and Practice.* London.

Spence, C. (ed.), 1994 (3rd edn), *Archaeological Site Manual.* London.

RECORDING SYSTEMS

English Heritage National Monuments Record Thesauri (on-line): http://thesaurus.english-heritage.org.uk/

Gawronski, J. H. G., 1986, *Amsterdam Project: Annual Report of the VOC-Ship Amsterdam Foundation 1985.* Amsterdam.

Milne, G., McKewan, C. and Goodburn, D., 1998, *Nautical Archaeology on the Foreshore: Hulk Recording on the Medway.* RCHM, Swindon.

RECORDING CONTEXTS

English Heritage, 2004a, *Geoarchaeology: Using Earth Sciences to Understand the Archaeological Record.* London. (www. english-heritage.org.uk/upload/pdf/Geoarchaeology-2007. pdf)

Hodgson, J. M. (ed.), 1997 (3rd edn), *Soil Survey Field Handbook. Soil Survey Technical Monograph No. 5.* Harpenden, UK.

COMPUTERS AS PART OF THE RECORDING SYSTEM

Conolly, J. and Lake, M., 2006, *Geographical Information Systems in Archaeology.* Cambridge.

Eiteljork II, H., Fernie, K., Huggett, J., and Robinson, D., 2003, *CAD: A Guide to Good Practice.* Oxford.

Fletcher, M. and Locke, G., 2005 (2nd edn), *Digging Numbers: Elementary Statistics for Archaeologists.* Oxford.

Wheatley, D. and Gillings, M., 2002, *Spatial Technology and Archaeology: the Archaeological Application of GIS.* London.

Historical Research 9

Contents

Ships usually represent a massive capital outlay to states or mercantile enterprises, and consequently generate an abundance of paperwork – building specifications; accounts relating to running, maintenance and repair; tonnage and capacity calculations; manifests of cargo, provisions, equipment and armament; crew and passengers lists and so on. From the 16th century, plans and construction diagrams begin to emerge . . . should a vessel be wrecked, yet more documentation may be generated by enquiries into its loss, the fate or survival of those on board, insurance assessment, and salvage. Such material may touch on matters far beyond the immediacy of the shipwreck. (Martin, 1997:1)

Historical research is a requirement of maritime archaeology, not an optional extra. Careful research, just like project planning and logistical preparation, can save a great deal of time. Like archaeology, to be most effective, historical research requires appropriate skills and experience.

The rewards of archival research can be great. At the very least, historical records can provide interesting background information about a site, which can enhance future visits. In some circumstances, historical records have helped to locate a specific site or vessel, or discover the identity of a vessel that has already been found. Any presumed 'fact' identified during archival research should, where possible, be supported by confirmatory evidence from a wide variety of different sources. Many different levels of research can a be undertaken into historical sources – whether one is seeking to identify a specific site or vessel, researching a known location, or merely looking into the history of a broader area.

Archival research also presents a series of challenges:

- Where is the required information held? This could be in an obscure archive anywhere in the world or an unexpected, even undocumented, section of an archive.
- How can the material be accessed? Look for published or unpublished guides, handbooks to the archive, computer or card catalogues. Above all ask the librarian/archivist for help.
- What skills will be required to read the original documents? Old handwriting, even from the nineteenth century, can be hard to decipher and foreign languages or dialects even harder. Language also changes markedly over the years and many documents, particularly official documents, were once written in Latin.

It is not the objective of this chapter to list all the different sources, archives and methods of historical research available. However, it will introduce some basic concepts of historical research and summarize the main types of archive that exist.

TYPES OF EVIDENCE

There are many types of written record, ranging from eye-witness accounts of sinkings to newspaper stories of chance finds by fishermen. The accuracy of recorded positions and the detail of accounts will vary but this is still a valuable, if time-consuming, source of information for locating or identifying underwater and foreshore sites.

When using historical data, it is essential to understand the difference between primary and secondary sources of evidence. Primary sources are original documents produced at the time of an event, such as ships' and lighthouse-keepers' logs. Secondary sources are analyses of an event or original documents, such as newspapers, books containing information on shipwreck losses, or publications based on primary or secondary research material.

Secondary sources can be much easier to read and very useful because they frequently include an index. They are often the best way to understand the basic history of a site or event, providing sufficient knowledge in order to research further by accessing primary data. However, secondary sources alone are frequently insufficient for the types of research required by maritime archaeology. Only by consulting primary sources, and doing it thoroughly, is it possible to avoid gross errors and the perpetuation of myths that surround some wrecks. A typical example of the latter is where a named ship carrying a cargo of perishable material like spices or furs is said to be lost 'full of valuables' in contemporary newspaper reports. This description can then be translated in succeeding generations of publications into a ship full of treasure without recognizing the true nature of the material and its likelihood of survival.

Even original historical documents (primary sources) can be partial or misleading in several ways. People only recorded what mattered to them at the time, and descriptions such as the nationality of a ship, or numbers of survivors, may not be accurate. In wartime, the stress and confusion of combat often meant that a vessel's location became confused. Conflict often affects the survival of documents, especially from the losing side. Records may not reflect the entire contents or character of a site or vessel. Sometimes shipping companies or masters intended to ship one thing but, in the event, sent another (especially in times of war). Mistakes during loading and unofficial cargoes (smuggling) can complicate the situation. In addition, cargo manifests and other historical data rarely cover the personal possessions of crew and passengers.

Records may depend on the subsequent history of the site. For instance, administrative boundaries change over time, as do record-keeping organizations. There can also be changes of ownership of documents, archives or even locations of archives. Vessels can also change hands owing to purchase, capture or theft. For example, the seventeenth century warship *Hazardous* was originally owned by the French before being captured by the British and put into use as a warship, so archives relating to the vessel exist in both France and England.

Factors such as the destruction or loss of documents, deliberate errors, lies and accidental mistakes further complicate the interpretation and understanding of primary sources. In addition, many sites are simply unrecorded, especially prior to the sixteenth century, and often much later in remoter areas.

LOCATING PRIMARY SOURCES

Local and national archives throughout the world contain vast amounts of historical information relevant to underwater and foreshore archaeology and there are also many documents in private hands. A major drawback is that many documents, even in major collections, may be uncatalogued. Even if they are listed by their title or main subject, the contents may not have been read since the writer completed the document. Of course, many archives are in good order and can be investigated by the diligent historian with great success. Ships' logbooks and cargo manifests may be relatively easy to track down, if they have survived, but important information about specific vessels, people or events can turn up in apparently unrelated documents. These unexpected discoveries are one of the joys of historical research but, until all archives are catalogued and accessible, tracking down primary information about a specific site will usually be a difficult task.

Archives can include a wide variety of different materials. While the most common of these are written or printed other sources include:

- maps and charts;
- ships' plans and models;
- iconography and imagery (such as carvings, paintings and even graffiti);
- oral history (such as recordings of survivors, or witnesses to an event);
- place-names;
- aerial photographs;
- memorial plaques and stones (figure 9.1), rolls of honour, etc. in churches, town halls and public buildings.

Each category listed above can include a variety of types and formats, depending on the country or area and its administrative history. Some examples of the most useful types of primary documentary evidence for maritime archaeology include:

Figure 9.1 Representations of ships and boats can often be found on stone, as monuments or graffiti. This eighteenth-century gravestone near the River Tay in Perthshire shows a salmon fisherman's square-sterned coble, complete with his catch. (Photo: Colin Martin)

- shipbuilders' plans and notes associated with the construction of a vessel or class of vessels;
- ships' logs, equipment and repair lists, lading bills (shipping receipts), muster rolls (list of names), passenger lists, and other materials produced during the life of a vessel;
- military archives, which can include combat reports or war diaries, ships' logs, regimental and vessel histories, etc.;
- lighthouse-keepers' logs, and, from the mid-nineteenth century onwards, lifeboat records;
- port and customs records, recording the vessels, cargoes, harbour dues and customs paid;
- contemporary accounts, private letters and papers, including travellers' or crew diaries and journals;
- ship-breaking records.

In addition, a wide variety of other records can contain information of interest to maritime archaeologists. Britain, especially England, is particularly rich in some of these types of record because of its geographical size, political stability and early form of centralized government. However, such records also frequently survive in other countries.

Most records are made for legal or administrative purposes, such as collecting taxes. They will be found under various headings, and may not tell us what we want to know. Customs records, for example, only record goods on which customs duties were charged, not all cargoes.

There is an extensive variety of different locations for all these historical data – too many to list comprehensively here. While most surviving historical documents are now held by major archives, many will still be found within libraries, museums, churches, and private companies.

Local sources (UK): The best place to begin research is locally, either near the researcher's home or near the site or location being researched. This research can be as simple as asking at the local library, museum, dive-club, or even the local pub. Established local residents may remember the loss of a vessel or how a site was used in the past. Local dive-charter skippers may also have undertaken extensive background research into wrecks they regularly visit.

Most libraries and museums have local-history collections and archives. Stray finds or artefacts may have been brought to local museums and the staff may have a particular local knowledge relevant to any given research topic. Many universities allow the general public to use their library facilities, providing access to more obscure publications.

Some local archaeological and historical societies publish journals or transactions. Some of the more active local societies are involved in survey and excavation projects and may also produce unpublished reports. These should be consulted in order to gather an understanding of both the site and the region in general. Once these local opportunities have been exhausted, regional, national or international archives can be considered.

Regional sources (UK): Regional archives include county, borough and city archives. However, it is worth remembering that local authority boundaries may well have changed over time and documents may be stored elsewhere. One of the best regional sources of information are Sites and Monuments Records (SMRs), also known as Historic Environment Records (HERs). Although primarily designed to list archaeological sites on land under the requirements of the *Ancient Monuments and Archaeological Areas Act* (1979), an increasing number of these also list sites in the inter-tidal zone or under water. A number are available on-line and can be searched from anywhere in the world with an internet connection.

National sources (UK): There are numerous national organizations which hold primary and secondary documentary material that may be relevant to a research project. As well as the National Archives, the National Libraries also hold manuscript material. Wales was annexed to England early and has the same legal and administrative systems. Northern Ireland has its own National Archives;

Wales its own National Library. Scotland was united with England in 1707, but retains its independent legal system. It has its own National Archives and National Library and separate records for all but military and naval affairs.

English Heritage's National Monument Record (NMR), based in Swindon, holds a considerable archive of archaeological and architectural survey material, including drawings and photographs. The aerial photograph collection includes 600,000 oblique and 2 million vertical photographs. Similar records are held in the National Monuments Record for Wales (NMRW), the Monuments and Building Record, Northern Ireland, and the Royal Commission on the Ancient and Historical Monuments of Scotland (RCAHMS). In many instances it is possible to search the records held in NMRs via the internet. (Contact details are provided in the Further Information section at the end of the chapter.) Other countries will often have similar organizations holding regional or national sites and monuments records.

The National Maritime Museum (NMM), Greenwich, London houses the Caird Library, which contains books, academic journals, a vast collection of letters, logbooks, folios and manuscripts. The ships' plans and historic photograph collection is held at the Brass Foundry site in Woolwich. There are over 1 million ships' plans, including the Admiralty collection of the sailing navy from 1700–1835. Approximately 1 million prints covering ships and maritime related subjects are held in the library. A collection of over 2500 model ships is also held by the NMM, though they are kept at an outstation at Kidbrooke. In addition, the National Maritime Museum's PORT website (www.port.nmm.ac.uk) is a subject gateway that provides access to searchable catalogues of internet-based resources. PORT is organized under subject headings, one of which is Underwater and Maritime Archaeology.

The British Library, apart from being a copyright library of printed books, has been the leading repository for private papers since its foundation in 1753. The collection includes a vast number of personal and estate papers, including manuscript maps and drawings. It also holds the most extensive collection of Ordnance Survey maps in the country. Comparable libraries are the National Library of Scotland in Edinburgh, and the National Library of Wales. The latter, founded in 1907, is the central repository for public and private archives in Wales and contains a large collection of estate and family papers, illustrative and cartographic material and a set of sketchbooks second only to the British Library.

International sources: Researchers may find it necessary to visit an archive in another country. Over the course of thousands of years of seafaring, ships have been built and travelled between every country in the world with a coastline or river access, bringing with them materials of every shape, form and description. As a result, even relatively small and simple maritime archaeological sites in the UK may have materials from abroad.

Material discovered on an archaeological site may be something as small as an imported wine bottle or fragment of ceramic; equally, it could be highly diagnostic evidence such as coinage or weaponry that, once identified, can indicate who owned or operated a vessel or which coastal sites she visited. Countless foreign vessels have been wrecked off the coast of Britain over the years, so it is even possible that the majority of research on a wreck-site located in the UK will have to be undertaken abroad. Two of the best examples of this are vessels of the Spanish Armada (sixteenth century) and the Dutch East India Company (VOC) for the seventeenth and eighteenth centuries.

It can be expensive to live in another country, or even town, for a long period of time, so careful planning in advance of any foreign visit is essential. This can include making a preliminary visit to gain access to, and assess the extent of, an archive; find out where materials are; and make any necessary arrangements in advance. The internet and e-mail can be a great help here.

Private sources: Private and site-specific archives are often of use to maritime archaeologists. The most famous of these are the shipping records of the insurers Lloyds of London. Other good examples include the records of major port authorities as well as the records of major shipbuilders. In a rather different way, the manufacturers of specific items like ceramics, porcelain, and ships' fittings often maintain archives of their products, which can be extremely useful when attempting to identify materials from an historic site.

THE INTERNET

There is a vast range of websites containing information of interest to maritime archaeologists, but it can sometimes be hard to judge the authenticity or merit of a site. Any research undertaken via the internet should consider the following criteria:

- Can the full citation information from the website be stated, including author, date of publication, edition or revision, title, publisher, the date the website was accessed, and the full website address (uniform resource locator, or URL)?
- What can the URL tell you? Is it a '.org', '.ac.uk', '.gov' or other official or semi-official site, or merely a commercial '.co.uk' or '.com'? The former are usually more

trustworthy as they are not motivated by private profit.

- Who wrote the page? Is he/she or the authoring institution a qualified authority?
- Is the page dated or current and timely?
- Is the information cited authentic? Does it include references to other sources? Is any of this information verifiable?
- Does the page have overall integrity and reliability as a source, including evidence of objective reasoning and fair coverage? Or is there distinguishable propaganda, misinformation or disinformation?
- Who is the intended audience of the webpage? Are there evaluative reviews of the site or its contents?
- What is the writing style? Could the page or site be ironic, a satire or a spoof?
- If a website user has questions or reservations about information provided on the website, how can they be satisfied? Are there contact details for the author[s]?

One of the greatest benefits of the internet is access to online catalogues and databases. All major libraries and museums, and many smaller collections, now have online access. This can make research much easier, faster and cheaper, as the locations of materials can be checked online, and in some cases documents can be pre-ordered so that they are waiting for the researcher on arrival. However, it must be remembered that few online catalogues are comprehensive. If the material sought is not listed in an online catalogue, it can still be worth phoning the archive in question or even visiting in person, as many collections have far more extensive hard-copy catalogues, particularly of more obscure primary sources. Another caveat with computer searches is that they can encourage people to focus on the particular and ignore the general background, and this can be a big mistake.

Most major academic journals are also now fully or partially available on the web. This makes research into more obscure secondary data much easier. University libraries will have access to the means to search academic journals electronically.

METHODS OF RESEARCH

When undertaking historical research, it can be difficult to decide what to record. Collect too much data and the search will take an unnecessarily long time and be unwieldy; collect too little data and there may not be sufficient information to proceed with a project without returning to an archive at a later date.

Perhaps the most important aspect is to ask the librarian or archivist for help on arrival. When visiting an archive it is essential to speak to one or more members of staff because, no matter how good the other searching aids, the archivist's knowledge of the collection is usually unrivalled. Most are happy to provide a guided tour of an archive and explain where things are and how their catalogues work. Many are also fine scholars in their own right and frequently offer a wealth of additional advice on any given topic.

When visiting an archive to collect primary evidence, it is advisable to use a large book or a laptop and be systematic. Start by recording the title, author and place of publication of a published document, or the reference number of the manuscript, together with page or folio numbers (useful if requesting photocopies at a later date, and essential if it comes to including the reference in a publication). Sometimes the quantity of information involved favours a pro-forma to help systematize data collection. Only once this key information has been recorded should the research continue.

It is vital to store records safely, ideally making a back-up copy for storage elsewhere. When a project is complete, all these historical data should then be placed in the project archive (see chapter 19) so others can access and study it in the future.

If time is short, it can be worth employing an archivist to do the research, although there are both advantages and disadvantages to this. While it may be cheaper and quicker in the case of a specific archive that an archivist may know well, archivists generally follow instructions to the letter and, as such, may miss important items because they lack first-hand experience of the subject, including passing references to other events that may be vital clues.

Archival research can at first glance appear to be a dull and thankless task, a waste of time and resources and the antithesis of all that is good about archaeology and 'getting out there'. In fact archival research can:

- save countless hours of searching for a site in the wrong location;
- provide a wealth of historical context for a location or wreck; and
- present opportunities for networking with other researchers, which in turn can lead to a range of long-term benefits.

Well co-ordinated archival research can result in original documents and records relating to a site or event, even materials that actually belonged to or were used by specific individuals. The research itself can take place in a variety of old and fascinating locations – not just dry, dusty libraries but museums, art galleries and churches. Archival research might therefore be considered an essential and potentially stimulating and rewarding aspect of archaeology.

FURTHER INFORMATION

English Heritage, National Monuments Record, National Monuments Record Centre, Great Western Village, Kemble Drive, Swindon, SN2 2GZ. Telephone: +44 (0)1793 414600; fax: +44 (0)1793 414606; e-mail: nmrinfo@english-heritage.org.uk; website: www.english-heritage.org.uk/nmr

Monuments and Buildings Record (MBR) – Northern Ireland, Environment and Heritage Service, 5–33 Hill Street, Belfast, BT1 2LA. Telephone: +44 (0)28 90 543004, website: www.ehsni.gov.uk/

National Monuments Record of Wales, Royal Commission on the Ancient and Historical Monuments of Wales (RCAHMW), Plas Crug Aberystwyth, Ceredigion, SY23 1NJ. Telephone: +44 (0)1970 621200; website: www.rcahmw.org.uk/

Royal Commission on the Ancient and Historical Monuments of Scotland (RCAHMS), John Sinclair House, 16 Bernard Terrace, Edinburgh, EH8 9NX. Telephone: +44(0)131 662 1456; website: www.rcahms.gov.uk

RECOMMENDED READING

Ahlstrom, C., 2002, *Looking for Leads: Shipwrecks of the Past Revealed by Contemporary Documents and the Archaeological Record*. Helsinki: Finnish Academy of Science and Letters.

Kist, J. B., 1991, Integrating Archaeological and Historical Records in Dutch East India Company Research, *International Journal of Nautical Archaeology* **19**.1, 49–51.

Martin, C. J. M., 1997, Ships as Integrated Artefacts: the Archaeological Potential, in M. Redknap (ed.), *Artefacts from Wrecks: Dated Assemblages from the Late Middle Ages to the Industrial Revolution*, 1–13. Oxford.

Photography 10

Contents
- Photographic theory
- Digital photography
- Surface photography
- Photographing finds
- Underwater photography
- Underwater photographic techniques
- Digital darkroom
- Mosaics – photo or video
- Video cameras
- Video technique
- Video editing

Photography, both still and video, is among the most useful recording techniques available to the archaeologist. Photography can be used for generating a record of a site at a known time, and is also an effective tool for education and public outreach. Technological advances in cameras and digital image processing, combined with cheaper, user-friendly equipment and software, has resulted in the increased popularity of both still and video photography. The internet and computer processing of images have led to faster and wider dissemination of information and 'virtual' access to archaeological sites. However, despite such technical developments, it is still the end result – a good illustrative photograph or piece of video footage – which matters most. Figure 10.1 is a good example of a photograph showing the frames of a ship and it is visually enhanced by the presence of a diver sketching them.

This chapter aims to introduce the reader to a number of techniques, disciplines, and items of equipment that will enable the photographer to achieve acceptable results during an archaeological project, both on the surface and under water.

Although some basic issues and techniques will be discussed here, complete newcomers to photography are advised to refer to specialized books and to consult experienced photographers for a better understanding of the subject. This chapter is mainly intended to illustrate the additional thought processes and procedures required of a project photographer who has to work on an archaeological site, part of which may be under water.

A project will require photographs and video footage for a variety of purposes. However, it is advised that all images are taken with each specific purpose in mind. Working closely with the archaeologist, the task of developing the photographic archive is the responsibility of the project photographer. Together they should generate a task- or shot-list. To manage the task-list efficiently, it is advisable to categorize the shots under headings – for example: activities, people, techniques, processes, structure, artefacts, education, interpretation, enjoyment and sponsors.

To this end, it is essential to have a good idea of the reason behind the photograph or video sequence that is being taken, though without suffocating the artistic merit and spontaneity of a shot. For example, is it for documentary purposes or publication? With these ideas in mind, the first step would be to work out a rough script and from this produce a shot-list of both surface and underwater sequences to be taken.

Before taking a photograph, always consider why the image is being taken and whether the main subject is clear. By this simple act, the value of the image can be improved and the time and money spent processing, recording and storing images can be reduced. Once the photograph or video sequence has been captured to an acceptable standard the photographer can move on to another subject.

Figure 10.1 A diver sketching a late nineteenth-century shipwreck in Dor, Israel. (Photo: Kester Keighley)

PHOTOGRAPHIC THEORY

The basics of photographic theory apply both to traditional film cameras and digital cameras. To get a correctly exposed image, the amount of light reaching the film or the digital camera's light-sensitive chip has to be controlled. This is achieved by using the right combination of lens aperture and the camera's shutter speed. The aperture (referred to as the f-stop) alters the size of the hole controlling the amount of light passing through the lens. The shutter-speed controls the length of time for which the film or chip is exposed to light. Together, they affect the total amount of light reaching the film or chip. If one is changed (for example, to increase the depth of field – see below), the other must be adjusted accordingly. If only one is adjusted, such that the film or chip receives too little light, the image will be dark or underexposed; too much light and the image appears light or overexposed. In either case information is lost from the image.

A third influencing factor is the sensitivity of the image-capture medium (for film, the speed of the film; for digital, the sensitivity setting of the chip – as an ana-logy to film, it is referred to as the ISO number). This can be changed by using film of different speeds, expressed by its ISO or ASA numbers, or by adjusting the digital camera's sensitivity or ISO. The most common types of film, from least to most sensitive, are 64ASA, 100ASA, 200ASA, and 400ASA (the ISO number is the same). This affects the quality or graininess of the image. The lower the sensitivity of the medium used, the finer-grained the image will be, with better definition, resolution, and clarity. As the sensitivity of the medium is increased, 'noise' in the form of film grain (film) and pixelation (digital) increases. This cannot be avoided; it is a fact of life. In underwater photography, higher, more sensitive ASA/ISO speeds are usually used to compensate for lower light levels.

In traditional photography, photographic film, which is sensitive to light, retains the exposed (or latent) image until developed. In a darkroom, once the film is developed, light is projected via an enlarger, through the film 'negative', onto photographic (light-sensitive) paper. This paper is then developed and results in a permanent photographic print. The principles of digital photography are fundamentally the same except that, rather than exposing film to light, a light-sensitive chip converts light into an electronic signal. This allows the image to be stored digitally on a computer.

Both film and digital cameras have built-in light meters, which measure the amount of light in the frame. This allows for the manual or automatic setting of aperture and shutter-speed, depending on the type of camera system being used.

The depth of field is the range over which the image appears in focus. This can be increased by reducing the size of the aperture (smaller hole) with a higher numbered f-stop, say from f4 to f11. There is, however, a trade-off. For example, in an underwater or low-light situation, in order to get a reasonable depth of field, a small aperture is used which, for a correct exposure, requires a slow shutter speed. To avoid camera-shake, a faster film or more sensitive ISO setting must be used, which in turn reduces image-quality. The alternatives are to use a tripod and/ or flash, as well as changing the focal length of the lens (i.e. changing to a wider-angled lens).

Most compact digital cameras have a macro setting, usually indicated by a flower icon. Traditional cameras do not have such a setting and when it comes to non-digital cameras, only SLRs (single-lens reflex cameras), with a macro lens, can be used for close-up photography – for example, artefact photography (see below).

DIGITAL PHOTOGRAPHY

Traditional photography has changed dramatically with the introduction of digital cameras, which offer, amongst other things, excellent image quality. There are similarities between film and digital cameras in that they both have lenses with apertures, viewfinders, shutter releases, shutter speeds, light meters, focusing mechanisms and a method of storing the image (i.e. film or memory cards).

Inside the digital camera, behind the lens, there is a light-sensitive electronic chip that converts light energy into electrical impulses. The electrical impulses are processed into an image by an image processor and saved on a memory card. The quality of the saved image is based on the number of pixels that make up the chip, and on the quality of the lens and the image processor.

Most digital cameras have a few useful extras like a liquid crystal display (LCD) screen. The instant feedback offered by the LCD screen enables better control of composition, exposure and lighting because images can be reviewed, camera settings changed and the images re-shot. The size of the LCD screen is important – larger screens are better for reviewing photographs, especially when under water.

Another feature of digital cameras is the control of 'white balance', which corrects colour for differing lighting conditions. This controls the camera's interpretation of the colour of light by correcting the image to make nearly any light look neutral. The camera defaults to automatic white balance (AWB), although there are presets for specific light conditions like sunrise and sunset. Video cameras also have similar white balance controls. The resolution of a digital image is defined as the number of pixels it contains. A 5-megapixel image is typically 2560 pixels wide and 1920 pixels high and has a resolution of 4,915,200 pixels (rounded off to 5 million pixels).

With digital photography, there are three main variables that can be altered by the photographer: the image size, quality/resolution and file size. Altering any one of these variables will have an affect on the other two. When deciding what settings to use, it should be remembered that the end result needs to be fit for purpose. For example, an artefact record photograph for the database can have a small image size and low resolution but for publication, the resolution will need to be high and the image size large (see chapter 8).

Most camera images can be saved in three formats: JPEG (Joint Photographic Experts Group), TIFF (Tagged Image File Format) and RAW. For most compact cameras, JPEG is the most popular digital image format, allowing images to be compressed by a factor of 10 to 20 with very little visible loss in image quality. In most cameras, images can be saved at three compression ratios. The degree of compression has a significant effect on the image quality/resolution and file size (a lower rate of compression results in a larger file size). Digital SLRs have two further formats: TIFF and RAW. Unlike JPEGs, the TIFF option supports 16 bits/channel multilayer CMYK (cyan, magenta, yellow and black – the primary colours in printing) images and compresses files with no loss of information, and it is therefore the preferred format for printing and publishing. RAW, as the name implies, refers to the raw unprocessed data. It gives very high image quality, and is the starting point for all other formats. A RAW image retains all of the image data available to the sensor that recorded it, allowing maximum manipulation of the image without degradation. The downside is that it needs further processing and the file sizes are large. Remember that a small image can always be created from a large file, but a large image cannot be obtained from a small file.

The master-list of photographs should be the best that can be afforded, including the price of storage (CDs or DVDs or an external hard drive). From these originals, all future copies can be made.

A drawback with most digital cameras, except digital SLRs, is that they suffer from shutter lag. This is the delay between pressing the shutter release button and the moment the picture is actually taken by the camera. It is occasionally advisable to use a tripod to hold the camera steady. There is also a further delay while the image is processed before the next shot can be taken, especially if the camera is set to the highest quality levels.

SURFACE PHOTOGRAPHY

The photographer's duty is to take a range of photographs to cover all aspects of the project, above as well as under water. This begins with the project mobilization and the setting up of the work site (e.g. mooring the dive support vessel, launching hardware like acoustic survey transponders, a grid, airlifts etc.). It then continues throughout the project as diving starts and other equipment and hardware is deployed. It should also include photographs of team members carrying out various aspects of project work, including diving operations, artefact lifts, artefact recording, data processing, wrapping and storage of finds. It finishes with putting the site to bed and demobilizing the project.

Remember, there is a distinct difference between archaeological shots of the site, the artefacts found, and the techniques like survey and excavation. The main difference is in how images are to be used. An image suitable for a popular journal may not be suitable for an academic lecture (see chapter 20) while other images, important for the site archive, will not always be of use for public presentation. In the past, it was traditional to take two photographs of everything – a black-and-white print and a colour slide – for the site archive, but nowadays this is not necessary, as a colour digital photograph can be saved in greyscale as a black-and-white image. It should be remembered that shots taken primarily for archaeological reasons should normally include an appropriate scale.

As artefacts are uncovered, recovered and registered they should all where possible be photographed, in some cases more than once from different angles. Any unique features or marks should be highlighted and a photographic scale and unique artefact number should be included in the shot. Photographs of artefacts should be undertaken *in situ*, before conservation, during conservation and after conservation. This can be achieved using a digital camera and the downloaded images can then be linked directly to the artefact database or using a thumbnail image. Database photographs can be taken at a lower image size and quality, resulting in smaller file sizes. Alternatively they can be taken at the best quality for the project archive and a thumbnail version can be used for the database.

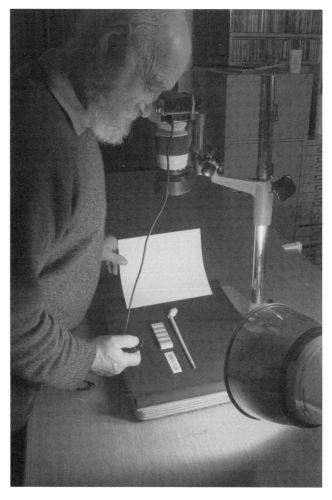

Figure 10.2 A simple set-up for photographing finds using a vertical stand. The white clay pipe has been set on a matt black background, with a scale and reference number appropriately placed. A single light source has been augmented by a white card reflector to 'kick' light back into the shadow side so as not to obscure the edge of the object. A remote release is being used to avoid camera shake. (Photo: Edward Martin)

PHOTOGRAPHING FINDS

A specific area should be set aside for photographing finds. Preferably, it should be outside and use natural light – but not direct sunlight, so as to avoid harsh shadows. Ideally a camera on a tripod should be used and shots taken against a suitable single-tone background that will contrast with the object (figure 10.2).

Wherever possible, the macro setting (flower icon) should be used, but be aware of the physical range of distance a camera requires to produce the focused image. Include a scale and artefact number, which includes a site code and year (see chapter 8). If carefully placed, scales and identification information can be cropped out when a 'glamour' rather than a record shot is required for use

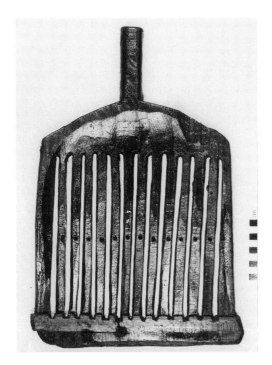

Figure 10.4 An obliquely photographed find. These wooden bellows from the Armada wreck *La Trinidad Valencera* (1558) have been placed on a light-neutral background – not white, which is too reflective. (Photo: Colin Martin)

Figure 10.3 A vertically photographed find. This wooden weaving heddle from the Armada wreck *La Trinidad Valencera* (1558) has been photographed on a translucent sheet, lit from beneath, to create a neutral white background without shadows. Careful lighting of the object has been used to bring out tool marks on the wood. Scale in centimetres. (Photo: Colin Martin)

in publications. A selection of chosen artefacts can also be photographed at a higher image size and quality/ resolution and even photographed using a traditional SLR film camera with a macro lens.

It should be noted that there are conventions for photographing certain types of objects (e.g. pottery sherds). For further details on site and finds photography refer to Dorell's book *Photography in Archaeology and Conservation* (figures 10.3 and 10.4).

UNDERWATER PHOTOGRAPHY

This section will give a brief overview of the various underwater camera systems – the Nikonos system, housed SLR cameras and digital cameras. It will also summarize some of the problems facing photographers under water and how to overcome them with the use of wide-angle lenses, flash equipment and some underwater techniques. Traditionally, underwater photography has been very expensive, but with the advent of mass-produced digital cameras in underwater housings, prices are dropping as the market expands.

Originally, dedicated underwater amphibious cameras, like the Nikonos system produced by Nikon, were the

most popular camera for underwater photography. This changed as underwater housings were developed for conventional cameras, but both systems have remained expensive. A number of underwater compact cameras have been produced which are cheaper but the results are generally not good. This is because of the difficult conditions typically found on underwater archaeological sites, such as poor visibility (caused by particles suspended in the water), low light levels, loss of contrast and loss of colour with depth.

To overcome these problems, underwater camera systems have to include:

- a wide-angle lens enabling the photographer to get close to the subject; and
- an underwater flashgun or strobe to overcome the loss of light and colour with depth and to improve contrast and resolution.

The best set-up for use in archaeology is a Nikonos V with a 15 mm lens, although there are cheaper wide-angle lenses and adaptors which attach to the standard lens. The 15 mm lens gives little distortion for such a wide-angled lens, but it is expensive. In clearer waters, however, the Nikonos 20 mm lens would suffice. A housed SLR offers more flexibility, but to photograph a wide-angle shot, not only a wide-angle lens but also a dome port for that lens is required. Housed SLRs require a different lens and port combination for each type of shot, including wide-angle, standard and macro. Both systems can use the range of flashguns available, but the flashgun must have at least the same angle of coverage as the lens used. For this reason, and to avoid shadows, some photographers use two flashguns.

If underwater housings are used it is worth noting that there are usually two types of port available – the dome port and the flat port. The dome port requires little focus adjustment once set, and the lens behind it retains its focal length under water. A flat port will need focusing all the time and it will act as a magnifying glass, increasing the effective focal length of the lens behind it by around 25 per cent.

Digital cameras can be put in housings with wide-angle adaptors and external flashguns as optional extras. Digital-camera housings usually have diffusers to tone down the light from the internal flash and alleviate problems of backscatter under water, due to the flash's proximity to the lens. They are ideal for use under water due to some of the advantages mentioned under digital photography, particularly the opportunity to review the images as they are taken. However, their main disadvantage under water is shutter-lag, which is more pronounced because of the difficulty of holding the camera still until the shot is taken. This becomes a particular inconvenience when trying to photograph an action shot. Lenses in compact digital cameras often suffer from severe distortion, especially at wide-angle settings. Wide-angle adaptors usually have to be re-fitted under water, to allow water in and air bubbles out, between the lens and the adaptor. The lens of a digital camera should also be inserted into a small tube of neoprene to mask light around the lens where it sits in the housing. The top end of the digital market is the digital SLR with associated expensive housing, interchangeable lenses, dome ports and powerful flashguns. For high-quality images, there is nothing better; they offer high-quality lens optics and minimal shutter-lag.

UNDERWATER PHOTOGRAPHIC TECHNIQUES

Diving technique: Perfect buoyancy and fin control are essential techniques to master so as not to disturb the visibility. As part of the daily project briefing, other team members should be informed of the intention to take underwater photographs. Ideally a period of the day should be devoted exclusively to photography, with no other divers in the water. This could either be first thing, before the silts are stirred up, at slack water, or around midday when natural light levels are at their best.

Equipment care: Before and after entering the water, check the equipment for bubbles and leaks. Refit wide-angle adaptors under water. After diving rinse the equipment in fresh water.

Photography techniques: Do not punch the shutter-release when taking a photograph – "squeeze" it gently to avoid camera-shake. Shoot within one-third of existing visibility.

Cleaning: Carefully clean ship's timbers and tidy up the excavation area before photography.

Composition: This is very important with respect to the type of shot – with and without a scale, with and without a diver. Try shots from different angles to obtain the best angles to show the subject. Distinguish between an artefact record and a public-relations shot, as each needs different treatment. Wait for the diver's bubbles to prove the shot was taken under water. Occasionally an obliging fish swims into shot, which can produce a dramatic shot of a diver working with bubbles and marine life.

Aiming the flash: The correct positioning of the flash is one of the most important factors for taking a successful underwater photograph. To avoid backscatter, move the flash further away from the camera, this avoids illuminating particulate matter between the lens and the subject. Hold the flash above the subject and to one side, as shown in figure 10.5. Think of the flash as mimicking the sun. One technique with a powerful flash is to put it on a pole to extend the flash-to-subject distance and better mimic the sun. When following the table of settings for the flash, remember that it is the flash-to-subject distance that is needed for the correct exposure, not the camera-to-subject distance.

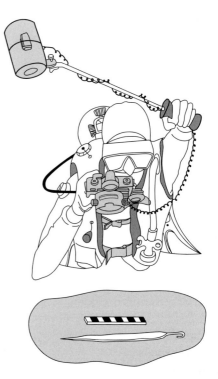

Figure 10.5 Lighting, composition and scale are all important considerations when making a photographic record of a site. (Based on original artwork by Ben Ferrari)

Scales: Underwater scales should ideally be matt yellow and black. These are used for a variety of shots but mainly for photographs of artefacts or a ship's structure. Place scales carefully so as to avoid masking any detail, and try to keep them parallel to the frame of composition and the artefact.

Using film: Shoot and process a sample film to test equipment and the local conditions before photographing in earnest. Use the technique of 'bracketing' by taking three shots, one correctly exposed, one over- and one underexposed.

Documentation: Write up a photo-log as soon as possible after the dive.

DIGITAL DARKROOM

One of the most significant developments in modern photography is the 'digital darkroom', where a computer can be used to improve the original image. The image can be digital or scanned from film. Ideally JPEG images should be saved at the highest quality, with the least compression. Once they are downloaded, a master copy should be saved as a TIFF file because every time a JPEG is changed and saved information is lost. The best practice is to work on the TIFF file then save a copy of the final version as a JPEG for export. Never work on originals in the digital darkroom – you never know when you will need to go back to the beginning.

There are numerous software imaging packages on the market and these are continually being developed. Whichever package is chosen, it is essential to make sure that the computer monitor is properly calibrated so that it represents colours accurately and prints match what is seen on screen. This is vital for colour management. Make a note of print settings because the output will change with the type of printer used, and also within a printer's quality options.

The image can initially be enhanced by adjusting density, contrast and colour, followed by a series of further refinements including reframing or cropping, resizing and making adjustments to isolated areas. A black-and-white or greyscale version can be saved for publication, but be sure to save the file at the highest resolution necessary for the final output, and at the final output size, especially for printing. The beauty of the digital darkroom is that it is possible to experiment quickly, cheaply and easily without being shut away for long hours in a dark room brimming with toxic chemicals.

When scanning photographs, the negative or slide should be scanned rather than a print, as film has the best quality in terms of sharpness, tone and colour details.

However, slides must be scanned at a resolution that will allow the image to be enlarged and still provide a good enough quality at the final output size. Remember that when the physical size of the scan is doubled, the resolution is halved. For the highest quality, shoot in RAW format and then process the images on a computer. This allows adjustment of exposure, white balance, hue, saturation and sharpness, with little or no degradation of the original unprocessed data. Some digital SLR cameras come with software to manipulate the unprocessed images. The alternative is third-party software, which can be downloaded from the internet.

This section will give an overview of some of the basic adjustments to improve an image, using Adobe Photoshop as an example. Some of these can be found under 'Image/Adjust'. One of the clever things about Photoshop is the use of 'adjustment layers'. All of the following can be done as an adjustment layer, which sits above the original image file, the advantage being that it is easy to experiment with changes without altering the original file.

- Start off by adjusting the overall exposure – brightness and contrast. This is best achieved by adjusting 'Levels', which allows adjustment of the highlights, mid-tones and shadows individually on a histogram. Move sliders in from the left (black) and the right (white) so that they are under the ends of the histogram. Move the middle grey slider left or right for the overall adjustment of grey or mid-tones. Alternatively use 'Curves' which permits very fine control of image density.
- The overall colour is controlled by adjusting the purity (hue) or vibrancy (saturation) of a colour with the 'Hue/Saturation' controls. Start off by boosting the saturation by +10 to +15 points. Hue can be used to make an overall change if the image has a colour cast (small adjustments); otherwise it is used to fix and adjust specific colours. 'Colour balance' can also be used to refine colour, particularly scenes with several different light sources that can cause problems in white balance.
- Isolate areas of the image for local brightness and contrast management and colour correction by selecting an area of the image and making changes inside it. There are three types of selection tools – 'lasso', 'automated', and 'defined shape' or 'marquee'. Once an area is selected, adjustments can be made within that area.
- Finally, sharpen the image, but only at the end and at the final output size. Use 'Unsharp mask' found under 'Filter/Sharpen', which looks for edges in an image and makes them stronger. For images under 10 megabytes in size, try setting the 'Amount' to 120,

'Radius' 1.2 and 'Threshold' 0. For images between 10 and 30 megabytes, 'Amount' 150, 'Radius' 1.5 and 'Threshold' 0.

MOSAICS – PHOTO OR VIDEO

Ideally, a mosaic should not be the primary means of survey but should be carried out to complement or assist it and fill in detail. If this is not possible, due to limitations of depth and dive time, then as a minimum some check measurements should be taken and used in the preparation of the final mosaic.

The main reason for producing mosaics is that often the lack of visibility and clarity under water and/or the scale of the site does not enable an overall picture of a site to be taken. Instead, a series of overlapping images must be taken which are then stitched, joined or merged together (figure 10.6). Traditionally this has been done with printed images but now images can be scanned (unless

Figure 10.6 A 5 metre square (264 sq. ft) photomosaic of ship remains on the Duart Point wreck, made up of 25 individually photographed 1 m blocks. (Photo: Colin Martin)

taken with a digital camera), and stitched together in the digital darkroom. The alternative is to use a digital video camera, with a diver or remotely operated vehicle (ROV) swimming over the site at a fixed distance, making a sequence of passes. The video camera could also be mounted on a neutrally buoyant underwater sled pushed along by a diver. Images are then captured, perhaps one per second, and stitched together in a similar fashion. The final mosaic can be adjusted to produce a balanced image in terms of contrast, brightness, and colour. Harsh lines and the like can be removed by using image-processing methods like cloning. However, it should be noted that mosaics only work well on flat sites.

There are various methods of collecting still images for a photomosaic: for example, setting up a lightweight photo-tower and systematically moving it along the wreck, at a pre-set distance from the subject (plate 10.1), taking photographs with a 50 per cent overlap. Alternatively, a diver or ROV can swim along a pre-set tape measure or acoustic survey line or track. Much of the success of the final mosaic is dependent on the care used in collecting the images (figures 10.7 and 10.8).

In the past, the Nikonos system has most commonly been used, often with a 15 mm wide-angle lens. However, the 20 mm lens is more suited as a compromise between optical distortion inherent in wider-angled lenses and the desirability of keeping the camera-to-subject distance short enough to ensure good image clarity. Only the

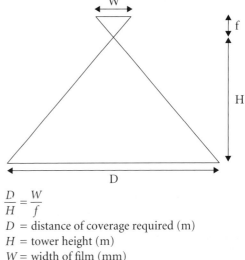

$$\frac{D}{H} = \frac{W}{f}$$

D = distance of coverage required (m)

H = tower height (m)

W = width of film (mm)

f = focal length of the lens (mm) multiplied by 1.33 for use under water

Figure 10.8 Photomosaics: formula to calculate lens focal length and camera height necessary to give the required coverage. (After Green, 2004:171)

central part of the image is used (the perimeter being discarded) in order to reduce optical distortion. This would similarly apply to the use of a digital video camera. A number of 'panorama packages' are currently available and the technique is described in greater detail in 'An underwater photo-mosaic technique using Adobe Photoshop' by Colin and Edward Martin (Martin and Martin, 2002).

VIDEO CAMERAS

Video footage can serve as a useful extension to still photography, especially under water (figure 10.9). It can be used to demonstrate techniques, assist with survey and to produce mosaics or a short documentary about the project. Too many projects have a video archive that never gets touched and the information never used or disseminated.

Video is only useful if it is properly thought out and edited. Once the purpose of the video has been established, then a rough script can be worked out and a shot-list of both surface and underwater images produced. One method is to produce a storyboard, which is a series of sketches visualizing each of the arranged shots. From this, a shooting plan can be devised and the images taken. These are then edited, which involves downloading them onto a computer and joining them together, allowing further refinement. The edited video can then be dubbed with a voice-over and/or background music. Titles and credits can be added to the introduction

Figure 10.7 A photographic tower positioned on a rigid site grid. (Drawing by Graham Scott)

Figure 10.9 Underwater use of a video camera (Photo: Kester Keighley).

and if appropriate a copyright statement can be added at the end.

This section concentrates on the use of camcorders. However, for more commercial underwater use, there is also a system using a small video camera sealed in a housing with a hardwire umbilical to the surface where the control unit and video recorder is located. These are mainly used on surface-supplied divers' helmets and on underwater vehicles, like ROVs. Their advantage is that they are powered and controlled from the surface, via an umbilical, so there is no problem of battery power for the camera and lights. ROVs usually carry an additional low-light black-and-white silicone intensified (SIT) video camera to give an overview and use the colour camera for detail.

For underwater use, a camcorder in a housing is ideal for archaeology, where the diver controls the camera. The disadvantage is the restrictions imposed by the limited battery life, especially when it comes to powering the underwater lights. The camera itself uses lithium-ion batteries that last up to 5 hours or more. Traditional underwater lights, however, have a low burn-time of less than an hour. The latest xenon high-intensity discharge (HID) lights (for example, Treble Lights) produce high-intensity white light, daylight balanced (colour temperature of 5600K), which burns six times brighter than a standard halogen bulb. These lights can be fitted with low-wattage bulbs, resulting in excellent burn times, but they are expensive. It should also be noted that rechargeable batteries lose their capacity with age as well as in cold water.

There are numerous different types of camcorders. Apart from the quality of the components, the main differences relate to the format, which affects the image resolution and quality. The basic functions are usually fully automated, like auto-iris for correct exposure, auto-focus, auto-white balance (to ensure correct colour reproduction) and auto-audio levels. Most of these functions have manual overrides. Like a digital still camera, the image from the lens is focused onto a single charge-coupled device (CCD) chip. More professional camcorders have three CCDs (referred to as three-chip cameras) and consequently produce a better quality image.

The most popular digital tape format is MiniDV. One of the main criteria is resolution: normal VHS is 200–250 lines, Hi-8 and Super VHS (SVHS) is 350–400 lines and MiniDV is 400–480 lines. Some three-chip camcorders have a semi-pro version using a format called DVCAM (500–650 lines), which can be recorded onto MiniDV tapes. The latest development is HDV (1080 lines), which is a new video format that records high-definition pictures onto either a standard MiniDV tape or an HDV MiniDV tape. The hierarchy progresses from single-chip to three-chip MiniDV to DVCAM to HDV. Sony's latest Pro-HDV camcorder is switchable and can record and playback in MiniDV, DVCAM or HDV formats. It can also record and playback in PAL (phase alternating line, the colour television coding system used for European broadcasting) and NTSC (National Television System Committee, the American colour TV broadcasting system).

Digital tapes can be copied without loss of quality and similarly they can be captured on a computer through a digital 'firewire' (IEEE 1394) connection. These can be edited and then exported back to digital tape without any loss of quality. Alternatively, a DVD can be made of the final edited movie, which will play on a domestic DVD player as well as on a computer.

VIDEO TECHNIQUE

Camcorders are able to operate in low light. Under water, they often 'see' better than the diver. However, the same problems that occur with underwater photography occur with video – poor visibility, loss of light, colour and contrast. To reduce these problems, the use of a wide-angle lens is advisable. This shortens the camera-to-subject distance and maintains a large depth of field. It is also best to shoot within one-third of the visibility and to use artificial light. However, if the light is held too close to the camera it may cause the problem of backscatter. It is best to shoot around the middle of the day when natural light levels are at their best.

On the surface, professionals use manual focus, refocusing for each shot. They zoom into the subject, focus manually, then zoom out and compose the shot. However, focusing under water is best achieved manually because on auto-focus the focusing system tends to 'hunt' or be forever adjusting itself as particles in the water column go past the lens. The lens should be set to a wide angle for maximum depth of field. The optimum method is to point the camcorder at something with some contrast (e.g. another diver) at the average distance required for the shot and to then press the auto-focus override button. The camcorder will then remain focused at that distance. It should not be necessary to adjust the focus during the shot, or even during the dive once the average camera-to-subject distance is set. If the distance is altered significantly, then refocusing will be necessary.

Sound recorded under water usually comprises the cameraperson's bubbles. On the surface, especially when interviewing, find a quiet spot. If this is not possible, try to use an external directional microphone with a windshield held close to the subject.

Many of the same rules apply as for still photography under water (see above). In addition, consider the following points.

Diving technique: Perfect buoyancy and fin control are essential to prevent disturbing the visibility. Other divers should be briefed about what the film crew are doing, especially if they are in shot, so they do not disturb the visibility.

Shooting technique: Under water, avoid the use of zoom; stay wide-angle and physically move in. This is required because of the limitations of visibility and in order to maintain depth of field and, therefore, focus. A test run can be done to check that the shot can be completed from the chosen dive position and that the planned camera movement can be achieved without snagging. To preserve continuity of screen direction, take consecutive shots from only one side of an imaginary 'line of action'. Do not cross the imaginary line. Use only one movement at a time (e.g. do not pan and zoom together). Record several seconds at the beginning and end of each shot, with the camera held still, before moving. By doing this three usable shots are created which are easy to edit – the hold at the beginning, the move, and the hold at the end. For each part of the sequence, take a series of shots that can be easily edited together to present that part of the story. For example, a wide establishing shot, to set the scene, followed by mid or closer shots to show more detail and identify the main character (or actions), and finally several close-up shots of some detail.

Taking additional shots that 'cut in' or 'cut away' from the subject is crucial in producing enough material to make a fluently edited final piece. A cut-in is a shot that shows the action, or a part of the action, in closer detail. Moving from a relatively wide shot showing a diver recording a grid square to a close-up of his/her pencil drawing a feature would be a good example of cutting in. Often, to minimize jarring, a cut-away is used as well. This could be a shot of the diver's face looking down, and when this shot is inserted between the wide and close-up shots of the action, a smooth and pleasing progression is shown. Cut-aways are also very useful when moving from one 'scene' to another. Shots of things such as fish, waving seaweed, divers' bubbles or sunlight filtering through the water can all be used to break away from one area of interest and move easily and naturally to the next. Shots like this can be collected at any time, and having a good stock of them will make editing much easier. A well-edited piece of video, like a well-written story, should have a beginning, a middle and an end, and should flow smoothly in a 'grammatically correct' way. When properly made, cuts should be virtually unnoticeable and the video should lead the viewer through the story without distraction.

Lighting: Under water, hold the light source away from the lens, at around 45 degrees to the subject, to minimize backscatter and flaring. Think of the light as mimicking the sun – 1 to 2 metres above and to the side. Alternatively use a lower-wattage lamp, mounted on the housing, and use another diver with a more powerful lamp as the lighting person. In this situation the lighting person points the light to where the light on the housing is pointing.

Documentation: It is very important to include this in the editing process, in particular to find good footage on a tape quickly. Shots can also be classified by their content and quality or usability (e.g. good, average or poor).

VIDEO EDITING

Editing is the skill that lies at the heart of video making. In its simplest form, it refers to the order and length of shots in a programme. At its most creative, it will determine the audience's response to the subject. The final video should consist of a series of linked shots, which will tell a story, so the sequence of images needs to make sense. A commentary or voice-over can be added, which helps to provide structure as well as continuity, and provides information that is not evident from the pictures alone. However, the test of an effective video is whether it could stand alone and tell a story without being enhanced by dubbing. Background music can be added in the later stages of editing. While most music is copyright protected, it is possible to buy royalty-free music to use for video editing.

There are several methods of editing. These include tape-to-tape, using a video editing controller and computer-based editing. Computer-based editing is currently the preferred method but it does require a high-specification computer with a large hard drive, as DV tape uses typically around 13 gigabytes of space per hour at full resolution. Chosen footage can be captured on a computer using the 'firewire' (IEEE 1394) connection on the camera, which enables fast data transfer, and many computers come with a firewire input.

Adobe Premier and Windows XP can be used for capture and editing, providing the computer has a capture card. There are, of course, other suppliers of capture hardware and software and editing software. Amongst the professionals, Apple Macs and Final Cut Pro are highly rated for all forms of editing and image processing. Windows XP Movie Maker is good for making quick movies but the editing suite is not as sophisticated as Adobe Premier. However, Movie Maker is quite good for saving the finished video to various formats, compressions and quality levels (e.g. to CD or for the internet). For optimum quality, export the video back to digital tape or use appropriate software to make a DVD. In this instance, the video files are transformed into a different format, which can only be played through DVD software on a computer or on a domestic DVD player.

FURTHER INFORMATION

Aw, M. and Meur, M., 2006 (2nd edn), *An Essential Guide to Digital Underwater Photography*. OceanNEnvironment, Carlingford, NSW, Australia.

Boyle, J., 2003, *A Step-by-Step Guide to Underwater Video*. www.FourthElement.com

Dorrell, P., 1994, *Photography in Archaeology and Conservation*. Cambridge.

Drafahl, J., 2006, *Master Guide for Digital Underwater Photography*. Amherst.

Edge, M., 2006 (3rd edn), *The Underwater Photographer: Digital and Traditional Techniques*. Oxford.

Green, J., 2004 (2nd edn), *Maritime Archaeology: A Technical Handbook*. London.

Martin, C. J. M. and Martin, E. A., 2002, An underwater photomosaic technique using Adobe Photoshop, *International Journal of Nautical Archaeology* **31**.1, 137–47.

Position-Fixing 11

Position-fixing at sea relies on the same basic principles as on land. Offshore, however, the environment often requires alternative methods for obtaining position. Position is found through the measurement of distances and angles.

Position-fixing is essential in archaeology for:

• pinpointing the exact location of a site;
• establishing relative locations of sites; and
• obtaining positional data during geophysical survey.

This chapter will outline some of the fundamental principles involved in position-fixing and give a summary of position-fixing equipment, including optical instruments, electronic systems and satellite-navigation systems. The first step is to establish exactly what a position is.

Position is normally expressed in terms of coordinates which can be depicted on a map, chart or plan. However, to use these numbers correctly it is important to understand how the numbers were obtained and this is more complex than it may at first appear. The map or chart is a scaled representation of the ground or the earth's surface. The earth has been identified, after much debate through the centuries, as being neither flat nor round, but an irregular shape resembling a rounded pear.

The problem faced by all map-makers is how to represent the shape of the earth on a flat piece of paper. To do this, the shape of the earth must first be defined mathematically and then projected onto a flat piece of paper or, more specifically, a flat plane. Mathematical approx-

imations of the earth's shape are known as spheroids (or ellipsoids). Different national and international mapping systems use different spheroids that best fit the area to be mapped. In the UK, the National Grid is based on an ellipsoid defined by Mr George Biddell Airy, the Astronomer Royal, in 1840, while the American satellite-navigation network known as the Global Positioning System (GPS – see below) calculates position on the World Geodetic System 1984 (WGS84) which was defined by the US military specifically for satellite positioning. There are many spheroids, each developed for a specific purpose by a different sovereign state. India, for example, was mapped by the British in the early nineteenth century on the Everest spheroid. Devised specifically for the task, this spheroid was named after Everest, the principal surveyor at the completion of the survey, and is still in use today.

It is essential to be aware of the spheroid on which any chart and set of positions are based. This is especially significant when plotting or using positions from an electronic positioning system (e.g. a GPS receiver), which gives a set of coordinates. Using the wrong spheroid can have serious consequences; including the loss of sites, ships running aground and even territorial disputes. A position in the WGS84 ellipsoid plotted on a chart in the UK based on the Airy spheroid, can be 164 m (533 ft) away from the intended position. There is software readily available for transformations between different systems, but it is essential to know which system provided the position and on which system the chart or site-plan is based.

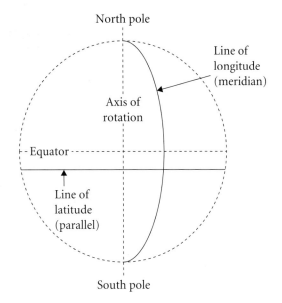

Figure 11.1 The earth, showing latitude, longitude and equator

GEOGRAPHICAL COORDINATES

Conventional marine chart coordinates are geographical coordinates and are described in terms of latitude and longitude. These are expressed as angular units (degrees, minutes, and seconds). A line of latitude is defined as a line or (more specifically, a plane) that bisects the spheroid at right-angles to the line between the poles. Latitude is measured as the angle subtended by an imaginary line (the normal) running perpendicularly through the point to be located to the equatorial plane (figure 11.1) and is expressed as a value north (N) or south (S) of the equator.

A line of longitude, or meridian, is defined as a line that describes the shape of the spheroid passing through the poles. Longitude is the angle subtended from the point to be located to the prime meridian in the equatorial plane. The prime meridian is an arbitrary meridian chosen as zero, in most cases the line that passes through Greenwich, UK. Longitude is expressed as either east (E) or west (W) of the prime meridian. On a marine chart these coordinates appear as a grid. This grid is not regular or orthometric (a grid formed of squares) but changes depending on the scale and the projection. This is because lines of longitude (meridians) converge at the poles.

The projection: This is the method by which positions on the spheroid are represented on a flat piece of paper. In simple terms, different projections are alternative ways of mathematically wrapping a piece of paper (or mapping-plane) around the spheroid. The projection is fundamental to mapping and caused the geodesists of the past great difficulties. It wasn't until Mercator devised his projection in the sixteenth century that mapping truly developed.

Figure 11.2 shows the basis of the Universal Transverse Mercator (UTM) projection. This is the most common projection used for nautical charts and mapping. It is a set of grid-projection parameters devised to apply to any and all spheroids. In general cases when working offshore with GPS position-fixing systems, UTM parameters are applied to the WGS84 spheroid. UTM was devised in an attempt to create a truly international mapping system. It divides the spheroid into six-degree segments and applies the grid parameters to each segment. These segments are called zones and in effect each zone is a

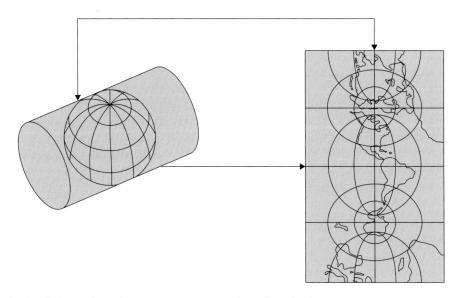

Figure 11.2 The basis of the Universal Transverse Mercator (UTM) projection

Table 11.1 UTM zone extents (Great Britain)

Zone	Western extent	Central meridian	Eastern extent	False easting	False northing	Scale factors
Zone 29	012 W	009 W	006 W	500,000	0	0.9996
Zone 30	006 W	003 W	000	500,000	0	0.9996
Zone 31	000	003 E	006 E	500,000	0	0.9996

different datum with a different central meridian value and there is no overlap between zones. It is important to note that Great Britain is covered by three UTM zones – zones 29, 30 and 31. The zone extents are as shown in table 11.1.

It is imperative that the zone is stated whenever grid coordinates are quoted and that online navigation or tracking systems are configured to the correct zone for the area of operation. The transverse Mercator projection has the advantage that distances and angles are represented by their scaled values; areas, however, are distorted.

Grid coordinates on land maps are often described in terms of eastings and northings. These coordinates describe a position on a flat plane, and orthometric grid. This is the case for both national and international mapping systems. The advantages of grid coordinates are that measurements, both distances and angles, can be related from the map to the ground in regular units, most commonly metres.

How the grid relates to the projection: The grid is overlaid onto the projection (discussed above). The method and parameters used to overlay the grid onto the projection are collectively called the map datum. There are many different map datums: for example the National Grid of the UK is based on the Ordnance Survey of Great Britain 1936 (OSGB36) datum. This datum uses the Airy 1840 spheroid and a transverse Mercator projection. It contains parameters defining the origin of the grid position on the spheroid and orientation of the projection. On an Ordnance Survey map this datum will be eastings and northings in metres. On old UK Admiralty charts, the datum may be quoted as OSGB36 but the coordinates will actually be longitude and latitude. In this case, the grid has not been applied; thus the positions have only been projected from the spheroid. It is critical when reading maps that the user knows which map datum is being used.

Vertical reference: So far only a two-dimensional position has been discussed. The third dimension that also needs to be considered is height or depth. Height can be expressed as a distance above the spheroid but it is more commonly related to a separate datum. On an admiralty chart the depths will be related to chart datum. Chart datum is a plane that defines zero height, therefore a distance below the plane is expressed as a depth and a distance above the plane is expressed as a height. Chart datum can change for each individual chart and is normally derived from tidal observations at a local point from which the lowest astronomical tide (LAT) is calculated. LAT is the lowest tide predicted from known tidal constants. The tide can go below this level on occasions when influenced by meteorological effects.

For national mapping systems a height datum is chosen that is constant for the total area mapped. For example, all the heights that appear on an ordnance survey map in the UK are related to Ordnance Datum Newlyn (ODN). This is a survey point at the Ordnance Survey observatory in Newlyn, Cornwall that has been assigned zero height. The point was derived through tidal observations in 1911. The difference between the individual chart datums around the coast of the UK and ODN are noted in nautical almanacs.

Archaeology and vertical reference: From the prehistoric period to the present day, there have been significant fluctuations in sea-level. Indeed, the process is continuing today. These changes have obviously had a pronounced effect on shoreline settlements, navigation, the viability of harbours, etc. The existence of extensive tracts of prehistoric forests, now only visible around coasts and estuaries at very low tides, and even earlier landscape features permanently submerged on the seabed, is testimony to the profound rise in sea-level relative to the land. In some coastal areas, for example, there has been rise of over 20 m (65 ft) relative to the land over the past 10,000 years. The chronology of these complex changes can be charted by the careful observation and recording of ancient features, whether natural or artificial, in relation to a fixed datum point. The absolute level of altitudinal benchmark used might be Ordnance Datum (OD) in the UK or Normaal Amsterdams Peil (NAP) in continental Europe. Such information can then be compared with the body of data being collected by institutions such as the International Geoscience Pro-gramme (formerly the International Geological Correlation Programme), which was established in 1974 to process such data globally.

ACCURACY

Accuracy provides an indication of the quality of measurements and hence a position. In chapter 14 (on Underwater Survey), accuracy will be discussed in terms of measurements and the quality of the site-plan. The same principles apply to any form of measurement and position derivation.

The concept of accuracy embodies the idea of 'absolute position', implying that there is a correct answer. The correct answer, however, is never realized because of intrinsic errors in any measurement or measurement framework. Schofield (2001) draws an analogy between survey measurements and target shooting. The centre of the target represents the absolute position of the point being measured. The skilled marksperson will produce a scatter of shots on the target. The degree of scatter represents the precision of measurement; the nearness to the centre of the target represents the degree of accuracy. Hence it is possible to have very precise measurements that are totally inaccurate (figure 11.3). An example of this would be a series of measurements taken to a point with a tape-measure to the nearest millimetre but the zero of the tape starting at 2 cm. Each reading of the tape could be made very precisely but the value itself would contain an error and hence be inaccurate. Accuracy is therefore expressed as bounds within which the measured value or the absolute position may lie. These bounds are expressed in statistical terms such as standard deviation (see below). The essence of achieving accurate survey data is to minimize errors. There are several definitions and terms that can be used when describing accuracy.

Errors: Errors come in many shapes and sizes. They can generally be split into two groups: systematic errors and random errors. Systematic errors are inherent in the instrumentation or measurement system as in the example above. Random errors are less predictable and include gross errors (e.g. from misreading a tape measure). The concept of accuracy is the concept of understanding and quantifying errors.

Standard deviation: This is the measure of variation of how close all the values are to the average value and is quoted as a single number. This is an evaluation of precision.

Measurement accuracy: This is related to the size of the error in measurement. Distance measurements can be expressed in terms of a relative error of 1 in 10,000. For example, this would describe an error of 1 mm over a distance of 10 m. This is a very useful definition and is often used when quoting equipment accuracies.

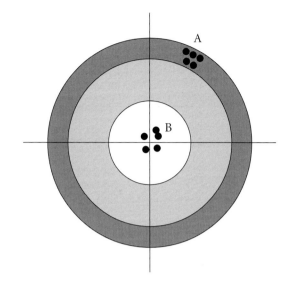

Figure 11.3 A scatter of shots showing precision (A) and accuracy (B)

Scale: The area depicted on a map or chart is subject to scale and represented as a unit of proportion. For example a scale of 1:1 indicates that 1 unit on paper represents 1 unit on the ground whereas a scale of 1:50,000 means that 1 unit on paper represents 50,000 units on the ground (i.e. 1 mm measured on a map would be 50,000 mm or 50 m on the ground).

Plotable accuracy: The plotable accuracy is a key consideration in any survey design. Surveys for the most part are represented as scaled drawings. When plotting at a scale of 1:100, 1 mm on paper will represent 100 mm (10 cm) on the ground. The width of a sharp pencil lead is typically 0.5 mm, equivalent to 50 mm on the ground at a scale of 1:100. It is therefore a waste of time and resources measuring to the nearest millimetre when it is only possible to plot to an accuracy of 50 mm.

Repeatability: This is a measure of consistency. In other words, can the derived position be relocated using an independent system or using the same system but on a separate occasion? Note that a position can be repeatable but not necessarily accurate. For example, using a tape-measure where the zero starts at 2 cm will give repeatable measurements but they will not be correct. A repeatable position can be derived despite the presence of systematic errors. It is very easy to become obsessed with accuracy to the detriment of producing a working plan or drawing. The key is to understand the concept of accuracy and the accuracy limits that are being worked to.

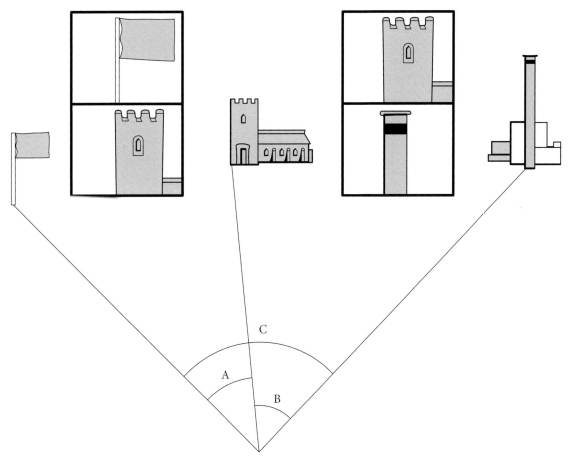

Figure 11.4 Taking horizontal sextant angles. Angle A is measured between left-hand side of the flagpole and the left-hand side of the church tower by lining them up in the split-viewfinder. Angle B is measured between the left-hand side of the church tower and the left-hand side of the chimney. Angle C, measured between the flag pole and the left-hand side of the chimney, should equal angle A plus angle B. (Based on original artwork by Ben Ferrari)

METHODS OF POSITION-FIXING

Horizontal sextant angles

The sextant is one of the most useful optical position-fixing tools for coastal surveying. It is basically a protractor and telescope linked by mirrors, which allow an angle between two separate features to be measured from the survey position (figure 11.4). Although superseded by electronic devices, sextants are still used in certain circumstances because they are accurate and portable.

Traditionally, two angles between three charted objects are used for position-fixing but, to increase confidence in the fix, it is preferable to take a third angle involving a fourth charted object, as this will give a check on the accuracy of the first two readings. Once a set of angles has been taken, the position can be plotted on a chart using a number of different methods. The angles can be drawn on a piece of translucent drafting film and then laid over the chart and moved around until the lines pass through the appropriate features, the intersection of the lines being the plotted position (figure 11.5).

Alternatively the plot can be constructed geometrically (figure 11.6), the simplest method being to draw a baseline between the left-hand pair of features and draw a line at an angle of 90 degrees, minus the measured angle, out from each end. The intersection of these lines is the centre of a circle that has a radius equal to the distance between the features and the centre. At any point on this circle the angle between the two features will be constant. A second circle must therefore be drawn, constructed in the same way on a baseline drawn between another feature and one of the pair already used. The intersection of these two circles is the plotted position. It is possible to construct a whole series of circles based on different angles, and these horizontal sextant-angle charts can be very useful if a lot of survey work is to be undertaken in the same area using the same charted features.

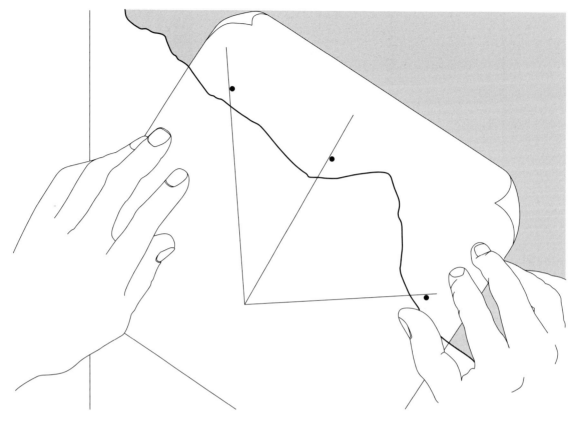

Figure 11.5 Sextant angles can be scribed on plastic drafting film to within about 20 minutes of arc (one-third of a degree). This can be sufficiently precise to plot a position on a large-scale chart. (Based on original artwork by Ben Ferrari)

Position-fixing using horizontal sextant angles can be accurate to within 1 m radius in ideal conditions, but practice is required to take observations from a moving boat. In the commercial world, sextant surveying is no longer used, having been replaced by cheap, versatile electronic position-fixing devices such as hand-held GPS units (see below).

Compass Bearings

A compass bearing could be used as a position-fixing device in the context of an archaeological survey but there are preferred alternatives. However, awareness of how a compass can be used is important because it adds to a general understanding of position-fixing.

To plot a position, a charted feature is aligned with the sights on the bearing compass and a reading taken. Bearings should be taken on at least two (but preferably three) separate features and, ideally, with a difference in angle between them of approximately 60 degrees. If two bearings are plotted on a chart, either by using the compass-rose printed on the chart or by physically measuring with a protractor from magnetic north, the two

lines should intersect at a point coinciding with where the readings were taken. A third bearing will act as an indication of accuracy and should pass through the existing intersection, creating what is often referred to as a 'cocked hat' (figure 11.7).

A prismatic or hand bearing-compass, or binoculars with a built-in compass, may be used from boat or shore. These traditional hand bearing-compasses employ a compass card rotating in a liquid, but hand-held electronic fluxgate compasses are now available which give a digital readout. These have a greater potential accuracy but the models so far encountered only give a bearing to within 1 degree.

The simple procedure of taking a fix requires practice to achieve consistent results in a moving boat, particularly with conventional magnetic compasses, as the compass card is normally moving continuously in response to the movement of the vessel. The major drawback with all magnetic compasses is their susceptibility to magnetic interference from electronic equipment, iron and steel. Great care must be taken to ensure the bearing-compass is not deviating. Unlike a ship's compass, it is not normally fixed in one position on the vessel. This means that its

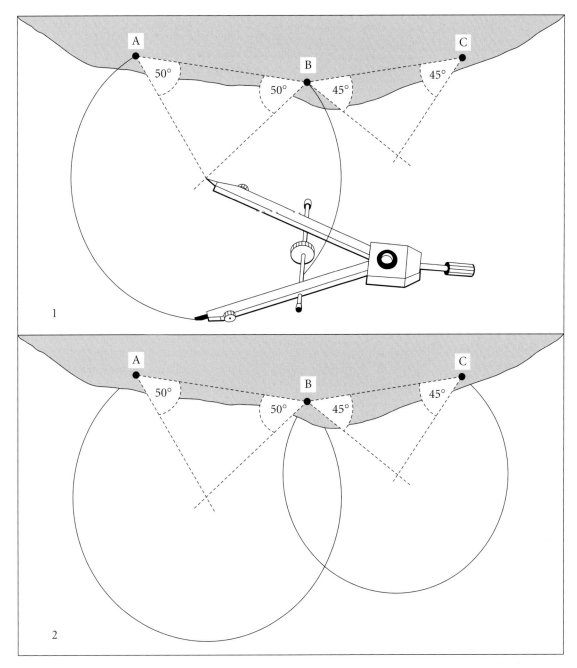

Figure 11.6 Sextant angles can be plotted geometrically from baselines between charted features. (Based on original artwork by Ben Ferrari)

relationship to potential sources of deviation often changes each time it is used, making correction factors very difficult to calculate.

It is important to remember that the bearings taken are relative to magnetic and not true north. The difference between the two slowly fluctuates over time and also varies with geographical location. Charts and maps usually have the relevant information printed on them and it is normally possible to calculate the difference between true north, magnetic north, and the north alignment of the reference grid of the map or chart being used (if that does not coincide with lines of longitude running true north–south between the earth's poles).

Transits

The visual alignment of two charted features establishes a line of sight that can be drawn on a chart; a second pair of aligned features, at approximately 90 degrees to the first and visible from the same position, will give an excellent

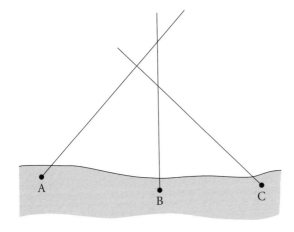

Figure 11.7 A large triangle of error, or 'cocked hat'. The smaller the triangle, the better the fix.

intersection when plotted (figure 11.8). A third transit will act as a check to see whether the observed features actually are the ones charted. This system is simple and potentially very accurate, particularly if the distance between the two features in alignment is a large proportion of the distance between the observer and the nearest feature (figure 11.9). It is also inexpensive given that all that is required is an appropriate map or chart, and perhaps a camera to record the transits.

Although the technique is extremely useful there can be problems. Archaeological sites under water and useful charted features are rarely conveniently positioned relative to each other. A choice may exist between features that do not quite line up and give an open transit (features just apart from each other) or a closed transit (one partially or totally hidden behind the other).

Often natural features have to be used and these can be difficult to equate precisely with what is drawn on the chart. For instance, the base of a cliff or the edge of an island or rock can differ as a result of tide, weathering or seasonal vegetation. In many instances the lack of suitable charted features will dictate that uncharted features will have to be used. Sometimes artificial transits, such as pairs of surveyor's ranging poles (figure 11.10) will have to be placed in appropriate positions. As with uncharted features, if these temporary alignments have to be used to plot the position of a site, each will have to be separately surveyed and marked on the appropriate chart or map. Establishing beforehand which features will be available for transits is not always practicable because they may not be visible due to poor horizontal surface visibility, or because they are masked by intervening landforms or vegetation.

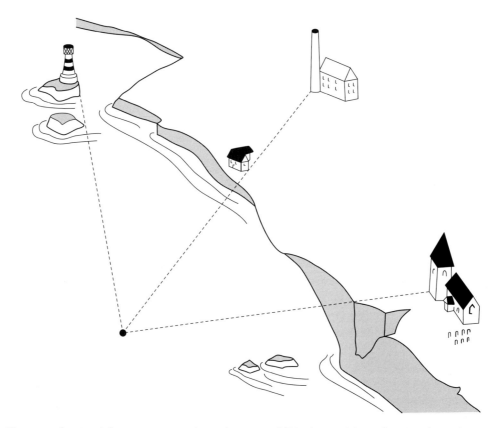

Figure 11.8 The use of coastal features as transit marks to establish the position of a site. (Based on original artwork by Ben Ferrari; after Oldfield, 1993:195)

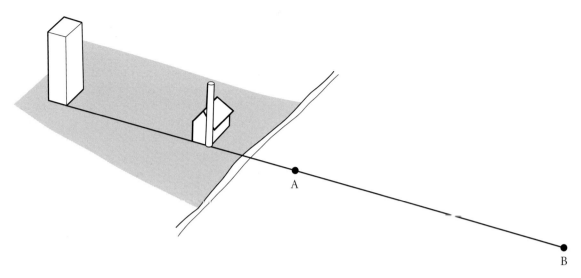

Figure 11.9 Accuracy of transits: a position fix at A will be more accurate than at B. (Based on original artwork by Ben Ferrari)

Figure 11.10 Temporary transits formed by setting up paired ranging rods along a shore baseline. When collecting data close inshore, or on inland waters, additional locational control can be provided by a tape measure or a distance line. (Based on original artwork by Ben Ferrari.)

Theodolite, electronic distance measurement, total stations

The theodolite is an optical measuring tool that measures angles in the horizontal and vertical planes. The theodolite was used in conjunction with chain and tapes for most surveying practices until electronic distance measurement (EDM) units were developed. EDMs use infra-red or laser light to measure distances and they can be attached or combined with theodolites, enabling distances and angles to be measured to the same point at the same time. From this, the 'total station' developed. The total station is a combined theodolite and EDM unit that processes and stores both sets of data in a microprocessor and data-logger (plate 11.1). The data is stored on an electronic medium, or downloaded onto a PC, or indeed both.

In essence a theodolite is a telescope that can be swivelled both horizontally and vertically against fixed scales for measuring the angle. The instrument is levelled, using spirit bubbles, so that the scales are precisely aligned in relation to the earth's magnetic field. Within the telescope there is a graticule along which sights are taken to the objects to be surveyed. In the case of the total station, a staff with a reflector for the electronic distance measurement is used to survey the relevant points. Like many electronic surveying systems, EDMs rely on the near-constant speed of radiated electromagnetic energy close to the visible end of the electromagnetic spectrum, commonly the infra-red wavelength. The time taken for a pulse of energy to reach a target and return to the instrument is measured and, using the known velocity of the energy, the distance is measured. With the combination of measuring angles and distances, several methods of determining position can be used. The use of a theodolite or total station requires specialist training but once mastered it is a simple and effective tool that can be used in many different situations. There are many texts available describing survey methods using the theodolite: *Surveying for Archaeologists* by F. Bettes is a good starting point.

Theodolites and total stations are used widely in land surveying and terrestrial archaeology for mapping survey detail on sites. Total stations have been used to measure boats and hulks on the beach and accurately to position survey datums for sites in the inter-tidal zone (figure 11.11). Millimetric accuracy is possible with these systems.

Angles and distances can be directly plotted on paper or, in the case of the total station, the instrument can be set to a local or the national grid. Theodolites are relatively cheap to hire and are readily available as they are used widely in the construction industry. The accuracy of these instruments is high. This is because a series of measurements is taken for each fix over a short period of time, usually between about 0.5 and 5 seconds depending on the accuracy required. A statistical average is then converted into a distance with an accuracy in the order of about 5 mm over an approximate maximum range of 4 km.

Electronic position-fixing and GPS

With the development of the offshore oil industry in the 1970s, electronic position-fixing (EPF) systems were developed to give adequate positional accuracies for survey and rig, and vessel and pipeline positioning. Some of these navigation systems, such as Decca Navigator and

Figure 11.11   Surveying a submerged site in shallow water using a shore-based EDM. (Based on original artwork by Ben Ferrari; after Morrison, 1985, fig. 5.2)

Loran C, were initially used in conjunction with GPS. They have now been almost entirely phased out and will therefore not be covered in this chapter.

The Global Positioning System (GPS) has revolutionized position-fixing at sea and on land. GPS is a satellite positioning system developed, owned and operated by the American military but available to commercial and leisure users worldwide. At the time of writing it is possible to purchase a hand-held GPS receiver that will give a position, anywhere in the world, to within 15 m. With the correct equipment, discussed later in this chapter, this can be enhanced, through differential corrections, to 1 m in 70 per cent of the world. On a local level, one person can set up and use a system that will give centimetric accuracy, in three dimensions. The ease of use and availability of equipment has led to the adoption of GPS as the standard form of navigation and survey position-fixing system both on land and at sea.

In the 1980s, GPS was developed on the back of an earlier system called Transit. Initially, GPS used a network of 18 satellites around the earth. Four of these satellites would be visible at any one time, from any location, and hence provide an instantaneous position fix. At the time of printing, there are 24 satellites in orbit plus a number of spare ones circling the earth in case of breakdowns.

Since the introduction of GPS, other interest groups have, or indeed are, developing their own satellite navigation systems, for the most part based on the same operating principles as GPS. Other commercial users and other military powers have felt that the control of the system by the US military could result in the service being withdrawn without prior notice. Not surprisingly, the Russians developed their own system called Glonass, which is operational and it is possible to buy dual Glonass and GPS receivers. There is a European system in development, known as Galileo, which should be fully operational by 2010. Galileo is being developed by commercial organizations for commercial and leisure users.

The derivation of position by GPS is relatively complex and beyond the scope of this book. It is based on the basic survey principles of trilateration discussed in chapter 14 (Underwater Survey). The positions of the satellites in space are known and ranges are calculated from the satellites to the receiver on earth. Each satellite transmits a signal containing information about the satellite position and the time of the signal transmission. The satellites are constantly updated with their positions from a series of ground stations that continually track and position the network. Within the receiver, there is a clock that measures the arrival time of the signal. From the known travel time of the signal, and assuming a known speed of travel, a range can be calculated. A position can be calculated using four satellite ranges instantaneously. The system broadcasts on two frequencies. This enables corrections to be made for atmospheric errors in the signal. The lower frequency contains the P code, which is the protected code for use by the US military and maybe her allies, but not released to civilian users. The higher frequency contains the C/A (coarse/acquisition) code, which is the code used in commercial and leisure receivers for position calculations. The C/A code is less accurate than the P code because of the pattern of coding used in the transmission signal. It was further degraded to reduce accuracy. This corruption of the signal was termed 'selective availability' (SA). At present, SA has been switched off with the result that a standard GPS receiver will give a position with an accuracy of between 2 and 20 m (6.5–65 ft), depending on the time of observations and the location. Prior to the relaxation of SA, an accuracy of between 25 and 50 m (81–162 ft) was obtainable but the system accuracy was quoted as being ±100 m (325 ft) 95 per cent of the time.

Enhanced accuracy methods

Several techniques have been devised to enhance the accuracy of the GPS C/A code. These techniques were devised when SA was implemented and have had a radical effect on the accuracy and reliability of the derived position. With the suppression of SA, they are still relevant today as they increase the accuracy still further.

Differential range-corrections / differential GPS

Differential range-corrections were initially used by the offshore survey industry and aeronautical community. Systems and networks that broadcast differential range-corrections have subsequently been developed for both commercial and leisure marine users on a national and international scale. The improved accuracy provided by differential range-corrections, otherwise known as differential GPS (DGPS), has had a profound effect on the importance of GPS as a surveying resource.

The principle of operation is that a receiving unit is set up on a known point and positions are derived. This receiver is known as the reference station. The difference between the derived satellite position and the known position of the reference station is calculated. The corrections between the observed ranges from the satellites and the computed ranges are calculated to derive the known position at the reference station. The stationary receiver is the key because it ties all the satellite measurements into an accurately surveyed reference point. This reference station receives the same GPS signals as the mobile receiver but, instead of working like a normal GPS receiver, it works in reverse by using its known position to calculate errors in the GPS signal. The receiver then transmits these errors to the mobile receiver in real time

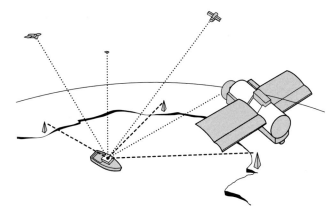

Figure 11.12 Differential GPS system with satellites, shore-based reference stations and in-boat mobile receiver. (Based on original artwork by Ben Ferrari)

so it can use them to correct its measurements. The corrections are only relevant to those satellites used in the position calculation (figure 11.12).

When this method was first developed, operators had to set up their own reference stations for their particular projects. Today many authorities have set up chains of differential stations around much of the coast and in many harbours. Corrections can also be received from satellites in a large proportion of the world. The International Association of Lighthouse Authorities (IALA) has set up a system that is free to users and extends from the northern North Sea into the Mediterranean. Corrections are received by radio link and the signal is decoded and sent to the GPS receiver. With the introduction of differential range corrections, positional accuracy with SA turned off is now of the order of <1 m to 4 m (3–13 ft) at a 95 per cent confidence level, depending on the time of day, location and the quality of the receiving unit.

At the time of writing, enhanced systems are being developed that are due to come online imminently. Termed augmentation systems, they will provide differential range corrections, more satellite ranges as well as system status information. This information will be broadcast from satellites and will be available in specific regions. The European system is known as EGNOS (European Geostationary Navigation Overlay Service), with the compatible system in the USA known as WAAS (Wide Area Augmentation System) and MSAS (Multifunctional Satellite Augmentation System) in Japan. These systems will be a significant step forward, as they will provide dynamic accuracies of 3 to 6 m (10–20 ft) from compatible standard hand-held GPS units. This will have serious implications for marine archaeology, as they will significantly bring down the cost of remote-sensing surveys and coastal-evaluation surveys, particularly in areas that do not benefit from coverage of national differential networks.

Real-time kinematic (RTK)

RTK GPS is similar to differential GPS in that a receiver is set up on a known point or reference station and corrections are transmitted to the remote unit. The nature of the corrections, however, is a little more complex as they are based on the phase, a property of the radio signal, not the derived ranges as in a differential system. The corrections are broadcast to the mobile receiver. The advantage of this system is that centimetric accuracy in three dimensions can be achieved in real time. The disadvantage of this system, at present, is the cost of the equipment. The system is also limited to a maximum range of approximately 40 km (64 miles) due to the changes in the properties of the atmosphere and the curvature of the earth.

EQUIPMENT

With recent developments in electronics, the cost (and, indeed, size) of GPS receivers has drastically reduced, making GPS one of the most cost-effective means of position-fixing. It is possible to purchase a hand-held unit with a differential receiver that is able to pick up the IALA correction service, or an EGNOS compatible unit, for less than the price of a medium-quality survey sextant. Most leisure craft and all commercial craft carry a GPS unit. Receiver quality and, therefore, price is dependent on the number of receiving channels, the quality of the internal clock and the sophistication of the algorithms used to calculate position.

It is important to stress at this stage that GPS undertakes all position calculations on the WGS84 datum. Most receivers have the ability to convert from WGS84 into other local datums. So, once again, it is vital to know which datum the receiver is set to and which datum is being worked in. Some receivers will also give an indication of the quality of the calculated position. This can be expressed as an accuracy figure or a DOP value (DOP being the 'dilution of precision') and it can be prefixed by H for horizontal or G for geometric. Basically, the higher the number, the worse the accuracy of the position.

Limitations of GPS

The fundamental principle of GPS is that all calculations are made on the WGS84 spheroid or datum and this in itself can cause problems to the unwary. If an alternative datum is selected in a receiver, a transformation is undertaken between WGS84 and the alternative datum. Few makes and models of GPS units have consistent transformation parameters (the method of converting from one datum to another between local datums and WGS84). Therefore, different coordinates can be derived for the same point, giving rise to low repeatability.

If several GPS points are laid over a site-plan, for example, errors may be encountered over and above the limits of accuracy. Again, this occurs when a plan surveyed using a flat grid, based on the assumption of a flat earth, is compared with GPS positions calculated on the spheroid. This is a significant factor when using highly accurate systems such as RTK GPS. If the distance between two stations is measured using a tape-measure, and then an RTK GPS unit is used to measure the coordinates and calculate the distance between them, the result will be two different values. This is because the RTK GPS unit is measuring the distance and taking into account the curvature of the spheroid. To obtain comparative measurement, the scale factor (a value representing curvature) must be applied to the direct distances measured by the tape-measure.

In summary, GPS is an extremely useful tool for position-fixing. It is relatively affordable and can result in sub-metric accuracy given the right equipment and conditions. It is essential, however, when using GPS, to be aware and have a clear understanding of the issues discussed in this chapter.

FURTHER INFORMATION

Ackroyd, N., and Lorimer, R., 1990, *Global Navigation: A GPS user's guide*. London.

Bettes, F., 1984, *Surveying for Archaeologists*. Durham.

Dana, P. H., *The Geographer's Craft Project*. Dept of Geography, The University of Colorado at Boulder. www.colorado. edu/geography/gcraft/notes/datum/datum.html (updated 2003).

Judd, P. and Brown, S., 2006, *Getting to Grips with GPS: Mastering the skills of GPS navigation and digital mapping*. Leicester.

Schofield, W., 2001 (5th edn), *Engineering Surveying*. Oxford.

12 Underwater Search Methods

Contents
- Underwater search methods
- Positioning
- Coverage
- Safety
- Diver search methods

Searches can be divided into two types: those deploying a diver or a submersible and relying on the human eye or hand-held equipment, and remote-sensing surveys usually employing acoustic or magnetic equipment and remotely operated vehicles (ROVs) deployed from a boat, or other craft, on or above the surface. This chapter covers the first type of search. For detailed information about remote-sensing techniques, refer to chapter 13.

POSITIONING

Whatever search method is used, it is vital to know where you are and where you have been (see chapter 11 for position-fixing techniques). This will save time in search operations and will significantly enhance the information recorded. A search only has value if the position of the area covered and the identified targets, together with other pertinent observations, are accurately reported and recorded. Divers should complete detailed dive-logs, recording all information of potential interest for subsequent analysis.

COVERAGE

Unless a well thought-out sampling strategy has been devised, a common objective is 100 per cent coverage of the selected area. Unfortunately 100 per cent is difficult to define in this context due to the varying degrees of efficiency for different techniques. For instance an extensive visual search may cover every square centimetre of the sea-bed but miss an object a few millimetres long, or a larger object camouflaged by a temporary dusting of light silt. Similarly, with remote-sensing techniques, magnetometer search-corridors separated by 50 m (163 ft) might be sufficient to detect a large wooden wreck containing iron cannon, but might not detect other smaller targets of archaeological significance, such as a single cannon.

It is important to realize that the fact that nothing has been detected during a search does not necessarily mean that nothing is there. The development of increasingly sophisticated remote-sensing equipment (see chapter 13) has made investigating the sea-bed a more reliable science, but there is still some way to go before it is possible to be certain of discovering all the available evidence.

SAFETY

Whichever search method is chosen, it is essential that team members are trained to undertake the technique safely. (Refer to chapter 6 for further information concerning safety on archaeological sites.) Many of the inherent problems associated with the following techniques, however, can be alleviated with planning and practice. This could include carrying out a dry run (plate 12.1) to help perfect diver positioning, communications and an efficient recording technique.

DIVER SEARCH METHODS

These mostly depend on visual observations, but hand-held instruments and equipment, such as metal-detectors and cameras, can also be deployed. The speed and efficiency of any search is proportional to the size of the targets, visibility and experience of the divers in such techniques. Small objects on a muddy sea-bed are a challenge to locate. The search organizers must consider which method is likely to achieve the best results, taking into account the nature of the material being sought, scale of the search area, depth, sea-bed type, number of divers and time available. The choice of the search team is also important. It should be remembered that a large variation exists amongst divers in terms of ability to notice things on the sea-bed. This ability is related to factors, such as:

- familiarity with the area to be searched;
- anticipated target material (e.g. small ceramic fragments or large ships' timbers);
- training in the search techniques;
- experience in the search technique to be used;
- apprehension caused by diving factors;
- level of diving experience;
- concentration;
- aptitude;
- commitment;
- diving conditions and diver comfort.

Towed searches

The equipment associated with this method ranges from the simple (a diver holding on to a weight on a line) to the relatively sophisticated (involving towed vehicles with moveable vanes capable of altering attitude and elevation relative to the sea-bed). A common and inexpensive system uses a simple wing or board capable of sufficient movement to 'fly' the diver over changes in the sea-bed topography (figure 12.1). All these systems rely on the surface crew controlling and recording navigation and making due allowance for 'layback' between the diver and the boat. Alternatively, an underwater vehicle can pull the diver along. Diver propulsion vehicles (DPVs) vary in complexity from those with a simple electric motor driving a propeller held in front of the diver to mini-submersibles. The major drawback of these systems, apart from cost, is the difficulty of position-fixing. Surface marker-buoys attached to the diver can be tracked from the surface or, alternatively, it is possible to

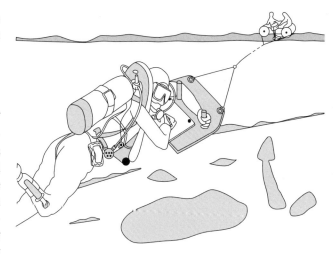

Figure 12.1 Towed diver search. (Based on original artwork by Ben Ferrari)

use through-water navigation systems based on acoustic transponders placed at known positions on the sea-bed.

The effectiveness of a towed-diver search will be dependent on visibility and speed of the diver over the sea-bed. At 1 knot the diver is covering just over 30 m (98 ft) a minute or 0.5 m (20 in) a second and this can be an effective way of covering relatively large areas of the sea-bed during one dive. In many circumstances, however, this will be too fast to allow observation to the required level of detail.

If the diver can control speed and even stop the forward movement to inspect potentially interesting sightings, the efficiency of the operation is dramatically increased. While rope signals are possible, diver-to-surface communications maximize the benefits of this method. Telephone-style (hard-wire) communications with the diver's microphone connected via wires to the surface tend to be clearer than through-water versions, which can be affected by water turbulence caused by the moving boat. The position of the towline relative to the propulsion unit must be taken into consideration when planning the search. With efficient communications it is possible for the surface team to log and plot observations made by the divers. Without communications, a less satisfactory alternative is for markers to be dropped on the sea-bed in the vicinity of any observation, which can subsequently be investigated further. These will then need to be accurately positioned so they can be relocated and assessed at a later date.

To reduce the inherent risks to the diver and to avoid pressure-related illness, very careful control of depth is important. This is often difficult to achieve when being towed. A solution is to use a diving computer that can accurately record the dive profile during the search.

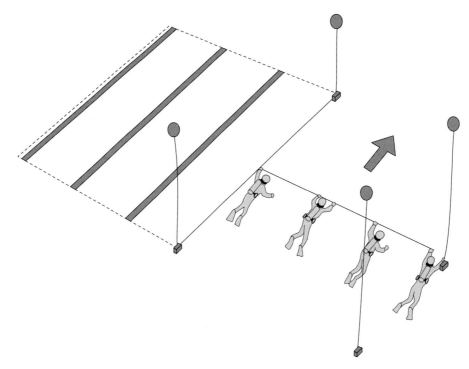

Figure 12.2 Swimline (freeline) search. (Based on original artwork by Ben Ferrari)

Many dive-computers have audible alarms to alert the diver in the event of a rapid ascent. In conditions of poor visibility, there should also be some capability for detecting potentially dangerous obstructions in the path of the towed diver. One solution is for the towboat to be equipped with forward-scanning sonar.

Swimline (freeline) searches

Practice, patience and perseverance are prerequisites for this method but it has been used with success on many archaeological projects in the past. The system relies on a string of divers, often between two and six, spaced along a tape-measure or graduated line at intervals equivalent to less than the limit of clear visibility, so that complete coverage is possible (figure 12.2). The search is usually undertaken with a ground-line to guide a controller. To enable positions of objects to be fixed, it is useful also to graduate the ground-line in convenient increments. This technique is in fact a large-scale offset survey (see chapter 14). These ground-lines can be as long as necessary (up to a kilometre has been known) and should be reasonably straight. To achieve this, they are probably best laid from a moving boat, keeping the ground-line under tension while it is being deployed. It can be difficult to lay long ground-lines precisely, but with practice it is possible. As long as the line is straight, the positions of the two ends can be fixed (see chapter 11) so that there will be a record of where the search took place.

The main difficulty with this technique is maintaining the line of divers at right angles to the ground-line, which inevitably becomes more difficult in poor visibility. Another difficulty encountered with this method is maintaining effective communication between the divers to allow the line to advance smoothly. This problem increases proportionally to the number of divers on the line, and it is also intensified by poor visibility. Usually, if a diver needs to stop to make an observation, or is ready to move off again, this is indicated to the others by a code of signals along the line. With practice, it is possible to communicate quite complicated messages along the connecting line using a set of bells (short tugs) and pulls (long tugs). Standard rope signals can be found in many diving manuals and additional ones appropriate to the individual operation can be devised.

Another difficulty is the variation in dive duration between individuals at the same depth and in the same environment caused by dissimilar equipment and breathing rates. Matching the contents of the divers' breathing-gas supply will help. It is considered best practice to terminate the search when one of the divers needs to surface for whatever reason, unless a buddy diver can accompany the diver to the shore or safety-boat.

Figure 12.3 The offset method can be used to record the position of features in relation to jackstays and ground-lines. In this example a hand-held metal-detector is being used to locate metallic features. (Drawing by Graham Scott)

Plotting the position of observations during the search can be achieved in the same way as described when using the offset survey method. All that is required is to record the distance along the ground-line and a distance out at right-angles to the ground-line. To achieve a detailed survey two divers can move slowly forward using a short rigid rule or tape, at right-angles to the graduated ground-line (figure 12.3). Alternatively, markers can be dropped at points of interest and the position fixed later. If the number of markers needed by each diver is excessive, an alternative search method, such as the jackstay system (see below) may be more suitable.

Swimline searches are often more effective in clear, shallow water, but this technique can be deployed successfully in deeper water if consideration is given to using a secondary pair, or team, of divers to record the points of interest located by the search team. With discipline, good communication and rehearsed procedures this technique can be very flexible, allowing the search-line to be stopped periodically to record the nature of the sea-bed, or count the number of surface-indicators (potsherds, for example) to compile a distribution-map.

Jackstay (corridor) searches

This is a useful system if total coverage of an area of sea-bed is required during visual or metal-detector searches, but it requires more pre-search organization than the two methods outlined above. The minimum requirement is usually two long ground-lines laid parallel at a convenient distance apart, often 10 m, 30 m or 50 m, and a further line, the jackstay, laid at 90 degrees between the two ground-lines. It is more efficient to use two jackstays to define a corridor because an area with defined edges can be more effectively searched. Once the area between the jackstays has been searched, another corridor is laid, usually by leapfrogging one line over the other (figure 12.4). It is important to take into account the potential safety implications of divers being separated during this part of the operation.

Fixing ground-lines to the sea-bed can be achieved with methods similar to those suggested for survey points (see chapter 14). In fact the ends of ground lines are likely to be survey points. One of the problems with ground-lines is keeping them straight on the sea-bed. In shallows, water-movement may mean the line has to be weighted or pinned to the sea-bed to prevent unacceptable lateral displacement. Even in deeper water, ground-lines may have to be placed along the line of maximum prevailing current to help prevent sideways movement; and even then, fixing to the sea-bed is likely to be required if the lines are going to be used as part of a site coordinate system for locating observations.

The jackstays themselves are less permanent features and may only be in position for as little as 5 or 10 minutes, depending on the size and intensity of search. The method by which they are anchored must depend on the nature of the sea-bed, and that might be totally different from one end of the line to the other. Like many survey points, line fixing will probably rely on either the weight of an object, such as a 25 or 50 kg (55 or 110 lbs) metal block, or on a pin or other fastening forced into bedrock, an immovable boulder, or sediment. Even heavy weights can be pulled across the sea-bed with surprising ease (except when you want to move them yourself, of course), so it is common for a weight to be pinned to the sea-bed for additional security.

If the ground-lines and the jackstays are graduated, one of the ground-lines should be considered as a zero axis. The zeros of the jackstays can then be positioned on that line. It is difficult to keep the distance between ground-lines constant and therefore the other end of the graduated jackstay will not always coincide with the second ground-line at the required distance. Rather than spend unnecessary time making everything perfect, and without sacrificing offset-survey precision, simply concentrate on the line of the jackstay passing across the appropriate graduation of the ground-line. To do this, it helps if the jackstays are over-length. While ground-lines are usually made from rope or line (leaded line is useful), the jackstays are often tape-measures or, better still, thin plastic measuring lines that are available in 50 m (165 ft) lengths. In many circumstances it can be advantageous to

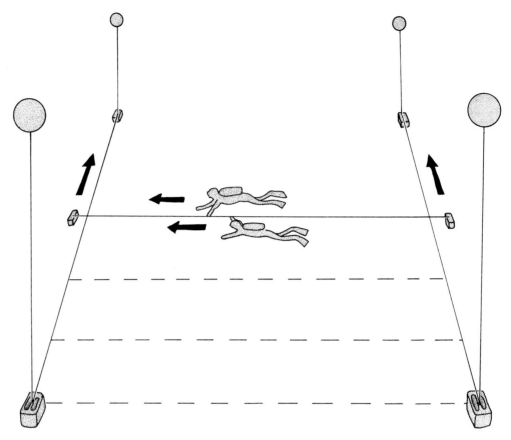

Figure 12.4   Jackstay (corridor) search. (Drawing by Graham Scott)

extend the length of these with about 2 m (6 ft) of elastic cord ('shock cord' or 'bungee-cord'). The advantage is that it keeps the line under permanent tension, therefore making it less of a problem should a diver accidentally apply pressure to it during the search. It also means that, provided there is sufficient elasticity, one end of the jackstay can be leapfrogged at a time with less expenditure of time and effort than if both ends have to be released before the line can be moved.

The normal width of the corridor between jackstays is between 2 and 8 m (6 and 26 ft), depending on the number of divers searching each corridor, visibility and the type and size of expected targets. Although searches with four divers can be organized between the jackstays, for intensive work a maximum of two divers for each corridor allows easier offset measurements to be taken from the jackstays, resulting in a precise location for each observation.

Experience has demonstrated that a 1 m (3 ft) wide strip for each diver in a 2 m (6 ft) wide corridor is suitable for very intensive sea-bed searches (perhaps using a metal-detector), almost regardless of visibility or nature of the sea-bed. In areas where the sea-bed is uniformly covered in fine sand or silt, the width of the search strip can be

greater. It can be as much as 6 m (20 ft) in certain circumstances, and even wider if the expected target is relatively large (e.g. a ship-sized ballast mound).

Grid searches

If an area needs to be searched thoroughly and features need to be located with precision, then the grid search has much to offer. The first step is to lay a series of ground-lines at 90 degrees to each other at an appropriate spacing to create a grid over the site. The size of the grid will depend on a variety of factors but is commonly between 2 m and 50 m (6–165 ft). Ground-lines left on the sea-bed for any time tend to get damaged or disappear. However, if the intersecting points or anchorage points have been adequately fixed, the grid could be reconstructed in future seasons, even if the actual lines have gone.

Once the grid has been established, the corridor-search technique can be easily deployed and the location of points can be readily identified by site coordinates with, conventionally, a series of numbers on both the horizontal and vertical axes with the zeros at the bottom left (south west) corner. Sometimes one of the axes is replaced by letters to help those inexperienced in the use of coordinates as a way of expressing a position. By using an appropriate number of digits it is possible to define a location down to the nearest millimetre, although that would not normally be necessary.

Circular searches

This is a simple search system that can be useful in poor visibility or when the need to conduct a search suddenly arises. The method can use equipment that is normally carried by divers, such as a diver's marker buoy and a weight to form the down-line and distance-line. Even the boat's anchor could be used as a starting point. The technique does not, however, easily or efficiently lend itself to total sea-bed coverage because of the elliptical overlap created by linked searches (figure 12.5). It is useful for trying to locate a known object whose position has not been accurately recorded, or to extend a search area after an isolated artefact has been found. It can also be of use as part of a sampling strategy to assess different areas of the sea-bed before intensive searching begins.

The system relies on a graduated line, often a tape-measure, being attached to a fixed point on the sea-bed. The divers swim round in a complete circle using either a compass (to take the bearing at the start and finish) or obvious marker on the sea-bed as a guide to when a circuit has been completed. A satisfactory solution is to set out a graduated, straight ground-line (running out from the centre of the search area) to act as a start/finish indicator. The distance between each circular sweep has to be related to the visibility and the type of target that the divers are expecting to find.

It is usual to begin in the centre and sweep at ever increasing diameters. Remember that the diver on the end of the line will be travelling further and faster than the diver closest to the centre-point of the search. However, starting at the maximum length of sweep can be more effective if a known object is thought to be upstanding from a flat seabed. One circuit should result in a snag if the target is in the circle. The distance line can then be followed back to the object. During normal archaeological searches such snagging is clearly not desirable. The problem of unwanted snags can be reduced if the distance-line is lightly buoyed at the mid-point. Some tension is usually maintained on the line to ensure that the diver keeps to the correct track. If the line catches on an obstruction, releasing the tension should allow it to rise and hopefully release itself from the snag. In very poor visibility, it is not always clear that a snag has occurred until the search pattern has been grossly distorted.

Once the position of the centre of the search area has been fixed, observations made during the search can be recorded by noting the distance from the centre and the magnetic bearing to it using a hand-held compass. In this way, each plot will be recorded but the level of accuracy will be as limited as surveying using the radial survey method (see chapter 14).

Metal-detector searching

Although these instruments are 'remote-sensing' devices and some types can be towed behind a boat, it is the numerous diver-held versions that have proved to be a valuable tool to many archaeologists. Unlike magnetometers (see chapter 13), they can detect both ferrous and non-ferrous metals, and those used under water usually work on the pulse-induction principle. Pulses of energy are emitted and produce a temporary magnetic field around the search-head. The rate at which this field decays is prolonged in the presence of metal. Comparison between the decay rate and the original pulses allows detection of metal objects of large mass to a maximum range of approximately 2 m ($6^1/_2$ ft), and objects the size of a single coin at a distance of approximately 10 cm (4 in).

Metal-detectors are used in three principal ways during archaeological work. First, during the pre-disturbance survey of a site, concentrations of metallic contacts and isolated responses can be mapped. The second way is to ascertain the approximate position of objects in a layer that is about to be removed. This can contribute to a very high recovery rate for metallic artefacts that might otherwise be overlooked because of their small size or poor visibility. The third use is to locate metal artefacts on bedrock which are either invisible because of a light dusting of silt or because they are hidden in holes and crevices in the rock.

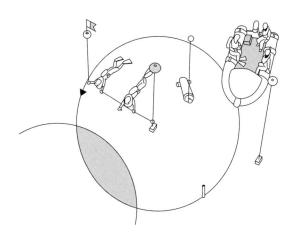

Figure 12.5 Circular search. (Based on original artwork by Ben Ferrari)

Hand-held metal-detectors must be used systematically if they are to be an effective search tool. This usually entails total coverage employing the corridor or grid-search techniques as shown in figure 12.3. It is important that the search-head of the instrument covers every square centimetre of the area of sea-bed under investigation. Investigating metal-detector responses by digging around an anomaly without regard to the other clues that might be disturbed is not a valid archaeological technique. Nor is any approach involving a diver with a metal-detector swimming in a random pattern over a site, looking for souvenirs.

Geophysical and Remote-Sensing Surveys

13

Contents

- Search patterns, navigation and positioning
- Acoustic systems
- Bathymetric survey
- Echo-sounders
- Multibeam swath systems
- Bottom-classification systems

- Sidescan sonar
- Sub-bottom profiling
- Magnetometry
- Integrated surveys
- Submersibles: ROVs and AUVs
- Aerial photography

Since the 1960s a variety of marine geophysical techniques have been used to investigate a range of submerged archaeological sites, with the principal aim of site prospection. More recently the emphasis of archaeological research has moved away from pure prospection towards a fuller understanding of individual sites in terms of detailed site mapping, the wrecking event, site-formation processes, and ultimately the development of heritage-management strategies for conservation and protection.

A significant advantage of geophysical surveying is the ability to collect large amounts of information quickly, often at some distance from the target. This allows search patterns to be much more widely spaced and undertaken at a greater 'speed over the ground' than could ever be achieved by divers. Some types of equipment can detect certain classes of information that are buried and out of the sight of divers. Furthermore, restricted underwater visibility and strong currents are less of a problem for geophysical survey instruments and, in many instances, they can be deployed in sea conditions worse than those in which divers can safely operate. However, these techniques should not be seen as ways of removing divers from underwater archaeological investigation, but as tools which can both enhance the effectiveness of diver investigation and extend the range of environments in which underwater survey can be undertaken.

Geophysical survey in archaeology is generally concerned with research and/or site management. This chapter summarizes a range of geophysical instruments and techniques that can be useful in an underwater archaeological project, including:

- acoustic systems (see below);
- magnetometers;
- submersibles – remotely operated vehicles (ROVs), autonomous underwater vehicles (AUVs).

SEARCH PATTERNS, NAVIGATION AND POSITIONING

The search patterns appropriate to diver investigations of the sea-bed (see chapter 12) are also appropriate to larger scale investigations, the differences being in the scale of operations and the method of location control. For a marine geophysical survey it is essential to plan the track of the vessel so that there is total coverage of the search area. This usually involves the survey boat following parallel lines at set distances apart, the distance proportional to the coverage of the instrument and any overlap required. For intensive work, the overlap is often more than 100 per cent and, to guarantee total coverage and provide a different viewpoint, two corridor search patterns at 90 degrees, or even three at 60 degrees, to each other may be considered. Any decisions regarding line spacing and line direction will depend on the type of equipment deployed and the nature of the sea-bed and environmental factors. For instance, in 10 m (33 ft) depth of water, an echo-sounder with a narrow-cone transducer might only cover a strip of seabed 1.8 m (6 ft) wide, considerably less than can be achieved with a sidescan system. If the sea-bed is fissured by deep gullies that need to be investigated acoustically, then the search pattern may need to be adjusted to allow transmitted energy to reach the bottom of the gullies. This consideration is not necessary with magnetometers (see below) as they are virtually omni-directional. Other factors that influence survey-line orientation include the proximity of shallows and obstructions, navigation buoys and fishing floats, shipping channels, the activities of other sea-users, the direction and strength of winds and currents, and variation in current direction and strength during tidal cycles.

To interpret any geophysical data acquired at sea, it is necessary to relate the observations to a geographical position; the more accurate the positioning, the more useful the data will be. For details of position-fixing techniques and technology, see chapter 11.

Currently, one of the most accurate ways of location control during marine surveys is for the position of the vessel to be displayed graphically on a computer monitor in front of the helmsperson. Specialist survey packages are often built into data collection software and allow an identified area to be quickly divided into survey lanes of appropriate separation and orientation. The helmsperson then 'steers' the cursor (representing the boat) down the selected line on the screen.

In good sea conditions, sidescan sonar can cover a swath up to about 1000 m (3250 ft) wide across the sea-bed whereas an echo-sounder with a narrow-cone transducer might only cover a strip 1.5 m (58 in) wide in 10 m (33 ft) of water. Such varying widths of sea-bed search highlight the difficulty of deciding what spacing is required between tracks to give a search pattern with 100 per cent coverage. It is important that the target type is known (or at least decided upon), the capabilities of the instruments deployed are fully understood and the environmental factors are considered – otherwise there could be gaps in the search area.

ACOUSTIC SYSTEMS

The most commonly used geophysical methods for marine archaeological survey are acoustic (sound or sonar) systems. These include echo-sounders, multibeam swath systems, sidescan sonars, sub-bottom profilers and bottom classification systems (see below). Which of the many available systems are chosen depends on the type of information required for a particular site. Important factors to consider are:

- Is it the morphology or the material make-up of the site that is important (or both)?
- Is qualitative or quantitative information required?
- Is the site of interest exposed and/or beneath the sea-bed?

No one system can provide all this information, and normally there will always be a compromise: in some instances it may be necessary to undertake multiple-instrument surveys to collect a wider range of information. Regardless of the methodology chosen, the survey should extend to include a meaningful proportion of the surrounding area so that there is the opportunity to put the archaeological site into its environmental context.

BATHYMETRIC SURVEY

An essential component of all investigations of submerged archaeological sites is the production of a detailed bathymetric (depth) chart. The degree of accuracy of the final presentation is dependent on both the geophysical technique used and the effective integration of a high-resolution navigation system. Attention must be paid to the coordinate system recorded by the navigation software. GPS data are conventionally output as geodetic coordinates (latitude and longitude) using the WGS84 datum. Where the survey is being undertaken as part of a seamless onshore–offshore investigation, it is common for the geodetic data to be converted to a metric coordinate system (e.g. UTM or OSGB36 for the UK). To ensure effective integration of the data, the archaeologist should always be aware of the vagaries of coordinate conversion (see chapter 11).

There are two primary systems for the acquisition of bathymetric data: narrow-track echo-sounders and wide-track multibeam swath systems. Whichever system is used, data quality is affected by the following factors.

Relative height of the sonar head: When taking a series of depth measurements with a transducer attached to a boat that goes up and down with the tide, the height variation has to be allowed for in the final bathymetric data. A simple way to allow for tidal variation is to check the depth reading over a fixed point on the sea-bed with the echo-sounder at regular intervals. Adjustments can be made to readings collected between checks to provide uniformity in the data. Often a tidal curve calculated from readings of a nearby tide-gauge is used to correct the depths. This works reasonably well if the tide-gauge is very close to the site but becomes less effective as the distance increases. It can be relatively simple to install a graduated board on a site as a tide-gauge, which can then be visually monitored and readings noted manually. Alternatively, sophisticated instrumentation can be set up which records the data automatically and transmits a correction in real time to the survey boat. Obtaining tide-gauge readings from a nearby source may work for some projects, but it is rarely good enough for high-quality, very high-resolution surveys.

While it is relatively easy to measure height differences caused by the tide, differences created by waves and swell are much more difficult to measure. It is for this reason that the highest quality surveys tend to use RTK systems (see chapter 11), which continuously and accurately monitor the relative height of the sonar head. This allows all vertical variations, regardless of their cause, to be compensated for automatically in the data set.

Roll, heave, pitch and yaw: Another factor which needs to be considered is the way transducers move about as they follow the motion of the boat on the water. Ground-swell can heave the boat up and down over considerable distances, which can cause problems with the depth readings if not taken into account. Similarly, waves can make the survey boat roll, pitch and yaw, which, in turn, can have a profound effect on the direction of the acoustic beam(s). For high-definition surveys, it is pointless to assume that transducers mounted on a moving boat always point directly down at the sea-bed. Most echo-sounders, except those specifically designed for professional surveys, do not have facilities for compensating for boat movement. As it is crucial to know exactly where the acoustic energy is directed when surveying, it is necessary to measure, to a very high degree of accuracy, movement in all four directions. This can be achieved with a motion-reference unit. While they are relatively expensive, they are an essential component of high-quality acoustic surveys in support of archaeological investigations.

Speed over the ground: A simple and effective way of improving the quality of geophysical surveys is to move slowly so that more data is collected in every portion of sea-bed. The biggest problem with adopting this simple technique is the difficulty of getting boats to steer accurately at slow speeds, but by heading into current or against the wind, natural forces can be used to help reduce the speed over the sea-bed. It is also possible to slow down survey boats by the use of drogues, but these can have an additional detrimental effect on steering.

ECHO-SOUNDERS

Conventional echo-sounder systems consist of a single, hull-mounted or pole-mounted transducer that acts as both an acoustic transmitter and a receiver (transceiver). These systems produce an acoustic pulse with a single frequency within a typical range of 100–300 kHz and a frequency-dependent, vertical resolution on a centimetric scale. The echo-sounder transducer produces an acoustic pulse with a cone angle normally between 5 and 45 degrees, oriented vertically downwards, so concentrating the acoustic energy in a small circular area of the sea-bed (the radius of this circle being dependent on the water depth). The horizontal resolution of these systems is controlled by a combination of source frequency, cone angle and water depth. For example, a 200 kHz echo-sounder with a 10 degree cone angle has a footprint diameter of 1.8 m (6 ft) in a water depth of 10 m (33 ft).

The echo-sounder system does not provide direct measurement of depth, but calculates a value from the recorded two-way travel time. The resulting depth information can either be recorded digitally or via post-acquisition digitization of two-dimensional analogue traces. Depths are conventionally recorded in metres, with the actual figures displayed representing the distance from the transducer to the sea-bed. For bathymetric analysis of data from a near-shore environment, all values obtained must be corrected for both tidal variation and the depth of the transducer beneath the water surface.

One major disadvantage of narrow-track systems is that the distance between the lines of a survey-grid controls the effective horizontal resolution of the system. In a tidal environment, the closest survey-grid spacing normally achievable is approximately 5 m (due to the limits imposed by the survey boat manoeuvrability). Therefore, the highest possible horizontal resolution for the bathymetric data is ±5 m. Bathymetric data are conventionally represented as profiles and/or two-dimensional contour plots.

Overall, the quality of echo-sounder surveys does not compare well with swath surveys and they take significantly longer to conduct, but they have the advantage of being less expensive and useful results can be obtained.

MULTIBEAM SWATH SYSTEMS

A development from echo-sounder technology is multi-beam swath bathymetry, which records depth measurements in a thin strip below and to the side of the boat, and repeats at up to 50 times a second as the boat moves forward. In one pass, this provides considerably more depth information about the sea-bed than could be achieved with a single echo-sounder. In the example given earlier, an echo-sounder at 10 m depth would cover a strip of sea-bed 1.8 m wide as it moves forward. A typical multi-beam swath system in a similar depth of water would perhaps cover a track 55 m wide. When coupled with the exceptionally good vertical accuracy, it is no surprise that multibeam swath systems are now the instrument of choice for professional hydrographic surveyors.

Many of these sophisticated systems have large sonar-head arrays and, although they can be built into the hull of a large boat, archaeological work is often done from smaller vessels of opportunity, which means that the sonar-heads need to be mounted on frames attached to the boat or on towed floating platforms. With care, it is possible with temporary mountings to get close to the theoretical resolutions available from multibeam swath systems, which can be of the order of 5 mm horizontally and 6 mm vertically (plate 13.1). Such precision is ideal for detailed archaeological site investigations.

As soon as the water depth increases, or the range when mounted on an ROV or AUV, the ping-rate has to be reduced so that returning echoes are collected before the next pulse is transmitted. This means that survey boat speeds must be reduced to maintain the highest resolution. This also has an additional benefit because sonar-heads, when mounted on frames attached to boats, tend to vibrate as speed increases, reducing data quality. As a general rule of thumb, when aiming for the best quality multibeam swath-survey data, try to keep the survey boat speed down to below 4 knots (c.2 m/s).

The data collected during multibeam swath surveys can normally be displayed in real time as a profile and as a colour contour plan, or as a complex three-dimensional image. After the survey, the millions of points of data will often include 'fliers' and 'spikes', acoustic returns caused by a variety of natural and physical phenomena. These are usually filtered out in post-processing but it is important that an archaeologist, or a surveyor with considerable archaeological experience, does this editing – otherwise archaeological features can be unwittingly removed.

The software available for viewing multibeam swath data usually has the facility to apply an artificial rendered surface to where the software thinks it should be. This is effective for normal sea-bed types (plate 13.2), but can be disastrous for archaeological evidence (plate 13.3). It is essential to view multibeam swath data of artificial material, such as wrecks, as point-clouds floating in space. Each point represents the x, y and z coordinates for each return for each beam at each ping and, with high-definition surveys, these can be very densely packed. Most of the proprietary multibeam swath data-visualization software packages allow these point-clouds to be looked at in three dimensions and rotated on the computer screen. This provides much more information than would be seen from stationary images because, currently, the human eye and brain are better than software for separating out and identifying features. It is also possible to produce profiles in any direction across a data set and take measurements between any two points, which is a tremendous help when trying to interpret and understand a site.

Multibeam swath bathymetry is a standard survey tool for both site-specific work and for coverage of the larger expanses of the sea-bed necessary for submerged landscape reconstruction (plate 13.4). It makes sense to undertake a multibeam swath survey first, to collect basic information about where the major components of the site are, before committing resources to diver surveys with tape-measures and drawing frames. As geophysical surveys are not constrained in the same way as diving operations by pressure of water, underwater visibility and currents, it is even possible to collect excellent data on sites where diving surveys would be either ineffective or impossible.

The advantage of multibeam swath systems is that they provide baseline surveys very quickly and, in terms of the overall site, at very high levels of accuracy. Basic site surveys can be accomplished at rates more than 100,000 times faster than can be achieved by even the most experienced diving teams (plates 13.5 and 13.6). However, tape-surveys by divers tend to be more precise where the measured distances are less than 2 or 3 m (6$^1/_2$–10 ft), and so are ideal for detailed archaeological surveys of small areas. For longer distances and relating various small areas of a site, there is nothing as quick and accurate as a high-definition multibeam swath survey, but it must always be seen as a complement to, rather than a replacement for, diver surveys.

Multibeam swath surveys are also very useful as a management tool because high-standard repeat surveys of a site are relatively simply achieved. This enables direct comparison between multiple surveys so that changes to sediments (plates 13.7 and 13.8) or changes to an archaeological site (plate 13.9) can be easily detected.

One perceived drawback of multibeam swath systems is the amount of data that such surveys can generate – up to 10 gigabytes in a day. Fortunately, advances in computer processing power and memory capacity have largely overcome this problem. Another potential drawback is cost, but systems are available for hire on a daily basis and it is not impossible to find a manufacturer or a grant-awarding body to pay for, or at least subsidise, the use of this important archaeological survey tool.

BOTTOM-CLASSIFICATION SYSTEMS

Multibeam swath and profiling echo-sounders were originally designed to give only quantitative data on the topography of the sea-bed. However, developmental work has resulted in attempts to extract proxy indicators of the material nature of the sea-bed from the returning echoes. There are a number of bottom-classification (sea-bed discrimination) systems. Most are based on single-beam 'profile' technology and so give an indication of the material type at individual points. These systems look at either the shape of the sea-bed return directly or in conjunction with a multiple reflection of the sea-bed to determine both the 'hardness' and the 'roughness' of the bed. Thorough ground-truthing of these parameters is then used to determine the grain size of the sea-bed. The available literature on bottom-classification systems suggests they are capable of delimiting broad sediment types (bedrock, gravel, sand and mud) and indeed they have been used for some archaeological surveys (plate 13.10). For these systems to become really useful tools for imaging archaeological materials, research needs to be undertaken into both their sensitivity to the rapid changes of material type common on archaeological sites and the actual acoustic properties of typical archaeological materials. Archaeologists are actively researching the potential for extracting such information from multibeam swath bathymetry data, but the extraction of non-normal incidence backscatter to material type is a major challenge. It should also be noted that those systems based on single-beam acoustic sources suffer the same horizontal resolution problems described above for the single-beam echo-sounders.

SIDESCAN SONAR

Sidescan sonar is a method of underwater imaging using a wide-angle pulse similar to those of the multibeam swath bathymetry systems. Rather than calculating depth information from the returning echo, the sidescan sonar system displays the intensity of the sound scattered back to the tow-fish from the sea-floor sediments and objects exposed on the sea-bed. Sidescan data can be processed to provide undistorted images of the sea-floor in real time. The area of the sea-floor covered in a single pass is controlled by the surveyor altering the altitude of the tow-fish above the sea-bed. This can be done by adjusting either the speed of the vessel or the length of the tow-cable.

Traditionally, a swath width equalling ten times the water depth was recommended but, while this may be adequate for basic sea-bed surveys, it does not always provide the optimum geometry for archaeological information. For detailed surveys, the angle at which the acoustic energy reaches the target is very important and running the same survey lines at very low and very high passes can often reveal very different, but complementary, information.

Sidescan sonar systems are available in a variety of types, depth ratings and operating frequencies. Recent developments that are now commercially available include 'Chirp' systems (so called because of the noise emitted),

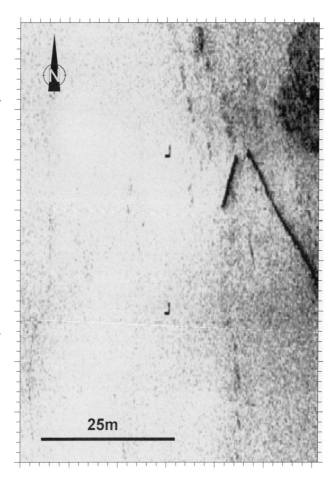

Figure 13.1 500 kHz sidescan sonar image of a v-shaped fish-trap from the River Barrow, County Wexford, Ireland. Data acquired using an EdgeTech Model 272-TD acquisition system. (Courtesy of the Centre for Maritime Archaeology, University of Ulster, Coleraine)

which utilize a sweep across a range of frequencies, and synthetic aperture systems, which provide excellent definition at longer ranges. Standard systems usually connect directly to a laptop computer to display the data. Such systems employ one of two industry-standard frequencies for imaging: 100 kHz and 500 kHz, although they may vary considerably from these, depending on the model and the manufacturer. In general terms, a 100 kHz operating frequency is chosen for regional surveys with swath widths in excess of 100 m (325 ft) per side. 500 kHz sources are generally used where a higher resolution is required, such as for shipwreck or waterside structure surveys (figure 13.1). Very high-frequency systems of up to 2.5 MHz provide even better definition but their effective range is limited, sometimes to less than 10 m (33 ft), and so are only useful on small, well-defined sites with excellent tow-fish positional control.

Material properties (primarily roughness characteristics) of the area being surveyed determine the strength of the echo (backscatter) from the sea-bed. Rock, gravel, wood

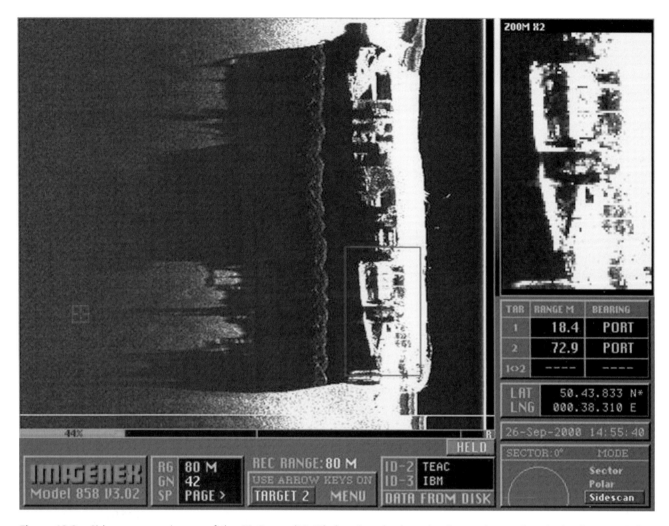

Figure 13.2 Sidescan sonar image of the SS *Storaa* (1943) showing shadow detail complementing the backscatter information from the wreck. (Courtesy of ADUS, University of St Andrews)

and metals are better reflectors than finer grained sediments and will therefore be recorded as darker elements on the sonar record. Target shape, including sea-floor gradient, also influences reflectivity and backscattering. Arguably the most important phenomena on sidescan records for archaeological purposes are acoustic shadows, which provide a three-dimensional quality to what is essentially a two-dimensional survey. Acoustic shadows occur alongside objects that stand proud of, or are partially buried in, the sea-floor. In sidescan sonar data, shadows can often indicate more about the shape and nature of a target than the acoustic returns from the target itself (figure 13.2).

The majority of sidescan investigations follow a predetermined survey pattern. Search patterns conventionally comprise a series of parallel survey lines with the lane spacing determined by survey requirements. The lane spacing must be less than the swath width of the sonar, thereby allowing for overlap between lines, ensuring 100

per cent coverage of the area. Although data has conventionally been displayed as a paper-trace, modern systems allow real-time data to be displayed through a computer and in conventional monochrome or almost any variety of colour mixtures the operator chooses. Software can then be used to identify the coordinates of any feature (geo-rectification) and, by measuring the length of the shadow and the height of the tow-fish, its projection above the sea-bed can be calculated. Geo-rectified sidescan images can be joined together in a mosaic to provide an image of large areas at the original resolution, thereby decreasing the trade-off between data resolution and survey coverage.

SUB-BOTTOM PROFILING

While sidescan sonar and multibeam swath bathymetry are the most effective techniques for finding and delineating archaeological objects exposed on the sea-bed, many archaeological sites coincide with areas of high sedimentation. This can result in the partial or complete burial of structures, features and artefacts. While wrecks with substantial iron content may be found using magnetic surveying (see below), the only technique suitable for detecting buried wooden artefacts is sub-bottom profiling. Furthermore, marine archaeologists are increasingly interested in the identification of the environment or landscape in which such artefacts were deposited, and to understand this it is essential to look at both the surface and the sub-surface.

Two principal types of systems exist: those that produce a single-frequency pulse (such as 'Pingers' and 'Boomers') and those that produce a swept-frequency pulse ('Chirp'). The single-frequency systems suffer from a penetration-versus-resolution compromise. Put simply, higher frequency sources give better resolution but can penetrate only a short distance into the sediment; conversely, lower frequency systems penetrate further but give poorer resolution. The development of Chirp technology in the early 1990s attempted to address this conflict by producing a pulse that can penetrate decametres into the sea-bed while still retaining decimetre resolution. It should be noted that the effectiveness of each system is dependent on the nature of the sediments being imaged, with coarser sediments (sands and gravels) being harder to penetrate than fine-grained sediments (silts and clays) (figure 13.3).

All sub-bottom profilers use a source that generates sound pulses that travel into the subsurface. These waves then reflect off boundaries or objects within the sub-surface and are detected by an acoustic receiver (or hydro-phone), which is usually mounted close to the source. Reflections occur where there are differences in density and/or sound velocity across a boundary. The returning echoes are then transmitted to a recording device, either a hard-disk or direct to paper printer (figure 13.4).

Sub-bottom profilers generate a data set that can be processed to give a cross-section in the direction of movement of the boat in two-way travel time (the time taken for the pulse to travel from the source to the reflector and back to the receiver). With additional knowledge of the speed of sound through the sediments present (obtained from *in situ* measurements of core

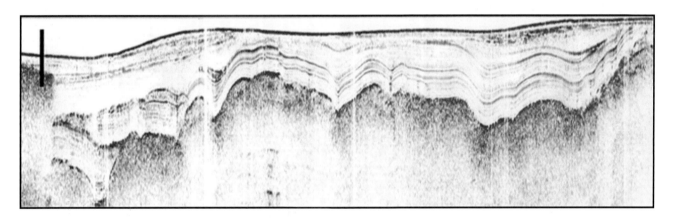

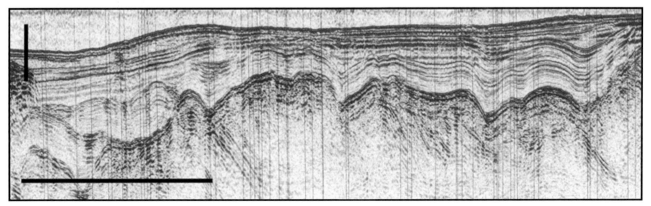

Figure 13.3 Chirp (top) and Boomer (bottom) image from the same location within Strangford Lough, Northern Ireland, showing the detailed layering within the upper, fine-grained sediments in the Chirp profile and the penetration through stiff glacial sediments into the basal bedrock from the Boomer section. Data acquired using a GeoAcoustics Chirp and Boomer system. Vertical scale bar represents c.15 m (49 ft). (Courtesy of the National Oceanography Centre/University of Southampton)

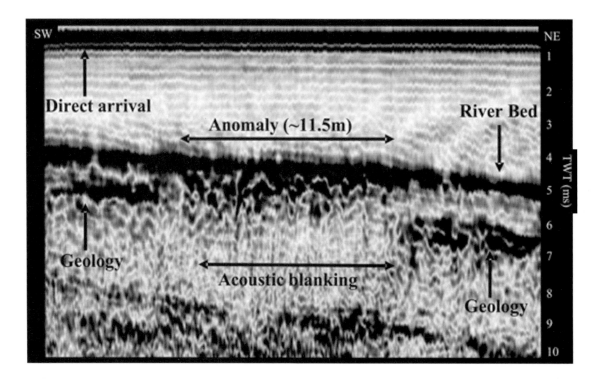

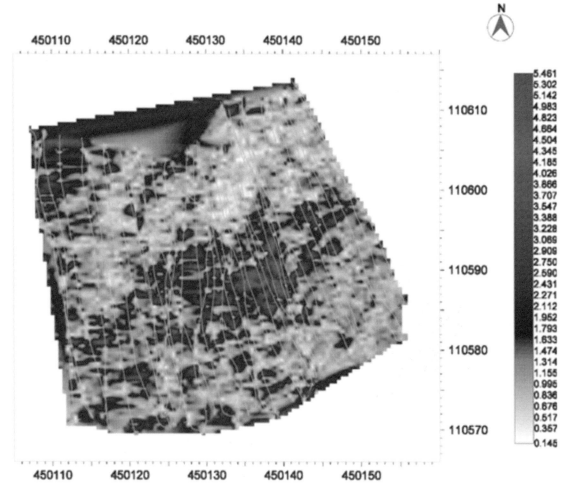

Figure 13.4   A two-dimensional profile and interpolated timeslice of the *Grace Dieu* (1439) wreck, River Hamble, Hampshire, UK. Data acquired using a 3-D Chirp sub-bottom profiling system, developed by the National Oceanography Centre and GeoAcoustics Ltd (UK). (Courtesy of the National Oceanography Centre/University of Southampton)

material or by comparison with standard empirically derived values), this cross-section can be converted to a depth section. Inevitably, the two-dimensional nature of the data results in sub-bottom data suffering the same coverage problems as described for echo-sounding. Recently, researchers at the National Oceanography Centre, UK, have developed a high-resolution 3-D Chirp system that is capable of acquiring a true 3-D volume of the sub-surface and has huge potential for the investigations of small-scale areas such as archaeological sites.

Pingers emit short pulses of a single high frequency (3.5 kHz for example), which gives a resolution of 0.3–0.5 m (12–18 in) and penetration of 20–25 m (65–82 ft). With such a system, the generation of the pulse and recording of the returning echoes are conducted within a single set of transducers, thus optimizing the horizontal resolution of the system. Boomer systems represent a single lower-frequency source, typically between 1 and 6 kHz. The boomer output pulse can be tailored to the survey's requirements, both in terms of the frequency of the pulse and the energy output. Boomer systems have a range of penetration of 50–75 m (165–246 ft) and an optimum vertical resolution of 0.5–1.0 m (18–39 in).

Chirp profilers are towed as close to the sea-bed as safety will allow, typically 5–10 m (16–33 ft) above the bottom, although if configured correctly they can operate in water depths as shallow as 2.5 m (8 ft). The frequency spectrum or bandwidth is wide for Chirp systems (typically between 6 and 10 kHz) and this is important as it controls the vertical resolution, with a practical vertical resolution of 20–30 cm (8–12 in) being obtainable at depths in excess of 30 m (98 ft).

For artefact identification, Chirp systems currently represent the best available technology, not only because of their good resolution but because suitable post-processing of data allows some degree of material characterization for buried objects. For landscape reconstruction, the Boomer system is considered the most reliable because it is capable of penetrating most sediment types found within the coastal zone and can thus guarantee some basic imagery of buried landscapes. However, in ideal circumstances both Chirp and Boomer should be deployed on a survey to ensure both detailed imagery of any fine sedimentary cover and penetration to bedrock.

MAGNETOMETRY

Magnetometers measure the strength of the earth's magnetic field and can detect variations in this field caused by the presence of objects containing iron and geological formations containing ferrous material. Modern sensitive systems can also detect weak magnetic signatures caused by ancient hearths and assemblages of ceramics. Marine

magnetic surveying is a well-proven technique and is often used for the location and detailed investigation of metal-hulled wrecks and wooden-hulled vessels that may have carried substantial ordnance or have some other ferrous component (plate 13.11). For inshore maritime archaeological research, three types of magnetometer are used. The proton precession magnetometer was the most widely used in the past, but is being replaced by the caesium (or optically pumped) and the overhauser magnetometer systems as the instruments of choice because they are both considerably more sensitive and so can detect smaller objects at greater range than proton systems.

The earth's magnetic field varies in intensity over the surface of the planet. At the poles, the field is concentrated and therefore has a high intensity – about 61,000 nT (1 nT = 1 gamma). At the equator, the field is quite weak, with a typical reading of 24,000 nT. In a localized area, the magnetic field tends to be even. If a ferrous mass or object is introduced into the area (e.g. an iron wreck), the lines of force are disturbed. Such local disturbances within the magnetic field are of potential archaeological interest and the amount of disturbance is a function of the mass of the object and its alignment.

Magnetometers for marine use are typically towed devices to avoid interference from the survey vessel. The minimum layback (distance between the stern and the tow-fish) can be ascertained by increasing the cable length until the boat stops registering as a magnetic anomaly. It may be necessary to carry out this procedure in more than one direction. The magnetic information is normally displayed as a numerical readout and as a graph, updated as the survey advances. The better systems generally use laptop computers to collect the data, rather than a dedicated logging device provided by the manufacturers. This has the advantage of being able to process the data quickly and, using appropriate software, produce informative graphical representations of the survey data, such as contour plots.

The proton precession magnetometer typically has a recording rate of 0.5–2.0 seconds and a sensitivity of 0.2–1.0 nT. Caesium and overhauser magnetometers have a faster measurement rate, typically 0.1 s, and sensitivities of at least 0.02 nT. These more sophisticated instruments can also be towed at higher speeds, tend to be more stable and are generally more effective for archaeological work than traditional proton precession magnetometers. The advantages of the proton type are a smaller relative size of tow-fish and they are cheaper to buy or rent.

One problem with magnetic surveying in coastal waters is the amount of detritus on the sea-bed from port developments and people's use and abuse of the coastal zone. Non-archaeological magnetic anomalies are abundant within developed areas such as ports. This is a particular problem at some sites, where objects such as anchors, chains, cables and ordnance with large magnetic signatures can effectively mask archaeological anomalies. A more widespread but localized difficulty is background magnetism from geological formations, particularly where these are in tight folds that reach close to the surface a number of times in a survey area. Various techniques are available to reduce this problem, including multiple magnetometers set in a fixed array, which allows differentiation between very large masses such as ferruginous rocks (containing iron) or nearby steel wrecks and smaller signatures of potential archaeological significance.

INTEGRATED SURVEYS

Integrated systems are often used in mapping the underwater cultural resource. Integrated survey systems may comprise a combination of two or more of the following: bathymetric survey equipment, sidescan sonars, sub-bottom profilers and magnetometers. This integrated approach can lead to the acquisition of a large but effective data set. Generally, data is acquired in digital format to facilitate offline processing and spatial integration (figure 13.5 and plate 13.12).

SUBMERSIBLES: ROVs AND AUVs

Remotely operated vehicles (ROVs) and autonomous underwater vehicles (AUVs) are examples of integrated systems. They can perform many of the tasks of a diver, including visual searches and photography, but are not limited in the same way by depth and time spent under pressure. These systems can host a suite of acoustic and video imaging data-collection devices with some underwater video systems incorporating image-intensifiers capable of showing greater detail on a surface monitor than the diver can see while under water.

ROVs are connected to a support ship via a cable and controlled from within the ship by an ROV pilot. Submersible technology is a field of rapid development but at the time of writing, deep-water ROVs can operate in depths of up to 6000 m (19,500 ft) while cheaper, shallow-water systems can operate in depths of around

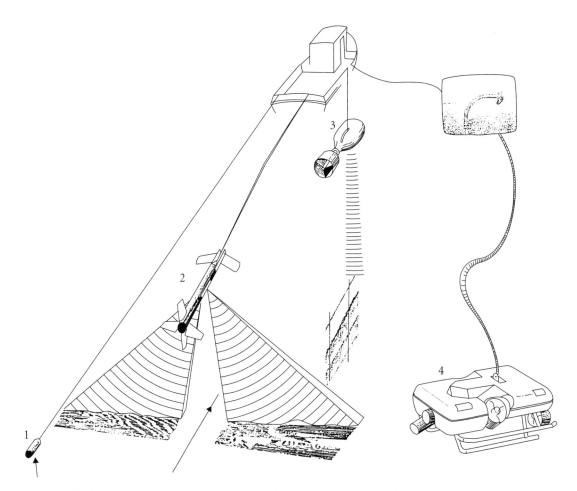

Figure 13.5 Geophysics equipment: 1) magnetometer, 2) sidescan sonar, 3) sub-bottom profiler, 4) ROV. (Based on an original drawing by Mark Redknap)

300 m (975 ft). AUVs, as their name suggests, are autonomous systems that function without remote control, tethers or cables. Their depth-rating at the time of writing is restricted by battery-life to approximately 3000 m (9750 ft).

AERIAL PHOTOGRAPHY

Aerial photography, particularly using modern digital cameras, is a successful method of investigation for inter-tidal and foreshore research but has limited success in underwater research due to an inability to penetrate the water column successfully beyond, typically, 1 to 3 m (3–10 ft). Modern high-resolution digital cameras produce images in either colour infra-red (false colour) or true-colour mode. The size of the ground covered by each pixel can range from 10 to 50 cm (4–20 in) and above, depending on flying height and speed. Although this pixilation does not give images as obviously detailed as conventional film, digital images are easier to rectify and analyse.

FURTHER INFORMATION

Boyce, J. I., Reinhardt, E. G., Raban, A. and Pozza, M. R., 2004, Marine Magnetic Survey of a Submerged Roman Harbour, Caesarea Maritima, Israel, *International Journal of Nautical Archaeology* **33**, 122–36.

Dean, M., 2006, Echoes of the Past: Geophysical surveys in Scottish waters and beyond, in R. E. Jones and L. Sharpe (eds), Going over Old Ground – Perspectives on archaeological geophysical and geochemical survey in Scotland, 80–87. *BAR British Series* 416, Oxford.

Fish, J. P. and Carr, H. A., 1990, *Sound Underwater Images: A Guide to the Generation and Interpretation of Side Scan Sonar Data*. Boston, MA.

Papatheodorou, G., Geraga, M., and Ferentinos, G., 2005, The Navarino Naval Battle Site, Greece: an Integrated Remote-Sensing Survey and a Rational Management Approach, *International Journal of Nautical Archaeology* **34**, 95–109.

Quinn, R., Breen, C., Forsythe, W., Barton, K., Rooney, S. and O'Hara, D., 2002a, Integrated Geophysical Surveys of The French Frigate *La Surveillante* (1797), Bantry Bay, County Cork, Ireland, *Journal of Archaeological Science* **29**, 413–22.

Quinn, R., Forsythe, W., Breen, C., Dean, M., Lawrence, M. and Liscoe, S., 2002b, Comparison of the Maritime Sites and Monuments Record with side-scan sonar and diver surveys: A case study from Rathlin Island, Ireland, *Geoarchaeology* **17.5**, 441–51.

Quinn, R., Dean, M., Lawrence, M., Liscoe, S. and Boland, D., 2005, Backscatter responses and resolution considerations in archaeological side-scan sonar surveys: a control experiment, *Journal of Archaeological Science* **32**, 1252–64.

14 Underwater Survey

Contents

- Types of survey
- An initial sketch
- Planning
- Setting up a baseline/control points
- Installing survey points
- The principles of survey
- Survey using tape-measures, grids and drafting film
- Vertical control (height/depth)
- Drawing/planning frames
- Grid-frames
- Processing measurements and drawing up the site-plan
- Three-dimensional computer-based survey
- Acoustic positioning systems
- Positioning the site in the real world

It has been shown elsewhere in this book that fieldwork should be undertaken with clear aims and objectives that are defined in the project design. It may be that the survey alone answers the questions posed by the project design, so an accurate site-plan can be the end product of the fieldwork rather than just one phase of a project.

The aim of this section is to give an introduction to basic survey techniques used in underwater archaeology. The techniques described here are the same as those used in archaeology on land, as well as in civil engineering and building work. Surveying is not the same as searching; divers looking for wreck remains are searching, while divers recording the positions of those remains are surveying. The purpose of a survey is to produce an accurate picture of the site, usually as a two-dimensional plan (figure 14.1) with supporting descriptions and measurements. In essence, this is an attempt to re-create, on paper or in a computer, the site as it exists now, and before it is disturbed. The site-plan must be an accurate representation, so it is not acceptable to simply sketch or guess where things are.

TYPES OF SURVEY

Surveys on archaeological sites can broadly be divided into four types:

1 assessment survey;
2 recording survey (including pre-disturbance and excavation surveys);
3 monitoring survey;
4 topographical survey.

The end product for each survey type is the same – the site is recorded to a known level of detail and accuracy.

An assessment survey provides information on which a fieldwork strategy can be based. It can be used to find the extent of the site before embarking on a more detailed pre-disturbance survey (see below). It can also provide information on the range, type and stability of archaeological material surviving on the site. This is important because such information can affect decisions about what action should be taken and when. Such a survey could help decide the location of primary survey

Plate 2.1 Forerunner to underwater archaeology: in the nineteenth century the Deane brothers worked as 'submarine engineers' in and around Portsmouth, UK. In this lithograph, Charles Deane is depicted in 1832 wearing his newly invented 'diving apparatus'. He is removing one of the bowsprit hoops from HMS *Royal George* (1782) in approximately 18 m (60 ft) of water in the Solent, UK. (Reproduced by kind permission of Portsmouth Museums and Records Service)

Plate 2.2 Ethnographic recording of a three-log kat in Edava, Kerala, India. (Photo: Colin Palmer)

Plate 3.1 NAS training project near Bristol, UK. These are activities the whole family can get involved in. (Photo: Mark Beattie-Edwards)

Plate 3.2 Archaeology on the foreshore in the UK. (Photo: Mark Beattie-Edwards)

Plate 4.1 Underwater sites, particularly within closed contexts such as this chest of longbows on the *Mary Rose* (1545), can result in excellent preservation. The diver is using a magnifying glass for closer examination. (Photo: Christopher Dobbs)

Plate 4.2 Site types: reconstruction of a crannog on Loch Tay, Scotland. (Photo: Colin Martin)

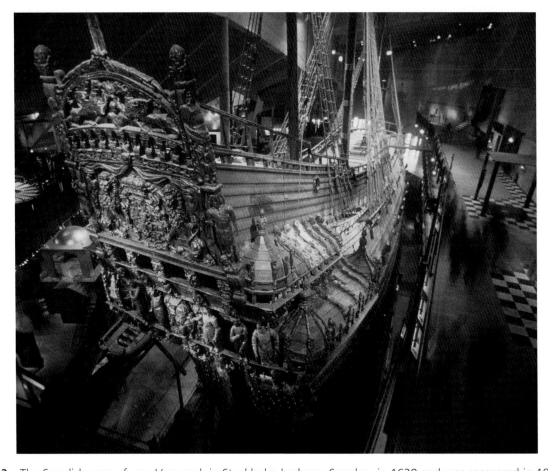

Plate 4.3 The Swedish man-of-war *Vasa* sank in Stockholm harbour, Sweden, in 1628 and was recovered in 1961. Low temperatures and low levels of salinity have resulted in spectacular preservation. (Photo: Colin Martin)

Plate 6.1 Archaeological team on surface supply preparing to dive. The supervisor (centre) reads out the pre-dive checks; the divers, assisted by their tenders (wearing life-jackets) confirm that each check is OK. The divers carry emergency bale-out bottles and have through-water communication with the surface. (Photo: Edward Martin)

Plate 6.2 A commercial archaeological diving unit working to UK Health and Safety Executive protocols for surface-supplied diving. The supervisor is in the cabin, where he is in voice communication with the diver, whose progress he monitors via a helmet-mounted video link. A tender monitors the diver's umbilical, while behind is the standby diver, fully kitted-up except for his mask. (Photo: Colin Martin)

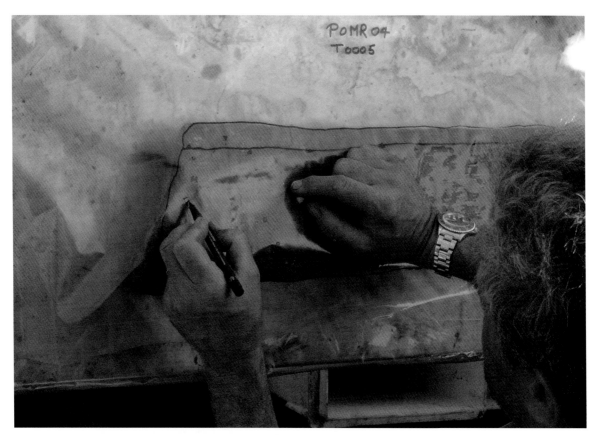

Plate 8.1 Recording timbers: 1:1 tracing of timber surfaces on polythene. (Photo: Kester Keighley)

Plate 8.2 Stratigraphy: an underwater excavation face showing several stratigraphic layers visible in the subtle changes of texture and colour. White tags highlight the different layers. (Photo: Kester Keighley)

Plate 8.3 Recording *in situ*: a slipware bowl during excavation on the Duart Point wreck (1653, Mull, Scotland), showing scale and finds number. (Photo: Colin Martin)

Plate 8.4 On-site finds processing: elements from a wooden barrel are numbered, identified, initially recorded and packaged prior to transportation to conservation facilities. (Photo: Kester Keighley)

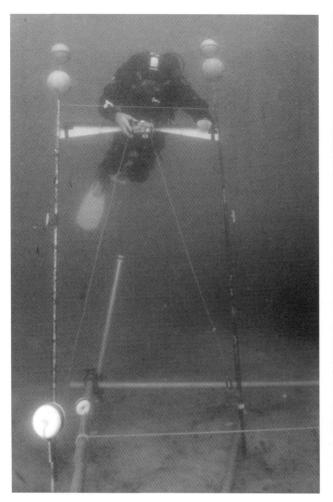

Plate 10.1 Free-standing photographic tower in use within a survey grid to record a photomosaic. For further information, see Martin and Martin, 2002. (Photo: Mark Beattie Edwards)

Plate 11.1 A 'total station' ready for use. (Photo: Kester Keighley)

Plate 12.1 A dry run of a search technique can prevent many potential problems. (Photo: Kester Keighley)

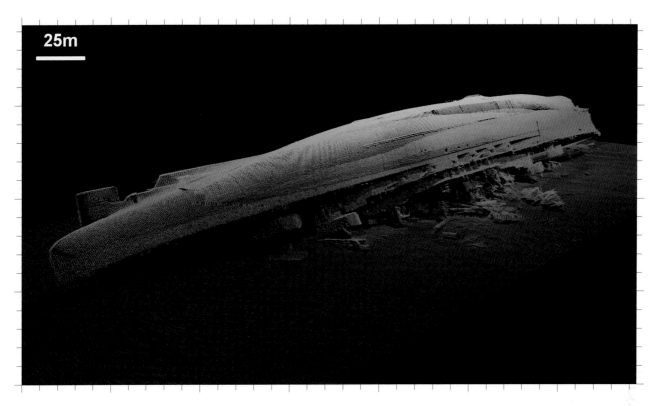

Plate 13.1 High-definition multibeam sonar point-cloud image of the 203 m (660 ft) long wreck of HMS *Royal Oak* (1939) in Scapa Flow, Orkney. (Courtesy of ADUS and the Department of Salvage and Marine, Ministry of Defence)

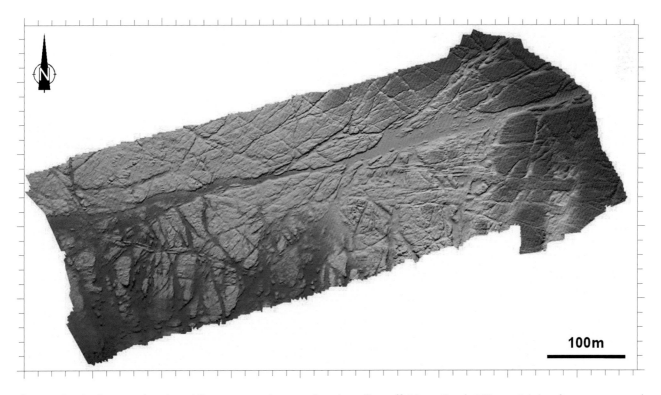

Plate 13.2 Surface rendered multibeam sonar image of rock gullies off Moor Sand, UK, containing bronze-age and seventeenth-century material. The 700 m (2275 ft) long area runs WSW to ENE and is 260 m (845 ft) at its widest. (Courtesy of ADUS, University of St Andrews)

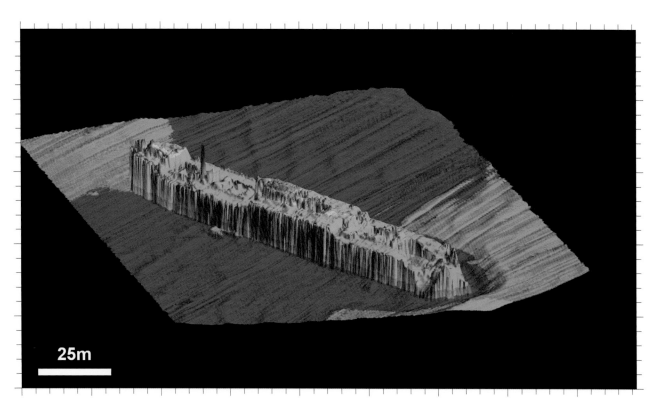

25m

Plate 13.3 Multibeam sonar image of the SS *Storaa* (1943) demonstrating the disadvantage of using a rendered surface, which can give wrecks a false slab-sided appearance. (Courtesy of ADUS, University of St Andrews)

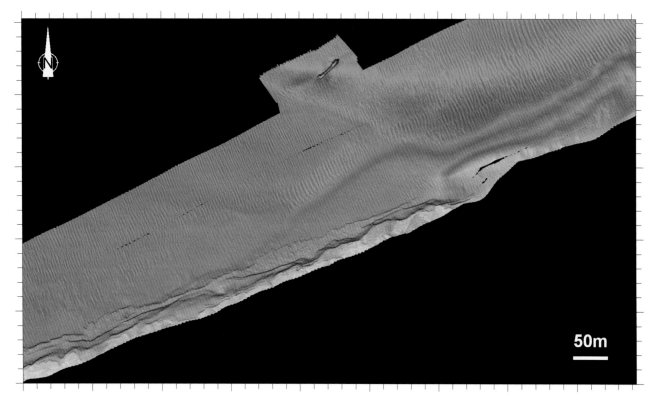

50m

Plate 13.4 Multibeam image of a prehistoric land surface at the base of the 8 m (28 ft) high underwater Bouldner Cliff in the Solent, UK. The wreck of the 44 m (143 ft) long dredger *Margaret Smith* (1978) is included for scale. (Courtesy of ADUS, University of St Andrews)

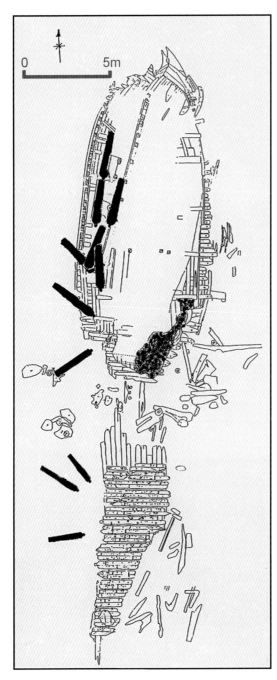

Plate 13.5 High quality, diver-recorded site-plan of the *Hazardous* (1706) wreck-site. (Courtesy of *Hazardous* Project Team)

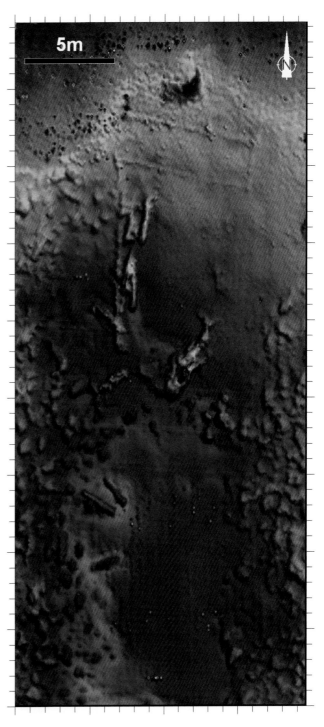

Plate 13.6 Multibeam sonar image of the *Hazardous* (1706) wreck-site for comparison with plate 13.5. (Courtesy of ADUS, University of St Andrews)

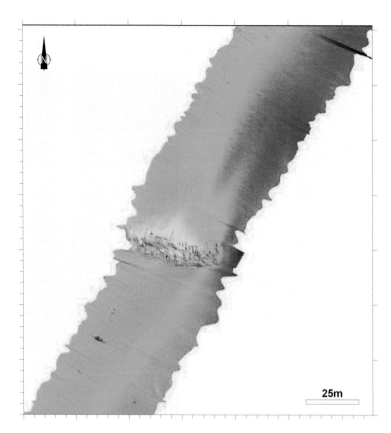

Plate 13.7 Multibeam image from a single pass in 2002 showing the sea-bed around the wreck of the *Stirling Castle* (1703). The wreck mound is 46 m (150 ft) long. (Courtesy ADUS, University of St Andrews).

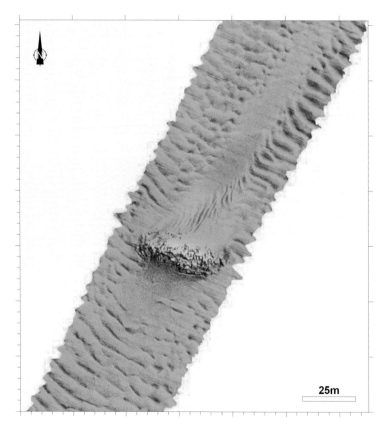

Plate 13.8 Multibeam image from a single pass in 2005 showing the sea-bed around the wreck of the *Stirling Castle* (1703). Well-defined sand waves have developed around the wreck mound in the 3 years since the previous survey shown in plate 13.7. (Courtesy of ADUS and the RASSE Project, University of St Andrews)

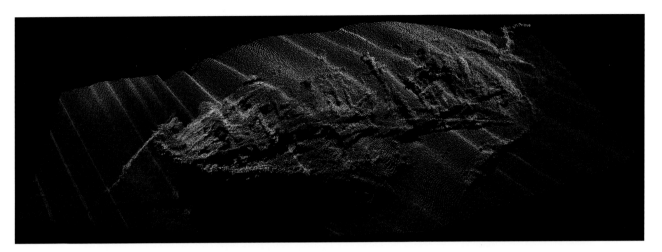

Plate 13.9 Multibeam image of a nineteenth-century wooden sailing ship on the Goodwin Sands at the entrance to the Strait of Dover. The end of the bowsprit now rests on the sea-bed but when surveyed 12 weeks earlier, it was in its operational position. (Courtesy of ADUS and the RASSE Project, University of St Andrews)

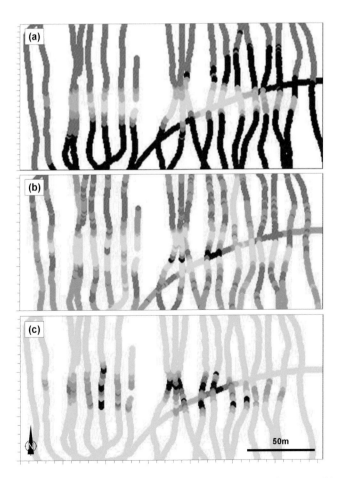

Plate 13.10 Ground-discrimination data collected from a single-beam echo-sounder showing (a) bathymetry, (b) hardness and (c) roughness of the wreck of the 178 m (580 ft) long *Markgraff* (1919) in Scapa Flow. (Courtesy of Mark Lawrence, ADUS, University of St Andrews)

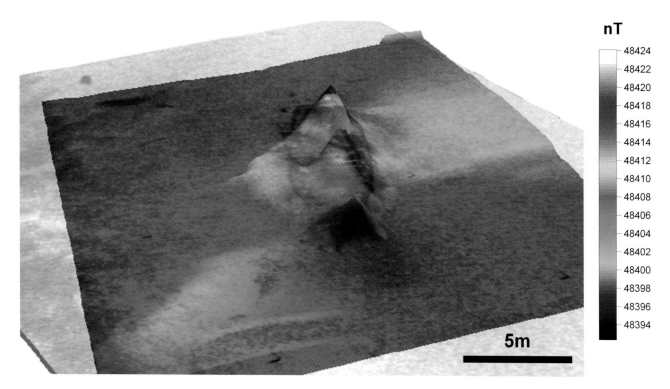

Plate 13.11 Three-dimensional plot of magnetic data acquired from the *La Surveillante* (1797) wreck-site, Bantry Bay, County Cork, Ireland. (Courtesy of the Centre for Maritime Archaeology, University of Ulster, Coleraine and the Applied Geophysics Unit, National University of Ireland, Galway)

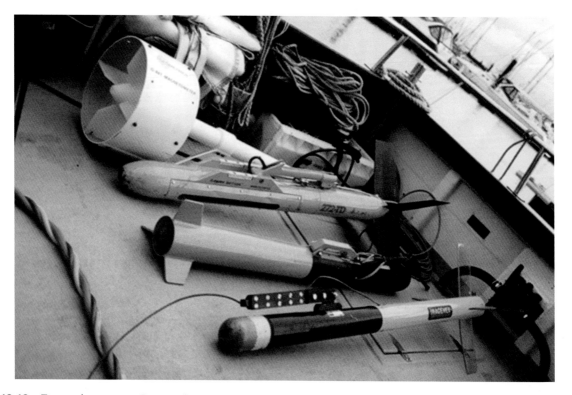

Plate 13.12 Top to bottom: a Geometrics G-881 caesium magnetometer (white tow-fish), an EdgeTech 272–TD side-scan, a GeoAcoustics side-scan and an Imagenex 885 side-scan sonar. (Photo: Rory Quinn)

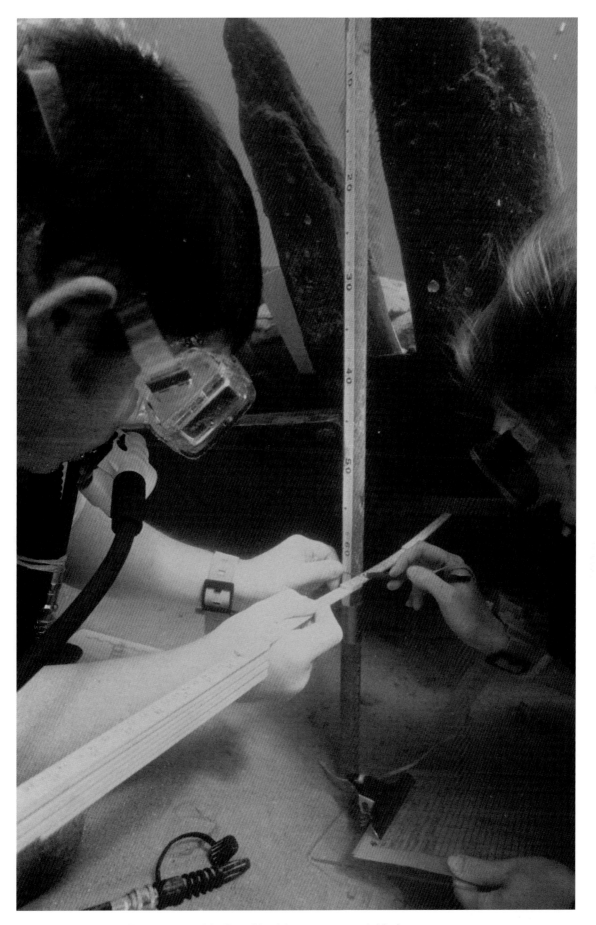

Plate 14.1 Using vertical offsets to record hull profile. (Photo: Kester Keighley)

Plate 14.2 An underwater board on which a sheet of drafting film has been secured with insulating tape. (Photo: Kester Keighley)

Plate 14.3 The clinometer in use measuring the angles of timbers on the Duart Point wreck. (Photo: Colin Martin)

Plate 14.4 A double-strung drawing/planning frame with levelled legs in use on the Duart Point wreck. Note that the frame has been positioned within a larger reference grid. (Photo: Colin Martin)

Plate 14.5 Drawing/planning frames being used vertically to record vessel remains on the foreshore. (Photo: Mark Beattie-Edwards)

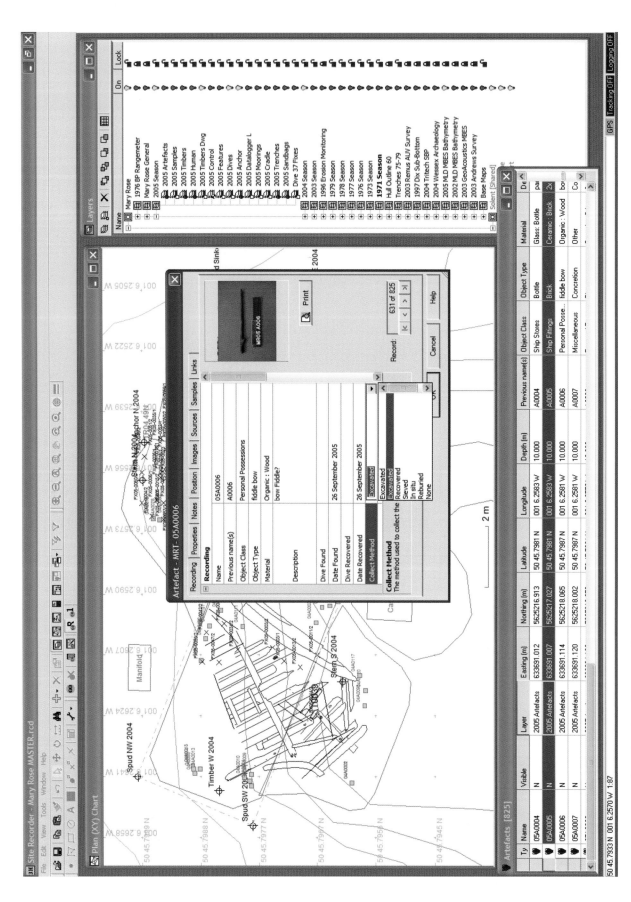

Plate 14.6 A screen-shot showing Site Recorder in use on the *Mary Rose* (1545) site. (Photo: Peter Holt)

Plate 15.1 The diver is clasping a site-grid with his feet. Maintaining neutral buoyancy enables the diver to keep clear of the archaeology while using one hand to excavate with a trowel and the other to remove spoil with an airlift. (Photo: Kester Keighley)

Plate 15.2   Excavating with a water-dredge in less than 5 m (16 ft) of water in Dor, Israel. The flexible tube attached to the suction end of the dredge aids manoeuvrability (Photo: Kester Keighley).

Plate 15.3 Excavating with a water-dredge. An air-filled plastic container keeps the dredge-head neutrally buoyant. The diver controls the dredge while excavating with a trowel. (Photo: Mark Beattie-Edwards)

Plate 16.1 Samples of oak from the *Mary Rose*. The one on the left is untreated and wet. Following drying, the sample on the right has suffered large volumetric shrinkage, indicating the need to stabilize with polyethylene glycol prior to drying. (Photo: Mary Rose Trust)

Plate 16.2 A gun carriage is lowered into a temporary polythene-lined tank pending dispatch to the conservation laboratory. (Photo: Edward Martin)

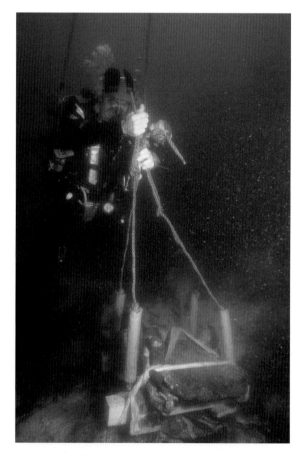

Plate 16.3 The base of a wooden gun carriage being prepared for lifting. The object has been placed on a wooden frame tailored to its measurements, ballasted with lead and cushioned with foam. Stretch-bandages are used to secure the object to the frame and the lifting strops are sheathed with pipe insulation to avoid rubbing damage. (Photo: Colin Martin)

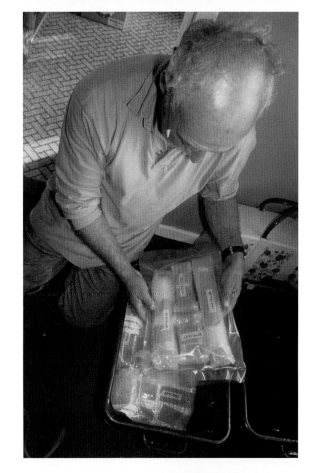

Plate 16.4 A batch of objects, wet-wrapped in kitchen towel, bubble-wrap and heat-sealed polythene bags, is packed for transport to the laboratory. Note the careful labelling on each package. (Photo: Edward Martin)

Plate 17.1 A gauge automatically logging the speed and direction of the current over a 1 month cycle on the Duart Point (1653) wreck, Mull, Scotland. (Photo: Colin Martin)

Plate 17.2 Monitoring of the underwater environment (both sediments and water column) on the *Mary Rose* (1545) wreck using an RBR data-logger. This records temperature, pressure (depth), conductivity, salinity, pH, dissolved oxygen and turbidity. (Photo: Mary Rose Trust)

Plate 17.3 Oak blocks placed on the *Mary Rose* (1545) wreck-site to study the activity of marine wood-boring animals. (Photo: Mary Rose Trust)

Plate 17.4 A conservator attaches an aluminium anode to an iron gun on the Duart Point (1653) wreck, Mull, Scotland, as part of an experiment to stabilize iron objects in seawater by electrolysis. (Photo: Colin Martin)

Plate 20.1 Archaeologists of the future? Public outreach is an important element of archaeology. Here, two children experience challenges presented by poor visibility as they try to identify archaeological objects while wearing blacked-out masks. (Photo: Hampshire and Wight Trust for Maritime Archaeology)

Plate 20.2 Many opportunities exist for publishing archaeological work, both electronically and in hard copy. (Photo: Kester Keighley)

Plate A1.1 A one-hole stone anchor found at Chapman's Pool, Dorset, UK. (Photo: Gordon Le Pard)

Plate A1.2 A two-hole stone anchor found near Golden Cap, Dorset, UK. (Photo: Gordon Le Pard)

Plate A3.1 Divers practise underwater survey techniques during a NAS Part I course. (Photo: Mark Beattie-Edwards)

Plate A3.2 Obtaining samples for dendrochronological dating during a NAS Part III course. (Photo: NAS Training)

Plate A3.3   Excavation and survey during a NAS Training project on the foreshore near Bristol, UK. (Photo: Mark Beattie-Edwards)

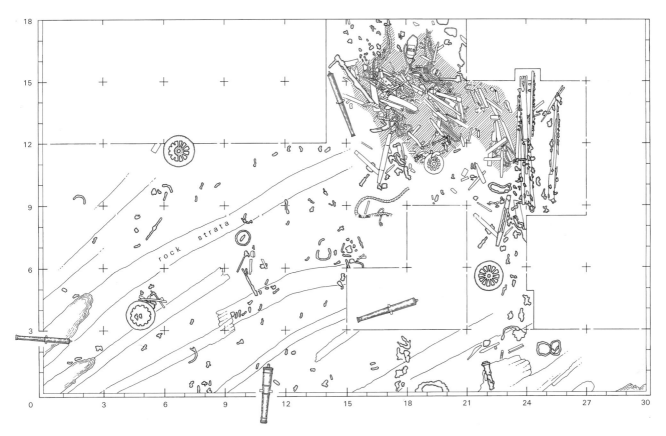

Figure 14.1 Excavation plan of part of the Spanish Armada wreck *La Trinidad Valencera* (1588), a scattered site on a flat, sandy sea-bed. The excavation limits are shown within a grid framework, and the locations of prominent objects shown. The extent of a major deposit of organic material is shown by hatching. (Colin Martin)

points. The aim of an assessment survey is to get a rough idea of the extent and layout of a site as quickly and efficiently as possible. Position accuracy is less important than speed, as the plan can always be improved at a later date. As a minimum, the assessment survey should result in a paper plan, drawn to scale, showing the significant features on the site, sea-bed type and an outline of the topography, as well as any information requested in the project design (figure 14.2).

Recording surveys include pre-disturbance and excavation surveys. These require high position-accuracy so careful planning, recording and processing are needed and they take time to do correctly. A common question regarding pre-disturbance surveys is, 'how much information should be recorded?' The simple but daunting answer is, 'as much as possible without disturbing the site'. The more information that can be collected through non-destructive pre-disturbance surveys, the more effective future work on the site is likely to be. This applies to both the planning of further work on site and avoiding unnecessary damage to archaeological material.

A pre-disturbance survey is an essential step in ensuring that a complete record of the site is made. A pre-

disturbance survey is not undertaken for the sake of it; the work will provide useful information on the condition of the site at that time. The results can be used as a benchmark for future monitoring (see chapter 17) if no further work is to be done, or can be a record of the original condition of the site if it is to be altered or destroyed by intrusive investigations such as excavation. The information collected during the pre-disturbance survey can help to ensure that the appropriate funds, conservation facilities and expertise are arranged well in advance of any disturbance to the site.

If intrusive investigation is considered to be the most appropriate way forward for an archaeological site, then recording during excavation is an essential continuation from where a pre-disturbance survey left off. Pre-disturbance survey results are developed during excavation as more artefacts, structure and stratigraphy are uncovered.

Monitoring surveys are designed to monitor how the site changes over time. Using an existing survey, a monitoring survey compares those parts thought most likely to indicate change. Typical examples might include monitoring the position, attitude and remains of a

(a)   **(b)**

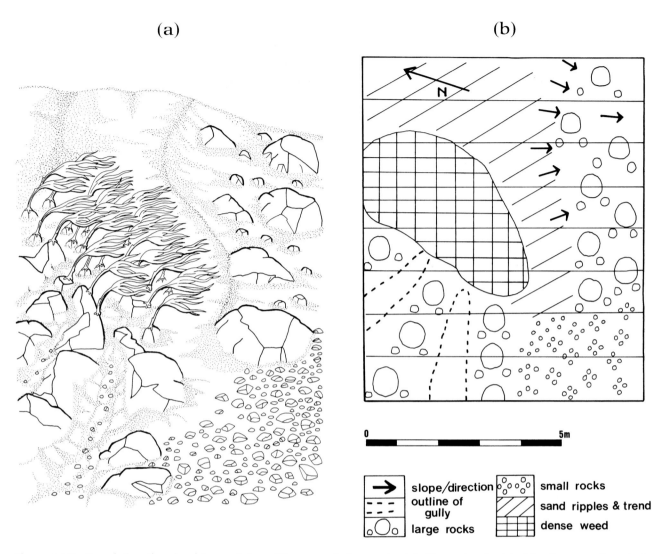

Figure 14.2 Translating the visual image into written record: the sea-bed (left) can be depicted by the use of symbols as shown on the right. (Based on original artwork by Ben Ferrari)

sternpost on an exposed site, or the depth of burial of a site under the sea-bed (see chapter 17).

A topographical survey aims to record the shape and characteristics of the sea-bed. The same principles are used; what makes this different is the need to cover a wide area. Recording the shape of the sea-bed usually involves recording the depth or height of the sea-bed at known positions. If the measurements are made at regular intervals over the site, then a plan can be produced showing the depths as contours. Details concerning the surface of the sediments and bedrock lying under water should include information about relative heights of features. Recording topography in detail is time-consuming if it is to be done accurately. Sometimes, more can be gained in a reasonable time by sketching the site and adding spot-depths at known features.

Accurate wide-area topographic maps and 3-D models are usually created using a multibeam echo-sounder (see chapter 13). There are standard ways of describing types of sea-bed and sediment. For further information, readers are referred to *Seasearch Observetion Form Guldance Notes* (www.seasearch – Seasearch Recording), which provide detailed practical information on how to record such information (see also table 14.1).

AN INITIAL SKETCH

The first step in any survey is to create a sketch of the site, as this will form the basis for any future work. A good sketch can very quickly provide a large amount of information about a site and this can be invaluable in the

Table 14.1 The sea-bed itself can be categorized by the size of individual grains within the sediment. (After Wentworth, 1922 or see http://en.wikipedia.org/wiki/ Particle_size_%28grain_size%29)

Sediment type	Particle size (mm)
Clay	up to 0.004
Silt	from 0.004 to 0.06
Fine sand	from 0.06 to 0.2
Medium sand	from 0.2 to 0.5
Coarse sand	from 0.5 to 2.0
Fine pebbles	from 2.0 to 8.0
Medium pebbles	from 8.0 to 16.0
Coarse pebbles	from 16.0 to 64.0
Cobbles	from 64.0 to 256.0
Boulders	over 256.0
Bedrock	N/A

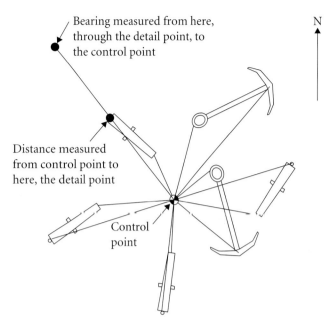

Bearing measured from here, through the detail point, to the control point

N

Distance measured from control point to here, the detail point

Control point

Figure 14.3 Radial method of survey, where the distances and bearings of features are measured from a central control point. Although accuracy diminishes with distance, the bearing shown here is measured from the extended point indicated because the effect of the iron gun's magnetic field on the compass has to be minimized

early stages of a project. Some form of sketch is essential before further work can be planned, as the size and shape of the site need to be known. The sketch should aim to give an overview of the main features, both topographical and archaeological, as well as any problems or potential hazards, and should help to decide where control points (see below) should be fixed. In later stages of a project, simple sketches can be extremely useful for recording the associations of objects (as can photographs). However, a sketch will have much more value if it is drawn at the site, and not later on dry land.

A checklist of information to record can be written on the slate before the dive. If the sketch was drawn under water on plastic film, it can if necessary be copied onto another sheet on land, either by hand or by photocopying or scanning. With an accurate sketch, future activity on the site can be thoroughly and efficiently planned.

If visibility is bad, then it may be easier to do a radial survey with a compass and tape-measure. The radial survey is a simple technique requiring a single diver with a tape-measure and compass. One end of the tape is attached to a control point in the middle of the site and the diver swims around the site recording the distance and bearing of each feature from the central point (figure 14.3). This, of course, works best on flat sandy sites with little weed-cover. It is simple enough for use on deep sites, provided that the tape is checked for snagging. Its main limitation is the inaccuracy of the measurement of bearing, especially as distance from the control point increases. Radial surveys are drawn up using a scale rule, a pair of compasses and a circular protractor. A point is marked in the middle of a piece of paper, which represents the central control point. After the selection of a suitable scale, a line (representing the recorded distance from the control point) is drawn from the central control

point at the appropriate bearing for each surveyed feature. For more about drawing up survey results, see below.

PLANNING

It is essential that all site work is well planned before it begins so that work on site is efficient, safe and problem-free. The actual amount of time the team will spend on site and under water is usually very short compared with the time spent planning and in processing the measurements. The first step is to define the scope of work, based on the project design (see chapter 5). The level of detail and accuracy required defines the type of survey to be undertaken and the techniques that can be used, so both should be specified in the project design. The scope of work will also depend on any work done previously. If the site has not been recorded at all, then an assessment is required; if an assessment has been done then subsequent work can be based on the previous results.

The level of detail and content of the survey depends on a number of related factors:

- Requirement – what needs to be recorded?
- Time – what is it possible to record in the available time?
- Equipment – what tools are available?

- Expertise – what skills are available?
- Environment – what can actually be recorded?
- Funding – what funding is available?

Requirement: What needs to be recorded depends on the aims laid out in the project design, which itself should take account of the other factors listed here and what information already exists (see chapter 5). For work where no change to the site is being made, a varying degree of recording can be considered because the work could be repeated to a higher standard if required. Where excavation is taking place, then 'everything' should be recorded, to a level determined by the archaeological director and set out clearly in the project design.

Time: How much can be recorded and the methods that can be used are determined by the time available on site, which itself depends on the time allocated to the project and working conditions on the site. It is important to plan how much time is likely to be spent working under water, and how much time might be lost due to poor weather or adverse tidal conditions. If working time is likely to be short, then the aim should be to do a simple survey using simple techniques – a little useful data is much better than no data at all. On average, it takes approximately 5 minutes to make a single tape-measurement under water. This time can be reduced or increased depending on variables such as depth, size of site, temperature, visibility, and experience.

Equipment: Most underwater survey methods use the same basic tools: the tape-measure and drawing board. Advanced tools such as acoustic positioning systems (see below) are not commonly available but they are occasionally used in support of underwater archaeological work and so should be considered in the project planning phase. It is important to ensure that the appropriate diving equipment is chosen to enable the diver to carry out the work safely and comfortably. A small rigid inflatable boat may enable the project team to get to the site quickly but a larger, hard-hulled boat may enable the team to spend all day on site in comfort. The ability to process results on site is a significant attraction of larger boats, as it allows for the correction of mistakes and collection of additional measurements there and then.

Expertise: It is essential to consider early in the project whether or not the team has the appropriate archaeological, surveying, and diving expertise to carry out the survey safely and accurately (see chapter 6). If additional skills are needed, they are readily available in the form of training, books, the internet, advice or recruitment of extra team members. The Nautical Archaeology Society provides training in many aspects of underwater archaeology to a range of skill-levels. If any team member needs a refresher, the survey techniques to be used can be practised on dry land beforehand.

Environment: The location and conditions on the site can limit how much can be recorded. Sites with features scattered over a wide area are more difficult to survey accurately. As a general rule, any group of features separated by more than 30 m (98 ft) should be treated as a separate site. In good visibility, it is possible to position control points further apart, therefore reducing the work involved in setting up control. In very poor visibility, there may be a limit to what divers can achieve (e.g. checking that the tape is not snagged takes longer in poor visibility). In strong currents the tape will bow, so control points will need to be located closer together to reduce the length of each measurement. More points will therefore be required to cover a site and survey work must be scheduled around periods of slack water.

Funding: Most of the limitations mentioned above can be solved by spending more money. However, most teams have a limited budget, which determines what equipment can be used and how much time can be spent on site. It should be remembered that an acceptable survey is normally achievable with limited resources.

SETTING UP A BASELINE/CONTROL POINTS

The basic principle of surveying is to be able to work out the position of a survey point from some other point or points, using known features to position the unknown ones. On land the positions of known points are often provided in the form of triangulation or 'trig' pillars, unfortunately these do not exist under water so they must be created. To start a survey, all that is required is to measure the distance between two primary survey points on the site. As yet, the location of the two primary survey points in the real world is not known. They are assigned arbitrary coordinates so they can be drawn (to scale) on a plan. These two points are now 'known' control points; with a 'baseline' drawn between them, they form the start of a site-plan. By measuring the distances and/or angles from both these points to other fixed points on the site, the other points can be plotted relative to the two initial control points on the plan. It is then possible to draw up a network of survey-points joined by distance measurements. The site-plan of points and distances is drawn to scale and represents the archaeological site lying on the sea-bed.

Most archaeological sites will require more than two control points, which will form a control point network. These points are the framework on which subsequent survey work is built. Permanent, fixed points are placed around the outside of the site and, if the site is large, possibly through the middle as well. A minimum of two points is necessary for manual recording (with a third advisable in order to make checks), and a minimum of three (or four with checks) for three-dimensional computer recording (see below), although more are usually used. The horizontal positions of control and detail points are defined using rectangular x, y coordinates, irrespective of the method used to survey them. The origin (0, 0 point) can be put anywhere on the real site, but it is easier if the origin is taken to be the extreme south and west of the area, thereby making the coordinates for all points on the site positive. If the site extent is unknown and it may extend further to the south-west, then it may be useful to put the origin a long way off the site. To achieve this, the coordinates of the extreme south-west point on the site can be given a large number such as 1000, 1000. This is known as a 'false origin'.

To avoid confusion, the survey team should standardize the units and conventions to be used. It is recommended that coordinates, distances and depths should be given in units of millimetres, metres and kilometres. Angles should be given in degrees. The way numbers are written down should also be standardized. It is recommended that distance and depth measurements are written in centimetres or millimetres, because distances in metres require decimal points, which can easily be lost when transcribing numbers from forms.

INSTALLING SURVEY POINTS

Once the position for each of the survey points has been decided, the next step is to install them on site. Where feasible, control points should be made as permanent as possible. Setting up new control points each diving season not only wastes time but may compromise accuracy.

Three types of survey point are used:

1   Primary control points are established in the planning phase and are the main reference points for the survey.
2   Secondary points may be added later to solve line-of-sight problems or to reduce measurement lengths.
3   Detail points are the points on artefacts and features used to position these objects.

Primary control points are the most important, and at least two primary control points must remain after the work has been completed so that any further survey can be related to the earlier one. Primary points should not be placed on the artefacts or structure of the site (both because archaeological materials should not be damaged and

because they might move), but should be fixed firmly and securely to the sea-bed. Some recent re-surveys of sites have re-used points installed more than 20 years ago. Secondary control points can be placed on rigid structures or artefacts that are unlikely to move. It is not so important to ensure the permanence of these points, as they could be re-established by measurement from the primary points.

For permanent fixings into rock, large galvanized steel or stainless steel coach-bolts can be used, or the bolts (spikes) used to hold train rails onto sleepers. For even more permanent attachment, the bolts can be cemented in using specialist underwater cement or a mixture of sand and cement held together with a little water and PVA glue (which stops the cement washing away while it is being applied). This mixture can be applied from a plastic bag like toothpaste. Climbing pitons, if made of a suitable metal, can also be used, driven into crevices in the rock. Enough of the bolt or piton should be left visible to provide a secure attachment for a tape-measure loop. If the bolts are too big then tapes can be temporarily attached using releasable plastic cable-ties.

Installing anything permanent on sand and mud can be difficult, especially if the sand itself moves, so a compromise is usually required. Long lengths of steel reinforcing-bar can be driven into the sea-bed and the longer the length of bar, the more stable it will be. Scaffold tubing can also be used (driven into the sea-bed with a stake-driver, sledgehammer or air probe), or the metal supports designed to protect and firmly locate the bases of fence posts. If there is a danger that the post will be knocked during diving work, then any excess length should be cut off. The actual point used for measuring on any post, bar or tube must be clearly identified; attaching a plastic hook to the bottom works well. It is better not to measure to the top of the post as any movement in the post will increase the further up you go. Clearly, it is important to check what is just under the sea-bed before the survey point is installed to avoid damaging the site.

Attaching control points to the wreck or structure itself is to be discouraged because it can easily damage the very object of the recording process. Where this cannot be avoided, it should be done with care and, ideally, should be attached to wood with no features (such as joints, treenails or decoration). Points attached to objects held in place simply by their own weight only work well on low-energy sites; any movement must be less than the intended survey accuracy. Extremely heavy objects do not always work well either; they move as they settle into the sea-bed or as scouring takes place. Brass screws are easy to fit and last a considerable length of time under water. Nails and cup-hooks are quick to install but tend not to last very long, and should be used only for temporary work. Large cable-ties or tie-wraps have been used to securely attach plastic shower-curtain hooks to cannons and anchors.

All survey points should be clearly marked in a way that will last for as long as the points are needed. One of the most common mistakes made when recording measurements is to measure to the wrong control point. The marking method chosen for primary control points should be designed so that it remains visible, even after a few years of marine growth has accumulated. Each point needs a label clearly identifying its name or number. For primary control points, the labels should be large and securely fixed on the point or very close to it. Labels made from plastic sheet with the point's name cut out seem to work well, as the name is still readable even if covered in marine growth. For secondary and detail points, less expensive garden-tags can be used; if the point name is engraved with a hot soldering iron then the label can be read after several years under water.

Finding the points can be a problem, especially on large sites or those with poor visibility. For points fixed to rock, the area around the point can be cleared of marine growth with a wire brush to make the point more visible. Additional ideas that have worked include bright plastic markers designed for survey points on land, or animal ear-tags, and using any durable tape or paint on the point itself or a float above it if there is weed cover.

Features and artefacts can often benefit from having a label attached, particularly cannons and anchors. When positioning any artefact, a record should be made of the point or points on the artefact that were actually positioned – a simple sketch is usually enough. Where a surveyed point on a feature or artefact is to be repositioned later, the point should be marked. Techniques that have been used include: white map-pins for temporarily marking points and defining edges or corners of objects in photographs; silicone sealant for attaching survey points to corroded metal structures in the inter-tidal zone; yellow crayon to mark crosses for points or for adding labels to iron or concreted items. Alternatively, the join in the cable-tie can be used as the point to be positioned.

THE PRINCIPLES OF SURVEY

Survey is about depicting features in a symbolic way, and showing the three-dimensional relationships between them (though usually the result is shown on a flat surface). This could, for example, involve the relationship between a natural feature, such as a gully, and the archaeological features and objects deposited within that gully, but equally the relationships between the various objects and features in the gully. The degree of accuracy required relates to what is being shown, and what methods are employed to show it (including the scale).

All survey is based on measurements and bearings, which can be combined to build up a complex picture. There are two basic methods: offsets and ties/trilateration.

Offsets

Offsets are measurements that position features relative to a tape baseline fixed between two control points. An offset measurement positions a feature using a single measured distance at right-angles to the baseline from a known point, and is simple and effective over small areas. Two divers are needed for these methods, one at each end of the tape-measure. Like radial surveys, the positions are usually only given in two dimensions although spot-depths can also be recorded. Offsets are most frequently used for assessment surveys, for recording detail or in confined areas such as rock gullies.

The first step is to set out a tape-measure baseline between two control points, which runs through the centre of the area to be recorded, and with a clear line of sight between the two points. Distances are measured from this baseline, horizontally or vertically, to features on the site using a tape-measure or measuring rod. Vertical offsets are often used for measuring profiles of ships' timbers (figure 14.4 and plate 14.1). To position using an offset, a measurement is made from the feature to the point where the offset measurement meets the baseline at a right-angle (figure 14.5(1)). The right-angle point can be found by swinging the 'offset' tape-measure and finding where the distance to the baseline is the shortest (figure 14.5(2)). In poor visibility it may be useful to have

Figure 14.4 Recording the profile of a wreck using vertical offsets from a horizontal datum. (Drawing by Graham Scott)

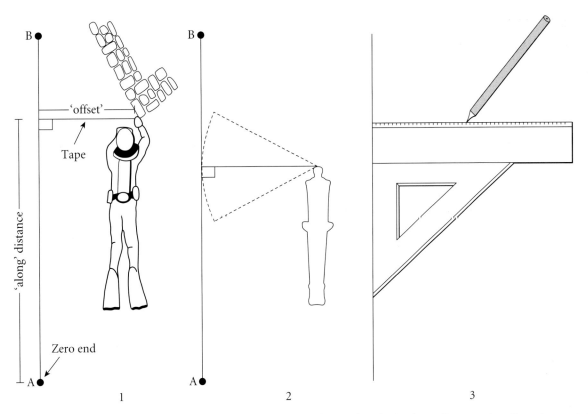

Figure 14.5 1) Offset method; 2) a method of establishing a right-angle, 3) plotting offset results. (Based on original artwork by Ben Ferrari)

a second diver swim up and down the 'offset' tape to ensure that it is not snagged.

In this case two distances are recorded – the 'offset' distance from the feature to the baseline and the 'along' distance from the zero end of the baseline to the start of the offset (figure 14.5). The position accuracy of offset measurements is limited by the fact that it is difficult to estimate correctly the right-angle with the baseline, and it is therefore only suitable for use in good visibility and when close to the baseline. Long tape baselines can move sideways and do not provide a stable reference for the offset measurement.

Ties/trilateration

This works by creating a triangle, taking two measurements from a feature to two chosen points (control points) on a baseline (figure 14.6). It is most accurate when the triangle is as close as possible to equilateral. Although this cannot be achieved precisely, the angle of the tapes at the feature should be between 30 and 120 degrees: this is known as the angle of cut.

As much as possible should be drawn up under water, so that details can be added and errors can be identified and fixed. Drawing up a plan-view on dry land, based on

either method, requires a scale ruler, a pair of compasses and a set square. For both methods, select a suitable scale based on the length of the baseline and the size of the paper, draw the baseline to scale, and label the control points. For offsets, mark the scaled measurements along the baseline (from the appropriate control point) using the scale ruler. A set square can then be used to mark the distances out from the baseline (offsets) (figure 14.5(3)). When drawing up ties measurements, set the compasses to the correct measurement and draw an arc from one control point and another arc of the right length from the other control point. Where the two arcs cross is the third point (figure 14.6). Once completed it will be clear why angles which are too acute or too obtuse do not produce an accurate result, as the point where the two lines cross will not be as clear as when they cross at closer to the ideal 60 degrees. Drawing up on a computer uses the same principles.

The trilateration method does not need to be limited to baselines. It can be carried out directly from any two control points on a site. It can also be used to tie in additional control points and is particularly useful for recording isolated finds. For an area with tightly packed archaeological material, a grid frame is usually used (see below).

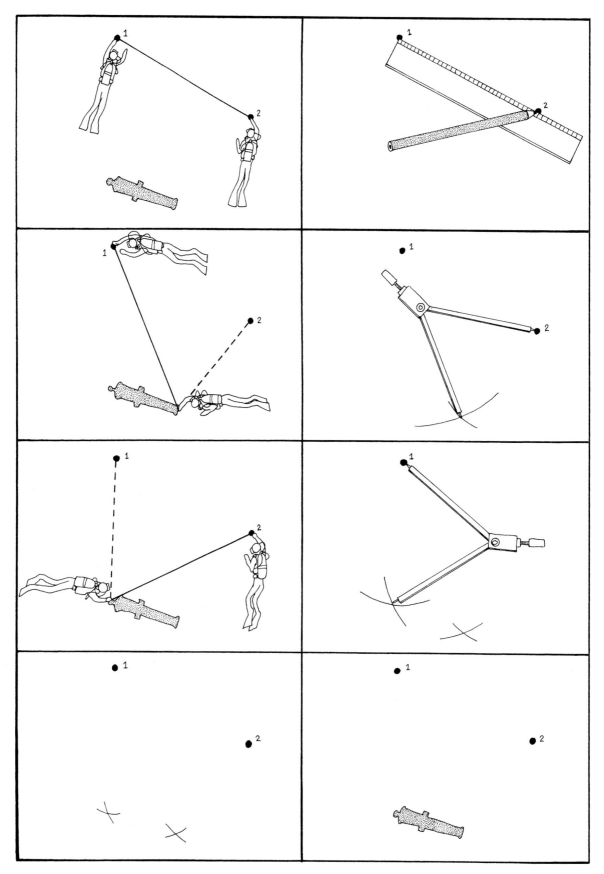

Figure 14.6 Ties/trilateration: survey and drawing up results. (Drawing by Graham Scott)

SURVEY USING TAPE-MEASURES, GRIDS AND DRAFTING FILM

With control points installed on the site, it is now possible to collect measurements that can be used to calculate the positions of the points. Each point on a site has a position given in three dimensions: *x* and *y* horizontally and *z* for depth or height. Positioning under water is largely a matter of measuring distances between objects. The most typical tools used for this are the tape-measure, the drawing frame and the rigid rule.

A small drawing board called a slate or clipboard is most often used for recording measurements under water. The slate should be slightly larger than A4 size. Plastic sheet of 3–5 mm thickness is usually sufficiently rigid. The slate should sink rather than float so, if necessary, a small strip of lead sheet can be attached to the bottom. It should have a short lanyard with a clip to attach it to the diver. If a slate is to be carried in a net bag, then the corners should be rounded. When carried in a bag with other equipment, the danger of figures smudging must be guarded against. A pencil should be attached to the slate with a lanyard long enough to allow it to be used comfortably anywhere on the slate. Propelling pencils made of plastic work well under water (soft (2B) leads are recommended). Ordinary pencils break easily when waterlogged.

The measurements are written on a sheet of plastic or drafting film attached to the slate with clips or electrical tape (plate 14.2). Standard waterproof forms can be made by photocopying a paper form on to drawing film (75 micron plastic paper will pass through most photocopiers and printers). Each form should have space for the site name, the site code, the date, a dive reference number, diver's name, estimated visibility and current, plus room for a small sketch. Numerous different types of form can be created – for distance measurements, offset measurements, depth measurements, blank forms for sketches and forms with square grids for drawing frames. Write clearly in capitals and avoid smudging work. Use standard units and conventions, which should be agreed beforehand. Any recording forms that have measurements on them should be well looked after. The forms are the primary record of any work and should be kept even after they have been processed. Keeping a notebook with day-to-day accounts, speculations and ideas about the site is often useful, as the notes can be helpful later when the measurements are being processed.

Many different varieties of tape-measure are available with different materials, lengths and designs. A tape-measure suitable for survey work under water is no more than 30 m (98 ft) long and has graduations as fine as the site-plan requires. Open-frame tapes are better as they can be easily rinsed after use and dismantled to remove silt and grit. Yellow tapes look good in photographs, whereas white tapes tend to flare in bright light and are harder to read.

The most common type of tape-measure is made of glass fibre-reinforced plastic and those with stainless steel or plastic fittings will last the longest. Fibreglass tapes are cheap and readily available from builders' merchants but they do stretch with use. Steel-cored tapes look very similar but stretch very little, although they are slightly more expensive and can kink if not handled properly. A typical steel-cored tape is accurate to about 6 mm at 30 m in ideal conditions whereas a fibreglass tape is only accurate to 30 mm at 30 m. The achievable accuracy does depend on the conditions under water, as any water-movement will tend to make the measurement less accurate. Measurements over 30 m are not recommended as this length of tape is difficult to manage and sag in the tape makes the measurements insufficiently accurate.

Given that fibreglass tapes will stretch, the correct amount of pull has to be used to ensure the correct distance is recorded. The weight of the tape itself tends to make it sag so measurements will tend to be slightly long; in pulling the tape straight the diver may end up pulling it too much and record a short measurement. Any water movement will make the tape bow outwards and this has the same effect as sagging, so where possible take measurements at slack water. It is possible to measure the correct tension but this is difficult to do under water and is very time-consuming. Getting the correct tension can only come with practice, but it is more important to be consistent with the amount of tension used. It is important to train divers in the use of these techniques on land before they use them under water.

The free (zero) end of the tape should be hooked on to a control-point and unwound in the direction of the point to be measured. Pull the tape tight over the point to be measured and record the distance at the centre of that point. The names of the relevant points should be clearly recorded on the form along with the measurement itself. Care is needed here, as a mistake can be made when the measurement is written down and when it is read by whoever is processing the measurements. The results should be recorded on a form clearly enough for someone who has never visited the site to be able to process the results. It is often necessary to refer back to the recording forms during processing to help decide whether a measurement should be rejected. A number of factors affect the quality of measurements: working in low visibility, strong currents and deep water all tend to produce more frequent mistakes.

Tapes are prone to being snagged on other objects between the two points being measured. Where both ends of the tape cannot be seen the tape should be checked for snags before the measurement is taken. Where practical,

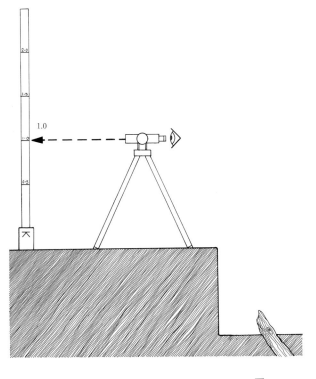

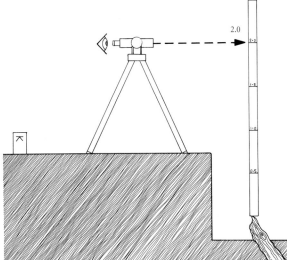

Figure 14.7 The principle of levelling to establish relative heights/depths. For example, if the benchmark is at 6 m: top image shows the level is 6 + 1 = 7 m above chart datum; bottom image shows the level is 2 m above timber; therefore, top of protruding timber is 7 − 2 = 5 m above chart datum. (Drawing by Graham Scott)

work in pairs where one diver takes measurements while the other ensures that tapes are not snagged. Avoid pulling tapes to free them from snags as this can damage them. Each tape being used for recording should be checked against an unused tape-measure at 15 m and 30 m (49 and 98 ft) lengths under normal tension. Any difference larger than the required accuracy for the survey indicates a problem and the tape should not be

used. The tape-measure used as the standard should be steel-cored rather than fibreglass and ideally should not be used under water to avoid damage.

VERTICAL CONTROL (HEIGHT/DEPTH)

The other component of a survey position is height or depth. In survey terms, this is called vertical control and the principles described here are similar to those used for levelling on land (figure 14.7). The most common tools for measuring depth or height are the digital depth-gauge and the dive-computer. These instruments measure the depth of water and display it on a screen, usually to a resolution of 0.1 m (4 in). A dive-computer takes time to settle to the correct measured value, so allow time for this before recording each depth measurement. The depth-sensor reading is affected by large changes in temperature, so measurements will change as the computer cools or warms up during a dive. If the temperature difference between air and water is very different, keep the computer out of the sun or put it in a bucket of sea-water for about 30 minutes before the dive. Some are more accurate than others, so try to use the same computer for all depth measurements. Waves and tides affect any instrument that measures depth. Underwater surveys are frequently done relative to a point on the sea-bed because this cancels out the effects of tides. A permanent feature on the site is nominated to be the depth reference or temporary benchmark (TBM) for all depth measurements (figure 14.8).

Other tools which may be useful in underwater survey include:

- rigid 1 m (3 ft) long rules and plastic folding rules for recording detail;
- a diver's compass for finding magnetic north;

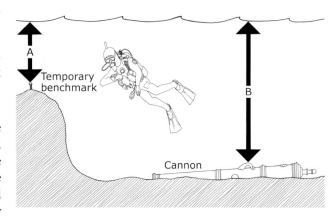

Figure 14.8 Depth of cannon (= B − A) relative to a temporary benchmark. (Drawing by Graham Scott)

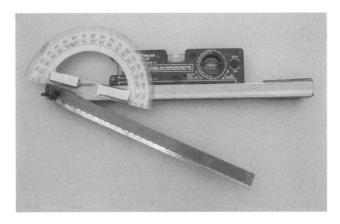

Figure 14.9 A simple clinometer made with items from a DIY store. (Colin Martin)

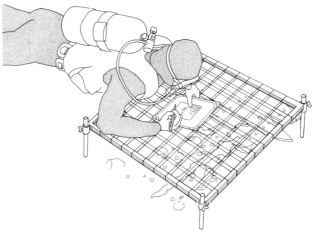

Figure 14.10 A double-strung frame helps counteract distortion caused by not being directly above the subject. (Drawing by Graham Scott)

- a clinometer (figure 14.9 and plate 14.3) for measuring the tilt of ships' timbers and structure;
- large callipers for recording the thickness of objects such as timbers or the diameter of cannons;
- plumb-lines for vertical offset measurements.

If divers are not carrying out scale recording under water, then the sooner the measurements can be transferred from the diver to the site-plan the better, so that dud measurements can be re-measured as quickly as possible, before they distort or hold up the overall plan.

DRAWING/PLANNING FRAMES

Drawing or planning frames are used to quickly and accurately record small areas of a site in detail, and can be used vertically as well as horizontally for recording sections and areas of standing structure (figure 14.10, plate 14.4 and plate 14.5). A typical drawing frame is a 1 m rigid square made from metal or plastic tubing subdivided into squares of 100 or 200 mm using thin cord or elastic. A 200 mm spacing (5 × 5 sub-squares) allows the user to judge which square an object is in by eye. A 100 mm spacing (10 × 10 sub-squares) involves counting to ensure the correct sub-square is being worked on. The diver should be directly above each square as it is drawn to avoid parallax error. Double-stringing the frame is an effective way to ensure that this happens, as when both sets of strings are in line, the diver knows s/he is in the correct position. Drawing is done directly onto a drawing board, with the shape of the frame and its sub-divisions drawn to scale on it.

Alternatively, if a sheet of transparent acrylic or PVC is laid over a drawing frame, the detail can be drawn directly on to the sheet using a wax crayon or Chinagraph pencil. Back on the surface, hand drawing, photography or scanning can be used to scale down the drawn squares. Or the squares can be traced off at the same scale onto drafting film.

Steel reinforcing mesh with 100 mm spacing can be used as a drawing frame in 1 m × 1 m or 2 m × 1 m sections. These frames have the advantage of being robust and can be left on site between dives. However, care is needed in using steel mesh frames because they are heavy and can damage delicate artefacts, structures and diving equipment. The use of a drawing frame is a simple technique and very accurate over small areas but the frame must be accurately positioned. Designs vary but the frame must be portable, must not distort, and the strings must be tight. In use horizontally, the frame needs adjustable legs and spirit levels on both sides.

Where an area larger than one drawing frame needs to be recorded, it is essential to position each frame accurately so all the separate drawings can be fitted together. If survey control is not used, then errors will accumulate through the recording process, leading to distortions in the final plan. Frames can be placed along a tape baseline if a strip of sea-bed is to be recorded. Sometimes positioning the ends of this baseline is sufficient to position the frames on the plan. For larger areas, the corners of each frame can be positioned relative to the control point network; however, this can be time-consuming. A third method involves putting labelled markers (control markers) over the site, positioned relative to the control point network – plastic survey markers are ideal for this. There must be enough control markers for at least two to appear in each frame as it is drawn and the markers, along with their names, must be shown on each drawn square.

To ensure that there is coverage of an area, it is recommended that some overlap is created between frame squares. The overlap can be useful in aligning adjacent squares when drawing up the survey but this should not be the only method for positioning squares when drawing up because this could lead to large distortions in the site-plan. Before moving a frame, markers can be placed to indicate the corners of the area just drawn so the next frame can be aligned accordingly.

The frame must be placed close to the subject so detail can be seen. In poor visibility the operation will be slower but should not be any less accurate. A set of symbols and conventions can be developed for each project so that various materials and features are represented in the same way on each drawn square. Recording using drawing frames produces lots of small squares to form a site-plan. The original drawings should not be inked over or stuck on to the main site-plan because the originals are the primary record and should be kept as a check on the final plan. Each original drawn square should have the following additional information marked on it: site code; date drawn; name of diver; magnetic north direction; individual or dive reference number; location description.

To draw up the plan by hand, a site-plan of the same scale as the squares is required, showing the position of the control markers. Each square can then be aligned on its appropriate marker and drawn onto the plan. To draw up a planning frame survey using a computer involves producing a computer-generated plan of the site. The plan should show the position of the control markers. The original drawings can then be scanned and overlaid onto the site-plan digitally. Although it is possible to fit the drawing squares together simply by lining up features on the edges, this is not recommended. It is important that control markers and/or survey points are included in the plan – otherwise the overall shape of the plan tends to distort and the accuracy of the plan will be unknown.

GRID-FRAMES

Another method for controlling positions on a site is to erect a grid-frame over the entire site, or the part of the site actively being worked on. The frame is usually made up of strong metal or plastic poles attached to each other to form a rectangular grid securely fixed to the sea-bed. Each square can be clearly labelled and colour-coded to indicate different areas on the site. Points on the grid itself then become the control points for the survey: placed high up above the sea-bed, they are ideally placed for making measurements. Control points and their labels are added as the frame is constructed. The frame has the advantage that control points can be put where they are needed, not just where the site dictates. Alternatively, a graduated

movable bar with movable vertical measuring rod can be moved up and down the grid frame and be used for recording the positions of points on the site.

A grid-frame works well if the site is small – otherwise the grid becomes too impractical to install. It can be used, however, on the part of a larger site that is actively being worked. The frame can be constructed on land and taken out to the site or it can be built from components on site. The site should be in a relatively protected place because storms, fishing activity or ships' anchors can move, damage, or even destroy the grid-frame. Because of the investment in time required, grid-frames are usually only used where a significant amount of work is to be done on a site over one long period.

A high degree of survey accuracy can be achieved using a grid-frame. It is likely that each point on it will need to be positioned under water, even if it has already been positioned on land, because the shape of the frame may distort during deployment on site. A distortion of 10 mm in any one point would be noticeable in a set of adjusted distance measurements. A few external control points should also be installed around a grid-frame so that its position can be re-established between seasons or if it gets moved accidentally. A grid must be carefully levelled before it can be used or the depths of each control point measured and dealt with in processing.

A rigid grid-frame of, for example, 5 m sides can be used as a framework within which to position 1 m square drawing frames. Rigid grid-frames are sometimes installed over trenches during excavations and these can be used simply as support for divers and may not necessarily be used for survey control. Rope grids can be used, but they suffer from poor positioning accuracy if they are larger than about 10 m square. These are useful for quick surveys but can be time-consuming to install and difficult to position accurately.

Site-plans can be drawn at any orientation but sometimes it can be more useful to align them so that north is upwards on the plan. To align the site, it is necessary to measure the bearing (angle from north) of something large on the site that appears on the site-plan: a baseline between two control points is ideal for this. The bearing measurements can be made using a diver's hand-held compass held alongside a tape-measure laid between the two control points. On completion, the plan can be rotated so that the baseline is at the correct bearing. Measuring the orientation of an iron or steel wreck is difficult, as the magnetic effect of the wreck will affect the compass measurements. In such cases, it may be preferable to use a heading derived from surface GPS positions (see chapter 11). Measurements from a compass will be given relative to magnetic north; if the magnetic variation is known at the site then the bearing can be corrected and the site-plan can be oriented relative to true north.

PROCESSING MEASUREMENTS AND DRAWING UP THE SITE-PLAN

Chapter 19 summarizes information that must be included on each survey drawing and site-plan. The site-plan is created using sketches and measurements collected from a site. The plan may be drawn by hand or on a computer. The advantage of computer-generated plans is that printed copies can easily be made, they can be printed at different scales, and the level of detail shown can vary as the work progresses. If the site has not been recorded before, the sketches or assessment survey results can be turned into a basic site-plan. In doing this, missing information is often identified and additional work can be planned.

If a recording survey is being undertaken, then the first step is to add the control points that were installed around the site to the site-plan. If a previous site-plan has been created from an assessment survey, then measurements may be needed from the new control points to a couple of the main features on the site, to align the previous site-plan with the control points. Once the control points have been established, the features can be added to the plan as they are recorded and processed.

The advantage of drawing up site-plans on paper or plastic film is that it requires few tools other than paper, pencils, scale ruler and dividers/compasses. Measurements made under water can be scaled down and drawn directly onto the paper. There are, however, a number of drawbacks with using paper:

- only one 'fair' copy of the site-plan exists so its loss or damage can be catastrophic (although this can be overcome by regular digitization);
- the plan has to be redrawn if the site changes due to environmental effects or excavation; unless a fair copy is made, this destroys the previous site-plan;
- large pieces of paper are usually involved and these require the use of large drawing boards or tables;
- if the site is extended the new drawing may be off the edge of the existing paper plan;
- paper shrinks and stretches as humidity changes, though drafting film is more stable.

The reasons listed above mean that it is preferable to draw site-plans on a computer. Personal computers are now very common, as are the computer-aided design (CAD) programs used to draw site-plans. These can be drawn in two or three dimensions, can cover any area and can be plotted at any scale. Layering facilities in these programs allow plans to be plotted with selected subsets of the information visible. Processing the measurements made on site also becomes easier and the positions of points

calculated by survey processing programs can be readily imported in bulk. A number of suitable CAD programs exist and deciding which one to use is largely a matter of personal choice. It is recommended that one of the more common programs is used or one that uses a standard file format such as 'drawing exchange format' (DXF). This allows the site-plan to be shared easily and minimizes the risks associated with 'future proofing' (see chapter 8).

Drawing the site-plan in one computer program separates it from the rest of the information recorded about the site. To keep all of the information together in one place requires a geographic information system (GIS). GIS programs run on personal computers and provide the combined capabilities of a CAD drawing program and a database. The GIS program can be used to draw the site-plan and to record information about finds, control points, measurements, dive-logs and anything else relevant to the site itself. As with CAD drawing programs, there are various GIS programs available. Any of these programs can be used to record information about a site but cost and complexity vary widely. To date there is only one GIS designed specifically for marine archaeology work and this is the 'Site Recorder' program from 3H Consulting Ltd. Like other GIS programmes, Site Recorder can record information about the site and be used to draw site-plans but it also includes archaeology-specific tools such as the ability to process survey measurements (plate 14.6).

It is essential that the site-plans and site information be copied and stored in a secure archive. If the site is destroyed accidentally or deliberately (by excavation), then the plans may be the only record of the site itself. Paper records should be copied photographically or digitally and deposited separately with team members and the authorities responsible for the site. Digital records can also be deposited with professional archive organizations such as the Archaeology Data Service in the UK (see chapter 19).

THREE-DIMENSIONAL COMPUTER-BASED SURVEY

Sites with little height variation can be recorded by adding contours to the two-dimensional plan. When dealing with a very three-dimensional site, such as substantial remains of a ship's hull, then three-dimensional recording is needed, and is probably best done by using a computer program. Three-dimensional (3-D) trilateration or 'direct survey measurement' (DSM) uses direct distance and depth measurements to position features on a site. Distances are measured directly from control points to features and any difference in depth is dealt with in the processing (figure 14.11). The 3-D trilateration technique has a number of advantages: it can be very accurate, it can

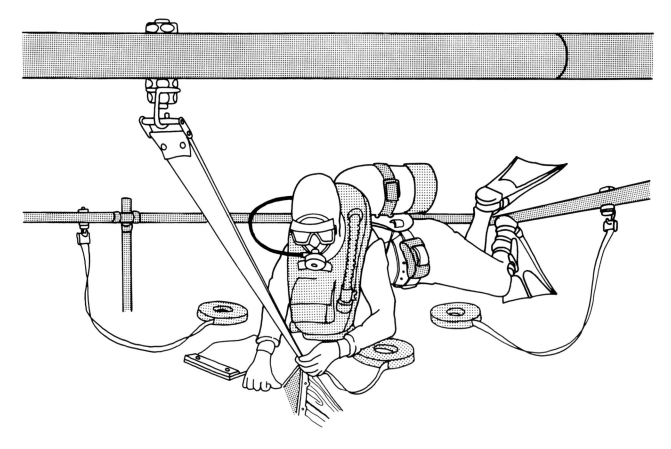

Figure 14.11   A diver taking DSM measurements. (Drawing by Graham Scott)

record positions in three dimensions and it is very easy for divers to take the measurements. In addition, standard survey processing techniques can be used to compute the positions of the features and also show how good those positions are. The main drawback with this technique, however, is that it requires a computer program to compute point-positions efficiently from the measurements.

The problem of positioning a point in three dimensions can be simplified by separating the horizontal position from the height or depth. Most underwater sites are three-dimensional and the tools available to survey them are usually limited. A tape-measure can be used to measure the distance between points so that those points can be drawn on a site-plan. But what if there is a big difference in height between the two points? In such cases, the distance measured is a slant distance and not a horizontal distance that can be drawn on a piece of paper. It is straightforward but time-consuming to correct each slant distance to a horizontal distance, so a technique that uses slant distances is preferable. A very high degree of accuracy can be achieved using 3-D trilateration, as this is the most accurate and reliable technique that uses tape-measures. A good survey point position computed using this method can be as accurate as 20 mm (95%), though this is dependent on the accuracy of the control point network.

Three-dimensional trilateration requires a network of survey control points to be established around and inside the site. This network of points is the framework on which the whole site-record is built. These control points are used for the same purpose as the concrete triangulation pillars (trig points/trig stations) found on the tops of hills and mountains. Trig points are used for providing survey control across countries and control points are used to provide the same control across a site. The main or primary control points should be established around the outside of the site. These points should be placed where they will not be removed if the site is excavated. If a network of survey control points is not used, it can be difficult to make the measurements fit together or to prove the accuracy of any survey. If the position accuracy specified in the project design is not achieved in the control network – or, worse still, not even measured – then the subsequent survey work can never achieve the required accuracy.

The first step in establishing the network is to plan the positions of the control points based on the current knowledge of the site. Planned positions for the control points are often based on the results of a previous assessment survey. Next, the positions of the control points are calculated using distance measurements between them, plus depths for each point. The calculated positions of the control points are then considered to be fixed, so the addition of subsequent detailed measurements cannot alter the positions of the control points. Measurements from four or more control points can be used to position detail points on features such as artefacts and structure.

Survey work on land is based on triangles because the tools used for land survey, such as theodolites, measure angles. Survey under water is largely done with tape-measures, measuring distances rather than angles. If three points are set out on land and the angles between them measured and added together, the total should add up to 180 degrees. If the total is not 180 degrees, then the difference is a measure of how good the angle measurements were. Unfortunately, this does not work with tape-measurements, as three distance measurements will almost always fit together, so a 'braced quadrilateral' is used instead. The braced quadrilateral or 'quad' is made from four control points in a square or rectangular shape, with the sides and both the diagonal distances measured (figure 14.12). If the positions of the points are plotted onto a piece of paper, the four points can be positioned using five of the six measurements and the sixth measurement is the check. In most cases the check measurement will not fit perfectly and the size of the difference gives an idea of how well the other measurements fit together. If the check measurement is significantly different to the value expected, there is a mistake in one of the six measurements.

A complete control network can be as simple as just these four points. However, if tape-measurements are kept to less than 15 m (49 ft), this can only cover a small area of a typical site. To cover a larger area, a number of quads joined at the edges are used (figure 14.13). Avoid placing control points less than 5 m (16 ft) apart, as this will not improve the survey but will waste a lot of time in positioning the extra points. Install the control points high up so they have a clear line-of-sight to as much of the site as possible but note that high points with a good line-of-sight are often the most easily damaged.

Once a couple of quads are joined then it is possible to measure between points in adjacent quads. The net-

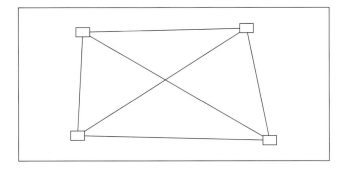

Figure 14.12 Braced quadrilateral (3-D survey)

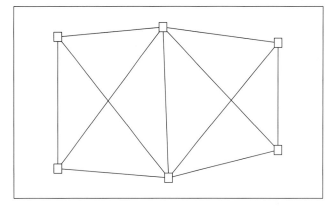

Figure 14.13 Joining quads (3-D survey)

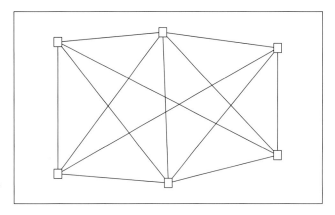

Figure 14.14 Measuring between quads (3-D survey)

work in figure 14.13 has the barest minimum number of measurements to be able to compute the positions of the points in three dimensions. If a mistake occurs in one of the measurements, it will be difficult to find and fix. By adding more measurements, as in figure 14.14, more information about the positions of the points is available, so any measurements that are mistakes should be easier to find. Sets of quads can be extended to cover large areas but, on very large sites, they can be time-consuming to set up. In some cases, it is easier to treat groups of features more than 30 m (98 ft) apart as separate sites.

The best network shapes for surveying are circles and ellipses (figure 14.15). Circular shapes can be used but not many sites are circular, so in most cases an elliptical network is required. The ratio between the length and width of the ellipse should ideally be less than 2:1 (less than twice as long as it is wide). To achieve this sometimes requires extra 'outrigger' bracing points to be placed either side of the site. It is essential that all control points for a site be connected together into one network. Separating parts of the site into smaller, unconnected networks will cause problems when drawing up. Where the extents of the site are not entirely known, the control point network may need

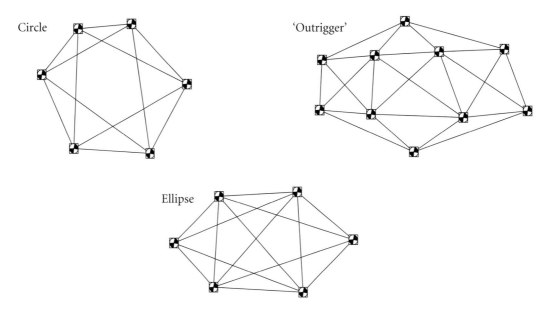

Figure 14.15 Good control-point network shapes (3-D survey)

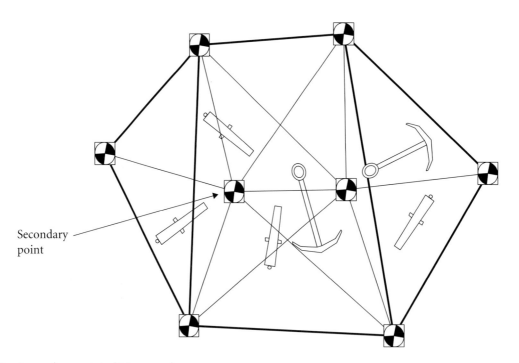

Figure 14.16 Secondary points (3-D survey)

to be extended as more of the site is discovered. The overall shape of the control point network should remain circular or elliptical after it has been extended. Where the site is large, secondary points can be placed within the site so that long tape-measurements are avoided (figure 14.16). On some sites, these points can be permanent and fixed to the sea-bed; however, this may not always be possible.

Secondary points fixed to structure or large artefacts can be used to span the gap between the two sides of the site

(figure 14.16). Long, thin networks or networks with a very pointed shape should be avoided because the position error will be large for the points furthest away from the main body of the network (as shown in figure 14.17, which illustrates poor configurations of control points). The angle between measurements to a point should be no smaller than 45 degrees and no larger than 135 degrees for a good control-point network shape.

It has been demonstrated that six distance measurements are used to position four control points, with one

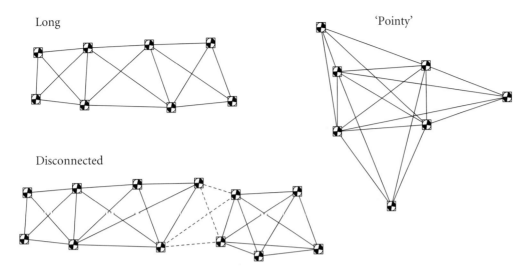

Figure 14.17 Poor control-point network shapes (3-D survey).

measurement used as a check. It is possible to plot the four points at the same height by hand and to determine how different the check measurement is from the expected value; this value is called the residual. If the points are at different heights then the depths can be measured and the measurements can be corrected for the difference using Pythagoras' theorem and the points can be plotted.

When there are more than four points, or more than one check measurement is made, then a problem occurs. The points can be plotted in a number of different places depending on which measurements are used. There are no rules to say which measurements should be used, so statistics are employed. A mathematical technique called 'least squares' can be used to compute the best answer for any given set of measurements. This is known as 'adjustment'. The technique is mathematically intensive and is best done using a dedicated computer program. Most sites using 3-D trilateration utilize the 'Site Recorder' or 'Site Surveyor' software from 3H Consulting Ltd or 'Web for Windows' (DSM), although other programs can also be used.

The computer program takes in initial guesses or estimates for the positions of the survey points, along with measurements between the points. From this, the software calculates new positions for the points on the site along with some position-accuracy information for each point. Based on the new positions for the points, the programme can also give an idea of the quality of the measurements or how well the measurements fit together.

Because many sites are essentially flat, the distance measurements tell us very little about the depths of each point. This is why it is essential to measure the depth of each control point and detail point, and to include these measurements when processing. When the calculated position accuracy for each of the control points is within the accuracy specified in the project design, this phase of the work is complete. At this stage it is essential to 'fix' the positions of all of the control points in the computer program so that the addition of further points and measurements does not affect the carefully calculated control point positions.

With the control points carefully positioned, they can be used to position detail points on the site. The detail points are added to features, artefacts and structure, so, by positioning the detail points, the positions of the features themselves can be calculated. To position detail points, a direct distance measurement is needed from each detail point to the four nearest control points. Ideally the measurements should be made from points all around the detail point rather than from one side only (figure 14.18).

Distance measurements between detail points on the same object can be used, such as the distance between the detail points at each end of a cannon. However, measuring distances between detail points on different objects should be avoided. This is because the objects themselves may be moved as the site is worked on and it also makes processing much more difficult because it is harder to find incorrect or 'blunder' measurements. Mistakes in measurements between detail points will affect the position of both of the points, so one incorrect measurement can alter the positions of a whole chain of detail points linked together by measurements. A blunder measurement between a detail point and a 'fixed' control point will only move the position of the detail point, constraining the effect of the blunder measurement to one detail point only. As with control points, it is essential that a depth measurement be made at each detail point.

The computer program used for processing or 'adjusting' the measurements will calculate the position of each

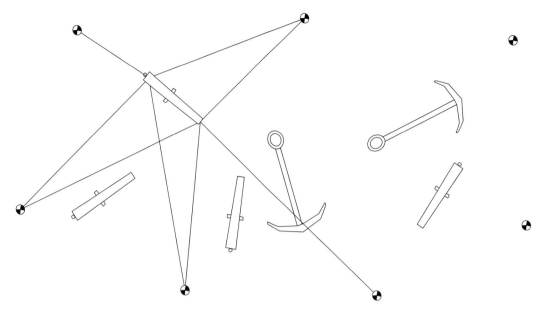

Figure 14.18   Positioning detail points on features using control points (3-D survey)

of the control points. These positions can be plotted on paper charts or imported into a CAD or GIS program. These adjustment programs show the positions of control points and detail points graphically on a chart, completing a significant part of the drawing up. Instructions for the use of any of these software packages is beyond the scope of this book but can be found in the manuals provided with each program.

ACOUSTIC POSITIONING SYSTEMS

An underwater 'acoustic positioning system' (APS) can provide positions under water like the Global Positioning System (GPS) does on land and on the water's surface. APSs are widely used for survey work in the offshore oil and gas industry, positioning ROVs, divers and remote-sensing tow-fish. These systems position objects under water by replacing distances measured using tape-measures with distances measured using pulses of ultrasonic sound-waves. They can be very accurate, can work effectively over very large areas and can continuously report the position of a diver under water. The main drawbacks of these systems are that they are expensive and the higher quality systems can be complicated to use.

APSs have been used for marine archaeology projects for many years. In 1972, the visible timbers of the *Mary Rose* were surveyed using an acoustic distance-measuring system. In recent times, the use of the APS has become more common and they are now in regular use by archaeology teams all over the world. Two types of APS are applicable to marine archaeology: 'long baseline'

(LBL) systems and 'ultra-short baseline' (USBL) systems. Both types of system use a personal computer on the surface to calculate and display the positions of the objects being tracked in real-time and in three dimensions. This allows the dive supervisor, archaeologists and ROV pilots to see the position of the divers and ROVs under water. By connecting the APS to a geographic information system, the position of the divers can be displayed live on the site-plan, allowing the site to be recorded in real-time.

An LBL system works in a very similar way to the way that 3-D trilateration is achieved using tape-measures. Four or more acoustic transponder beacons are deployed on the sea-bed for the APS and these do the same job as the control points installed around the site for 3-D trilateration. Acoustic signals measure the distance from a transceiver unit on the diver to each of the beacons. The diver's unit can also measure its depth so this and the distance measurements are used to compute the diver's position using exactly the same mathematics used for 3-D trilateration. LBL systems can be used in depths between < 5 m and 1500 m, and provide the same high accuracy whatever the depth. As LBL systems require a network of beacons to be set up on the sea-bed, the area that can be covered in one deployment depends on the size of the network and this is itself dependent on both the system being used and the depth. Typical sizes for high-accuracy work in shallow water (< 50 m) would be 100 m × 100 m, but in deeper water this can be increased to 1000 m × 1000 m. Position accuracy depends on the quality of the system and how well the positions of the beacons have been calculated. Typical position accuracy for a low-cost system can be 500 down to 100 mm, while the best-quality system can reliably position to within 30 mm. The object being positioned has to be connected to the surface by a cable, as the transceiver needs to communicate with the computer system on the surface. This is less of a limitation than it may appear because divers using these systems need to be in voice communication with the surface and this usually requires a cable.

A USBL positioning system can position divers and ROVs relative to a boat on the surface. An acoustic transceiver is fitted to the end of a rigid pole and deployed in the water over the side of the boat. The transceiver on the pole receives acoustic signals from a transponder beacon on a diver and these can be used to calculate the distance and direction of the diver relative to the boat. By combining these measurements with the boat's position, provided by a GPS receiver, the position of the diver in the real world can be calculated. The position accuracy of USBL systems is dependent on the distance between boat and diver – the positions get less accurate the further the diver is from the boat. As the USBL system is attached to the boat, there is no limit on the area that can be covered because the boat can simply be moved. Often these

systems are used to track ROVs, allowing the boat to follow the ROV anywhere, similar to having a dog on a lead. Because the position accuracy is dependent on the distance between boat and diver, accuracy is usually specified in terms of percent slant range (distance); however, a typical working figure would be 1 m accuracy at 100 m. Unfortunately, the accuracy of the system is very much dependent on the quality of the GPS receiver and compass and motion reference unit, which also have to be fitted to the vessel. USBL systems are seen as being easier to use than LBL systems, but they require careful installation and calibration to get the best results.

POSITIONING THE SITE IN THE REAL WORLD

For simplicity, sites are often recorded using local coordinates that are not referenced to points in the real world. For projects where the site itself is associated with features on land, or where remote-sensing data have been collected, it becomes necessary to determine where the site is in the real world so as to relate the survey data

Figure 14.19 Surveying a submerged site in shallow water using a shore-based total station. (Drawing by Graham Scott; after Morrison, 1985, fig. 5.2)

to the National Grid or latitude/longitude. The methods of obtaining a position on the surface of the sea have been described earlier in the book (see chapter 11); this leaves the problem of accurately relating a point on the sea-surface to a point on the sea-bed. In shallow water, poles long enough to reach the surface can be placed on control points and their positions fixed using a 'total station' (figure 14.19) or DGPS ('differential global positioning system' – see chapter 11). In deeper water, a large buoy on the surface tied to a control point or heavy artefact can also work well. If the buoy is large, the rope can be tensioned, keeping the buoy above the point to be positioned; this method works best at slack water on a flat calm day. Any methods more elaborate than a simple surface buoy may be wasted effort, unless the surface positioning system is very good, because the amount the buoy moves from a position directly over the point is likely to be less than the accuracy of the surface position. To obtain a very accurate (300 mm/12 in absolute) position-fix in deep water, an acoustic positioning system can be used. This involves placing a transponder beacon on the point to be positioned. The APS positions the beacon by combining acoustic range measurements, depth measurements and positions from a DGPS receiver.

FURTHER INFORMATION

Atkinson, K., Duncan, A. and Green, J., 1988, The application of a least squares adjustment program to underwater survey, *International Journal of Nautical Archaeology* **17**.2, 119–31.

Bannister, A., Raymond, S. and Baker, R., 1992 (6th edn), *Surveying*. New Jersey.

Cooper, M., 1987, *Fundamentals of Survey Measurement and Analysis*. Oxford.

Cross, P. A., 1981, The computation of position at sea, *Hydrographic Journal* **20**, 7.

Erwin, D. and Picton, B., 1987, *Guide to Inshore Marine Life*. London.

Green, J., and Gainsford, M., 2003, Evaluation of underwater surveying techniques, *International Journal of Nautical Archaeology* **32**.2, 252–61.

Historic American Buildings Survey/Historic American Engineering Record, 2004 (3rd edn), *Guidelines for Recording Historic Ships*. National Parks Service, Washington.

Holt, P., 2003, An assessment of quality in underwater archaeological surveys using tape measurements, *International Journal of Nautical Archaeology* **32**.2, 246–51.

Holt, P., 2004, *The application of the Fusion positioning system to marine archaeology*. www.3hconsulting.com/downloads. htm#Papers

Holt, P., 2007, *Development of an object oriented GIS for maritime archaeology*. www.3hconsulting.com/downloads.htm#Papers

Howard, P., 2007, *Archaeological Surveying and Mapping: Recording and Depicting the Landscape*. London.

Rule, N., 1989, The direct survey method (DSM) of underwater survey, and its application underwater, *International Journal of Nautical Archaeology* **18**.2, 157–62.

Smith, S. O., 2006, *The Low-Tech Archaeological Survey Manual*. PAST Foundation, Ohio.

Spence, C. (ed.), 1994 (3rd edn), *Archaeological Site Manual*. London.

Uren, J. and Price, W., 2005 (4th rev. edn), *Surveying for Engineers*. London.

Wentworth, C. K., 1922, A scale of grade and class terms for clastic sediments, *Journal of Geology* **30**, 377–392.

SOFTWARE
3H Consulting Ltd (www.3hconsulting.com/)
Sonardyne International Ltd (www.sonardyne.com)

Destructive Investigative Techniques   **15**

Contents

- Probing
- Sampling

- Excavation

Throughout this book, stress has been laid on the importance of survey and recording, not because excavation and intrusive techniques in general are less important but rather the opposite. Just as on land, underwater archaeological sites cannot be un-excavated, so the process is inherently destructive. That destruction can only be mitigated by careful planning, pre-disturbance survey, comprehensive recording, and publication. The decision to excavate under water is additionally onerous because the conservation of materials recovered from aquatic environments is often problematic and expensive (chapter 16). So although for the general public 'digging' has long been regarded as the quintessential activity of archaeology, these days a great deal of fieldwork takes place without it, not least because of the dramatic advances in the technology of remote sensing (chapter 13).

To take a purist stance, excavation could be regarded as a last resort in the investigation of a finite, non-renewable resource and this is reflected in policies of heritage management worldwide, which has been adopting the principle of preservation *in situ*. In essence, therefore, excavation is primarily justified in two ways: when research questions cannot be answered any other way and/or the site is under some sort of threat (Adams, 2002a:192). In practice, a third reason is training, although this is slightly different from the other two in that it should never be the sole justification for digging. That is the reason why 'excavation' is the only subject group in the NAS Part III syllabus that is not compulsory. Examples where training runs alongside research and rescue imperatives include NAS projects, university training digs or some of the excavations funded by government bodies such as English Heritage.

When excavation is justified, various strategies can be employed, and each must be evaluated in terms of the balance between information retrieval and impact on the surviving remains. The most destructive option, total excavation, might not be necessary. Many excavation strategies involve some form of sampling, using test pits, trenches or larger, more open areas (figure 15.1).

As well as ethical constraints there are also practical issues to consider before excavating under water. It is often (though not always) more costly than other investigative techniques but it always requires a wide skills-base in a team with well-organized logistical support. This chapter briefly outlines the three basic methods that will disturb a site in search of clues: probing, sampling and excavation.

PROBING

The principle of probing is fairly obvious. It is an attempt to locate sediments or structures beneath the surface layers, but in practice it is not always as simple as it might seem. Systematic probing of a site may assist the evaluation of its extent, state of preservation and depth of burial (figure 15.2). However, since its operation relies on feel, the results of probing can be very difficult to measure and interpret. It is best used to answer only very simple questions, such as the depth of sediment over a buried land surface, or perhaps the extent of a buried wreck structure. As with core-sampling, because of the potential danger to fragile archaeological material, it should only be used after careful consideration of the consequences.

Probing will only be of lasting value if it is carried out systematically to answer particular questions. The nature

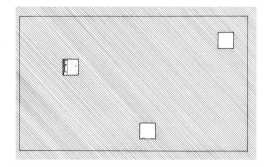

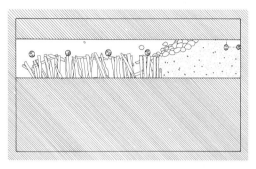

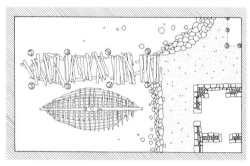

Figure 15.1 Excavation strategies: an example of the way different strategies – test pits (top); trench (centre) – will provide varying levels of information about the whole site (bottom). (Drawings by Graham Scott)

of the questions will dictate the probing strategy (e.g. readings taken at measured intervals along a line or at the intersections of a grid).

Types of probe: The simplest probe is a metal rod, thin enough to be pushed into sediment and thick enough to withstand bending. In practice, the resistance of the sediment imposes its own depth limit, beyond which it becomes increasingly difficult to distinguish a real obstruction. In these situations, or where the sediments are compacted, a more efficient probe can be made from tubing (e.g. 25 mm (1 in) bore steel pipe) down which water is pumped (figure 15.3). Only low water-pressure is needed to penetrate all but the most compact material and high pressure will cut through almost anything, including archaeological material. One of the drawbacks is that the water from the surface is often oxygen-enriched and this may upset the anaerobic environment in which fragile archaeological material survives.

SAMPLING

A sample is a representative amount of material that has been collected from an archaeological or natural context. Sampling for environmental or scientific analysis is relevant and appropriate for all sites. Samples may be taken for numerous reasons, ranging from dating to the identification of organic remains. There is a difference between collecting a sample of a material or deposit 'to see if there is anything in it' and taking a sample to answer a specific question.

Figure 15.2 Probing to record sediment depths and obstructions can be an effective method of assessing the extent of some sites. (Based on original artwork by Ben Ferrari)

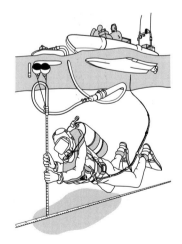

Figure 15.3 Air or water probes can be used to explore a site but are potentially destructive. In this example, measurements (distance along tape and depth of probe) are being relayed to the boat via diver-to-surface communications. (Based on original artwork by Ben Ferrari)

Samples should only be collected if three basic criteria can be satisfied:

1 There should be evidence that the sample will contain traces which will provide valuable information concerning the past. This is best checked by examination of a pilot sample either on site or in the laboratory.

2 There must be a sound reason for collecting the material. Specific questions should be asked, post-excavation analysis is made easier if objectives are clearly stated.

3 There should be a clear prospect that the material will actually be studied. This should be established by consultation with specialists before, or at an early stage of the archaeological investigation. However, important material should be sampled even if no scientific programme has been pre-arranged: a specialist can usually be found to work on material of significance.

These three criteria can best be satisfied if a clear strategy is agreed between the archaeologist and relevant specialists beforehand, or as soon as the problems become apparent during excavation. It is important to understand, however, that even after carrying out a detailed examination a scientist may be unable to provide a simple unqualified statement. Every method of examination has its own limitations. Often one analytical method must be employed to examine one group of phenomena and a second, quite distinct, method used to examine other aspects of the same sample.

It may be useful to attempt to divide non-artefactual or environmental archaeological remains into broad categories. For example:

- *Economic* – Environmental archaeology can make considerable contributions to our understanding of the economy of a site or period. At its simplest level this may relate to what was eaten on the site. At a more complex level the environmental information can be used to reconstruct the contemporary agricultural economy or used to illustrate differences (such as social, religious or racial) across a site or between sites.
- *Environmental* – This refers to the sampling of deposits that may yield information on the general climatic, environmental or ecological conditions prevailing on or near a site. With respect to underwater sites, this may mean samples that can generate information about the formation of the site or, perhaps, data on the chemical and physical characteristics of the site and particular preservation conditions.

- *Behavioural* – The biological remains contained in certain contexts and/or their distribution across a site can relate to various aspects of human behaviour. At its most obvious, the threshing and winnowing of cereal crops on submerged settlements could produce recognizable patterns among botanical assemblages. The practising of crafts or commercial activities on board ship may also reveal itself in characteristic groups of animal bone or other materials on shipwreck sites. It may also be possible to interpret the function of specific areas or determine the original contents of containers.

An important part of any archaeological investigation is stratigraphy: the study of the various sediments, embedded structures and features that comprise the site's stratification (chapter 4). Apart from visual methods of characterization it may be necessary to take samples of the various layers present for laboratory analysis. Sedimentology – looking at particle size and composition through the depositional sequence – helps determine the changes that have taken place over time (see column sampling below).

The principles of radiocarbon and dendrochronological (tree-ring) dating have already been introduced as the two main techniques of absolute dating (chapter 4). There are many factors that critically affect how viable samples are for particular types of dating analysis.

Radiocarbon sampling: It is recommended that contact should be made with a radiocarbon laboratory at an early stage, if possible before any samples are taken. The following points should be taken into account:

- Never submit a radiocarbon sample unless the archaeological problem it is intended to solve is clearly identified. It may have nothing to do with chronology. Dates well-related to the span and significant events of site chronology should always be sought. Do not be wooed by potential samples simply because they are there. Always try to form an opinion on the chances that a sample will actually date the human activity or natural phenomenon for which a date is sought. In most cases there is no absolute certainty of association or contemporaneity.
- Before taking a radiocarbon sample from an archaeological deposit, section or core, study the nature of the deposit or layer and the stratigraphical conditions (such as geological complications, possibilities of humic contamination from higher levels, root penetration, visible animal activity from other periods).
- Collect more samples or a greater amount of sample than required for one dating because a later

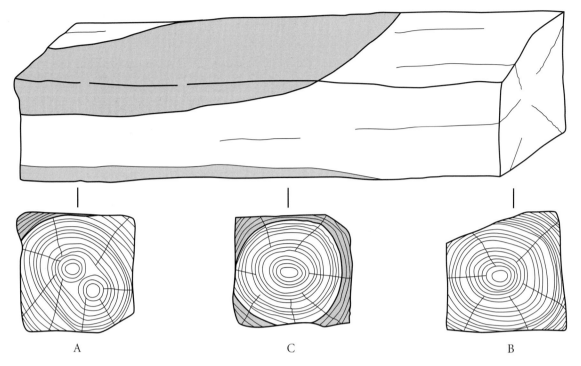

A   C   B

Figure 15.4 Optimum place for tree-ring sampling: A) branching is distorting the ring pattern, B) no sapwood is present, C) there is undistorted ring growth and good sapwood survival. (Based on original artwork by Ben Ferrari; after Nayling, 1991:47)

check may be required or the sample may get lost. Estimate whether the amount of sample available is sufficient to provide the required precision in age. Realize that botanical, zoological or chemical identification is possible only before treatment in the radiocarbon laboratory.

- Pack the samples in plastic or aluminium foil or in glass bottles and write immediately the name of the site, sample number (depth and horizontal position) and the name of the collector on each packed sample. Send the sample to the laboratory as originally packed, together with full documentation.
- If samples have to be stored before submission to the laboratory, keep them in a cool, dark and dry place. Don't use organic preservatives and, in the case of shipwreck material, don't submit samples contaminated by waterproofing agents such as tar.

Sampling for dendrochronology: It is recommended that contact should be made with a dendrochronology laboratory at an early stage, if possible before any samples are taken (English Heritage, 2004b). The following points should be taken into account:

- The determination of a date for a structure will require a sample or samples of wood cut from

well-preserved and long-grown timbers, preferably with sapwood surviving. Tree-ring sampling ideally involves selecting the widest part of the timber, free of branches and knots, and the sawing out of a 50–100 mm (2–4 in) thick slice. In general, ring sequences of less than 50 years will not date reliably. Size will not equate with ring-width as the growth rates of the original trees might vary considerably.

- Long ring-pattern sequences out to, and if possible including, sapwood are ideal for dating. Such complete samples will give the dendrochronologist the opportunity to sub-select the most suitable and informative samples for dating purposes, and also offer the option of characterizing the complete wood assemblage in terms of species, age and growth rate.
- A complete cross-section cut perpendicular to the grain is preferable, but a v-shaped piece from the back might be sufficient if the piece has been selected for conservation and display. In some cases, sections sawn from timbers can be tree-ring counted and then joined back to the main piece.
- Coring is also possible although it is recognized that it may cause compression and distortion of the ring sequence. An increment corer has been used successfully in dating boat-finds (for example, Tyers,

1989). A careful assessment should be made in choosing the optimum coring location with reference to the extent of the ring-sequence and the danger of damage to the outer (and perhaps most decayed) rings.

- Samples should be taken to provide the longest ring-sequence possible, including the latest surviving ring on the timber. For the greatest precision of dates, samples that include some sapwood, preferably complete, need to be identified on site and sampled preferentially (figure 15.4).
- Wood that is split or cracked may require support and strapping before sawing. Great care should be taken with the outer surfaces of the wood. This area may consist of sapwood that dries at a different rate from heartwood and it tends to become detached.
- Package the samples to prevent either physical damage or dehydration, and label them so that each is cross-referenced to the records of the original timber.

The decision regarding how much of the deposit or material to recover must be taken in the light of the answers to questions like:

- What is the deposit made up of? What is the principal component? Is it uncontaminated? The material that it is intended to sample for must have survived; the context must be well stratified and dated; and it must be possible to take a large enough sample to yield the required minimum of identifiable material in a manner unlikely to produce a sample bias.
- What is its potential? What can it reveal about the archaeology of the site?
- What will it entail, in terms of the excavation budget and resources, to recover all or a part of the deposit?
- What is the opinion of the specialist who will be providing the identification and interpretation?

Speculative sampling could be employed, provided that it formed part of a coordinated sampling programme (e.g. to provide test samples). Such samples should be processed and investigated with the minimum of delay and information about the quantity and range of evidence present can be quickly relayed to the archaeologist who has the opportunity to modify the excavation strategy relating to the original deposit. The planning stage of the excavation should include an assessment of the likely potential for scientific studies before and during the excavation itself and in post-excavation work. Account should be taken of the:

Figure 15.5 Taking a spot sample. (Based on original artwork by Ben Ferrari)

- time needed to carry out scientific work;
- cost (to include the cost of site visits and meetings as well as laboratory time);
- likely importance of the analysis, both absolutely and in relation to cultural archaeological studies; and
- intrinsic importance for the development of the discipline of archaeological science.

Spot sampling: These may be small local concentrations of biological materials (excluding wood). Examples are groups of fruit-stones, insect remains or small bones. Do not attempt to clean or separate the materials until they are transported safely to a specialist or more suitable processing conditions. General samples for biological analysis can be examined for the presence of many different kinds of remains (insects, fruits and seeds, parasite-eggs) depending on the nature of the material and the archaeological questions posed (figure 15.5).

The following is a basic procedure for the recovery of a sample illustrating some fundamental points:

- Have a suitable, clean container with a close-fitting lid ready.
- Identify the extent of the deposit from which the sample is to be taken.
- Record all locational details (such as relationships with other contexts, orientation of sample) on an underwater recording form/sheet in the form of measurements, sketches and notes.

- Make sure that the surface of the deposit is cleaned in the proposed sample area to reduce the risk of contamination.
- If a 'total sample' cannot be recovered, cut or gently separate a proportion of the whole deposit with a clean spatula or trowel. (Tools should be cleaned between samples to prevent cross-contamination.)
- Place the sample in the container and tightly fit the lid.
- If a number of samples are to be recovered on one dive, make sure that they do not get mixed up. Use pre-labelled containers or bags if necessary.
- Bear in mind at all times the possibilities of contamination. Note down any doubts.
- Record the quantity of the material or deposit that was actually recovered and how that compares with the original total deposit (even if it is just an estimate, such as '10 kg of an estimated 100 kg').
- Raise the closed container to the surface for prompt examination by the relevant specialist.
- Ensure that all notes are accurately transferred to the 'sample record form' (or equivalent documentation).

Coring and column (monolith) sampling: Chronologically stratified sequences, such as those found in naturally accumulated deposits (e.g. peats or lake sediments), should be collected in such a manner that will not disturb the sequence. Two possible methods are coring sediments down from the surface and the cutting out of a monolith of sediment (perhaps from a section face).

Column samples and monoliths are taken by inserting a channel or tube (of stainless steel or rigid plastic) into the sediment and then extracting it using the most convenient method to achieve an undisturbed sequence. Monoliths, or column samples, can be made up of a sequence of separate samples (figure 15.6). If the containers are pushed into the section, it is possible to remove large blocks of undisturbed sediment. Care should be taken to avoid contamination of the contents by smearing or the introduction of extraneous matter. All the containers should be carefully labelled with orientation, sample number, and location.

Further examination can take place under more suitable conditions, where contamination can be minimized. Sub-sampling of cores or monoliths for further analysis (e.g. for pollen) should be carried out by specialists or under their direct supervision. X-radiography of the undisturbed column may be useful for identifying layers or structures invisible to the naked eye.

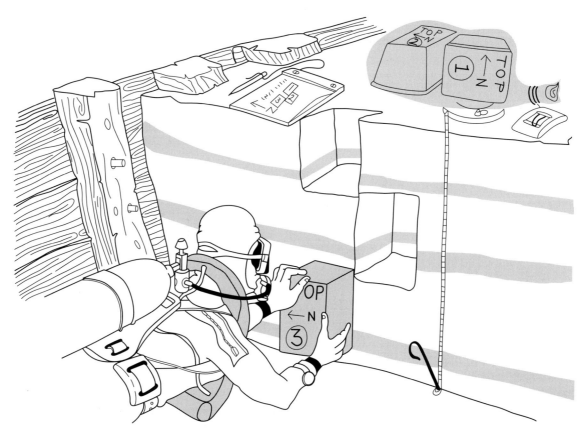

Figure 15.6 Column or monolith sampling from a section. Note that a sketch has been made of the work. (Based on original artwork by Ben Ferrari)

EXCAVATION

Archaeological excavation might be defined as the controlled dismantling of the contexts that form a site – sediments, surfaces, structures, objects and materials relating to past human existence – in order to understand their temporal, spatial and social relationships.

The aim of this section is to set out various methods and processes of excavation which have been shown to produce satisfactory results. This book cannot include every variation available, as each problem encountered prompts a slightly different improvisation. Nor can the use of any piece of equipment automatically lead to high standards of excavation. However, a sound understanding of the requirements of archaeological excavation, coupled with experience in the use of the equipment described below, will allow good work to be done in even the most unpromising circumstances.

A high standard of excavation is difficult to achieve without the necessary experience accumulated on a range of sites. Gaining excavation experience under water can be a long, drawn-out process because of the limited time that can be spent working under water at depth. It is further compounded by the relative difficulty of learning from others around you when under water. Another major constraint is opportunity. During the 1970s and 1980s in particular, several major excavations around the world involved large teams for several months, season after season. As there were far fewer professional diving archaeologists at that time, teams comprised professionals, students and amateur volunteers. Such projects are now few and far between, partly due to the factors discussed above and also because, in general, less underwater excavation is done these days in proportion to surveys. Even on the few developer-funded sites that involve underwater excavation, it is carried out by relatively small teams working within those contract archaeology companies which undertake maritime work (relatively few at the time of writing, though this varies between countries). It can therefore be hard to accumulate the skills required for a high standard of excavation. Yet training needs must be met and the onus rests with professional associations, organizations like the NAS and, hopefully, governments.

A concrete example, of course, are the NAS Part III courses (see appendix 3) which certainly provide a faster learning curve than is possible on most sites. The other valuable, some would say indispensable, way to gain meaningful excavation experience is to become involved in excavations on land sites. The principles and basic methodologies are exactly the same under water as on land; it is just that the environment differs.

It is important to define the area to be excavated and work to those planned limits. A disciplined approach is necessary for the following reasons:

- It increases the efficiency of the project by concentrating effort on the areas selected on the basis of their potential to answer the questions posed in the project design (chapter 5).
- Working in distinct, regular areas allows more efficient planning of subsequent investigations. The project will know where it has been because the precise limits of the excavation already undertaken will be recordable.
- Distinct boundaries of investigations help workers to be more thorough in retrieving all the elements of evidence necessary for interpreting the site. The inevitable damage caused by excavation is also then limited to distinct areas rather than spread over a larger area by wandering excavators.
- Disciplined work also has practical benefits, such as straight vertical edges at the limits of excavation to aid recording of stratification and the ability to concentrate site facilities such as airlifts and site-grids.

It should be impressed upon those carrying out the work that they must confine their attention to the defined area. This should be achieved in two ways:

1 A method of physically marking the work area will be necessary, such as rigid grids (which have the advantage of protecting the excavation edges) or line (which must be firmly anchored if it is to provide a permanent marker).
2 Effective briefing on the physical limits of the investigation, and the reasons for them, for those who will be carrying out the work will also be required. Without a reasoned explanation of the need for discipline, no amount of physical markers will produce a systematic excavation.

The diver's hand remains the most sensitive, accurate, and useful tool for fanning away or scooping silt towards the mouth of the airlift or dredge (figure 15.7). However, at intervals the working area may need cleaning or skimming with another tool to ensure stratigraphic features or other relationships remain visible. That tool is likely to be the 'mason's pointing' trowel, a fundamental tool on any archaeological excavation, whether on land or under water. The small 75–100 mm (3–4 in) bladed tool can be used either delicately or strongly, as circumstances dictate, scraping with the edge of the blade towards the body (figure 15.8) or, less frequently, using the point. In softer, less-compacted sediment, larger trowels can be used, especially for cleaning or scraping sections. Whatever size is used, they should have welded blades, as riveted blades tend to break at the attachment after prolonged use and exposure to seawater.

Figure 15.7 Excavating a wooden weaving heddle on the Armada wreck *La Trinidad Valencera* (1588). Note the delicacy with which the sediments are being removed – the archaeologist is using only an index finger to tickle away the spoil. The mouth of the water-dredge can also be seen – its only purpose is to carry away spoil, not to dig into the archaeological deposit. (Photo: Colin Martin)

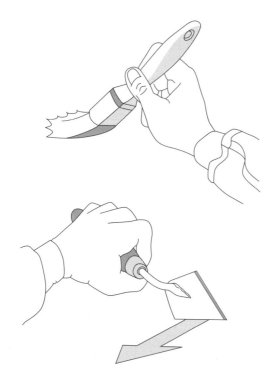

Figure 15.8 The trowel and the paint brush, along with the hand, are the most commonly used tools for excavation. (Based on original artwork by Ben Ferrari)

As indispensable as the trowel on underwater excavations is the paint brush. Larger ones are used like hand-brushes on land to clean surfaces. Under water they are particularly useful to clean timber surfaces prior to recording or photography. Smaller brushes (40–60 mm in width) are often the best tool in soft unconsolidated

sands, silts and clays, especially when excavating organic and other fragile materials because it is almost impossible to do this with a trowel alone.

As well as trowels and brushes, many other hand tools and utensils are suitable for excavation such as teaspoons, dental probes, spatulas, knives and the like. Non-metallic tools are particularly useful where it is important not to damage delicate organic surfaces. Small tools are best kept in some form of submersible container to avoid accidental loss.

When excavating, the diver must be aware of the need for care in defining the contexts making up a site (see also chapter 4) and the nature of stratification. An excavator should aim to remove the layers in the reverse order of deposition. Deeper deposits should not be touched until overlying contexts have received adequate examination and recording.

Some sediments may not allow clear layers to be excavated sequentially. In this situation, control can be maintained by excavating in measured spits (e.g. removing an arbitrary layer 10 cm (4 in) deep). The exposed surface is then cleaned, recorded and the sequence repeated until a recognizably different layer is reached. Later analysis of these apparently homogeneous spits may allow useful evidence to be extracted. Lack of apparent layering is not a justification for uncontrolled excavation technique; neither is the use of spits an excuse for ignoring context differences. Depth of excavation, known as arbitrary excavation, alone is not a reliable method of relative dating (see Harris, 1989:119). Where layers are difficult to distinguish, it is more important than ever to keep the work orderly and neat. Allowing an excessive amount of loose sediment to build up may mask subtle changes and also small finds.

Contexts and stratification should be recorded, in the first instance, from above as they are noted or uncovered. During survey, this is all that will be available. However, the opportunity should also be taken during excavation to record them from the side as they are cut through. This can give added information about the relationships between the various contexts (see chapter 8). There are three ways of achieving this, all of which may be used during the same excavation:

1 Permanent sections will exist at the edge of the site where the sediments have been left unexcavated. These sections are termed permanent because they are unlikely to be removed as the excavation progresses.
2 Temporary sections can be used to record contexts not represented fully enough in the permanent sections. During excavation, part of the context or contexts is/are left unexcavated while a side view is recorded. The rest of the context can be removed

and excavation continued. Since context differences tend to be more distinct when cut through, so temporary sections can be used as an aid to excavation. If a vertical face is maintained during the excavation of a context, any intrusion into underlying layers is more easily noted and stopped before it goes any further.

3   In mobile or deep sediments where these standing walls of sediment would be impossible or unsafe, cumulative sections can be recorded. This is simply the process of recording the section of each individual context before it is obscured by shoring, sandbags or a sloping excavation face. With accurate measurements along the same line, the section drawing will build up layer by layer throughout the excavation, resulting in a picture of the sediments cut through at that point.

Ideally the sides of the excavation should be as vertical as possible so that a true stratigraphic sequence can be recorded in one plane at 90 degrees from the horizontal. It helps considerably with stratigraphic analysis if plans and sections are relative to the natural horizontal datum. If for some reason it is not possible to compile a cumulative section, an alternative is to record information from a sloping or stepped 'section'. This is less satisfactory because relationships may be distorted by variations in the layers either side of the line of the section. Objects and structural remains should be left in place if they are sticking out of the section. Burrowing in after the object will only weaken the section and obscure the layers. However, when the section has been recorded it does provide an excellent source of samples of the various layers.

The positions of sections should be marked out. In the case of temporary and cumulative sections, which may not be immediately obvious, they should be clearly explained to other divers working in the area lest they unknowingly cause damage. Safety is, of course, paramount in these matters and unstable excavation faces can be a serious hazard. Sandbags and shuttering should be used where necessary.

Careless removal of objects will seriously compromise the results of the excavation so careful attention must be paid to the ways they are excavated and recorded prior to recovery. It is not acceptable archaeological practice to pull objects from the sediments that surround them for a number of reasons including:

- the risk of breaking the object;
- the risk of damaging other items close or attached to it which have yet to be exposed;
- failure to record the association of nearby objects; and

- failure to recognize which archaeological context it is associated with.

With objects that are reasonably robust, excavation involves systematically reducing the surrounding sediments until the object is sufficiently exposed for recording. The object is then lifted in an appropriate container or cradle, etc. However, as they become more exposed, objects are increasingly susceptible to damage, either from the activities of divers or from environmental factors such as current, water-borne abrasives (sand) and burrowing fauna. It may be necessary to physically protect and support exposed objects during excavation. Mechanical strength can be added by splints and padding, but delicate objects will always need a skilful excavator, and the co-operation of nearby divers, if they are to survive in one piece (figure 15.9). This is why, on sites where safety factors allow, excavators remove their fins, as these can cause extensive damage to both stratification and other archaeological material.

Figure 15.9 A conservator removing the surviving section of a gunpowder barrel, excavated from the Spanish Armada wreck *La Trinidad Valencera* (1588). The object had been secured with bandages before extraction and then placed in a container that was filled with sand and then lidded before being raised to the surface. (Photo: Colin Martin)

Even these measures may be inadequate for the most fragile objects such as textiles, leather and other very friable materials. These are best recovered with some of their surrounding sediments, usually by transferring them into a suitable container. This also avoids the object collapsing under its own weight, which would be the likely result should it be lifted from water into air. The whole procedure takes practice and if possible should be done by someone who is familiar with the type of object being recovered.

Some material will not need to be recovered after it has been exposed and recorded (e.g. ship's structure). Provision must be made to protect it in the long term. This may involve reburial, usually involving lining the trench with a protective membrane, held in place by sand-bags. The trench is then 'backfilled', either with some of the sediment removed during the excavation or, alternatively, with a sediment of specific characteristics brought in for the purpose.

Underwater excavation, just as on land, consists of two distinct procedures, each of which has its associated tools: 1) the actual digging and 2) the removal of 'spoil' (the unwanted sediments loosened in the process of digging). Although excavation can often be carried out by the same tools used on land, spoil-removal under water is very different. Occasionally the current alone is sufficient but normally some form of suction device is used, adapted from those first developed for industry. One of the mistakes of early excavators was to use these suction devices as the means of digging, in effect conflating the two activities. If control is the most important aspect of responsible excavation, then maintaining this distinction is vital.

The act of excavation is one of constant decision making – how deep to dig, what to cut through, where to cut to, etc. The by-product of this activity is loose sediment. Some of this may be recovered with an object or it may be recovered as a sample. The rest is 'spoil', having been judged not to contain sufficient information to keep. This judgement is not only the province of the individual excavator but of the trench supervisor and ultimately the site director. Once created, spoil needs to be removed. Just as spoil produced by the use of a trowel on land can be removed by shovel, bucket, wheelbarrow or conveyor belt, so various tools are used for the same purpose under water.

The airlift and the water- or induction-dredge are suction devices originally invented for other industries but which have become ubiquitous in underwater archaeology. In the past, they have been described as the underwater equivalent of a shovel, but their real function is to move unwanted excavated material (spoil) away from the excavation like a wheelbarrow or bucket. In this sense, they are one of the advantages of underwater excavation because once in operation they are virtually automatic.

Figure 15.10   Excavation using airlift. (Drawing by Graham Scott)

Airlift

The airlift is a simple device consisting of a rigid tube into which air is injected at its lower end, usually from a compressor on the surface (figure 15.10). As the air rises towards the surface, it lifts the tube to near vertical and creates a suction effect at the bottom. Water and any loose materials are pulled in and up. The power of the suction is dependent on the difference in depth-related pressure between the top and bottom of the tube, and the amount of air injected.

Airlifts can be run from any source of compressed air but are best powered by the sort of road compressors used for pneumatic tools such as jackhammers and rock drills. These provide large volumes of air at low pressure (between 7 and 10 bars (100–150 p.s.i.)). Neither high- nor low-pressure diving compressors provide sufficient volume. The size of compressor required depends on the depth of the site and the size and numbers of airlifts that will be used simultaneously. The smallest air-tool compressors deliver 2 cubic metres (275 ft^3) of air per minute and this is adequate to power two airlifts. In no circumstances should such a compressor (or any other air source) be used to supply air to divers and airlifts at the same time.

The limitation of these compressors is their weight; hence they need a sufficiently large boat or floating platform unless the compressor can be sited on land nearby. Another limiting factor is often the cost of such units although, for short periods, hire charges are not excessive, especially in relation to the total costs of excavation.

On small operations, the hose can lead direct from the compressor to the airlift. However, it may be necessary to secure the air hose to somewhere convenient on the sea-bed so that there is no additional pull on the airlift once the hose fills with air and becomes buoyant. On larger, long-term projects where more than one airlift is in operation, hoses led directly from the surface will inevitably become tangled and constitute a safety hazard. In this situation, it is better to have one delivery pipe from the compressor to a multiple take-off arrangement (manifold) fastened to the sea-bed. If this manifold is made from a long steel tube with a selection of take-off points (either manually or automatically valved), individual airlifts can be connected at convenient points by the divers. In many cases where the tidal flow will change direction by 180 degrees, it may be more convenient to have a manifold both up- and downstream of the site.

Prior to their first use on site, the airlifts need to be positioned and secured. The air hoses can then be connected or, alternatively, airlifts can be taken under water with the hoses already connected but this must be done prior to the air supply being turned on. If possible, they should be arranged on the sea-bed with the exhaust uppermost. Start up the compressor and allow it to reach working pressure. The excavator can now open the inlet valve. As long as the airlift is lying nozzle down, air will flow up the pipe and the airlift will slowly become vertical. If air merely bubbles out of the nozzle and the airlift remains stubbornly recumbent, simply lower the nozzle or physically lift the pipe slightly. Where it is difficult to do this, another method is to place a hand over the nozzle end so that the air cannot escape and so fills up the pipe. The danger with this technique is that the airlift becomes extremely buoyant. It should only be done if the nozzle is under a grid pipe and securely tied down. Once in operation, carefully adjust the air valve so the airlift is at a near neutrally buoyant state. If the airlift is too buoyant in operation, it needs to be ballasted with a little lead. Just as awkward is an airlift that cannot be made neutrally buoyant except on full power. In this case remove weight.

Careful positioning of the discharge is required so that the spoil does not cascade back down on top of the previously excavated or other sensitive areas. If the site has a constant tidal stream or current running across it, then the discharging spoil can be carried clear of the work area. If there is little or no current, the airlift can be restrained at an angle ensuring spoil drops outside the excavation area. If this is not far enough, the choice is between moving the spoil again (a measure to be avoided) or using a water-dredge. The problem with tying the airlift down is that it both restricts freedom of movement and reduces the ease of use.

If the discharge end of the tube projects out of the water, the weight at the lower end should be adjusted. Air in the tube when in operation gives more than enough buoyancy in most conditions and it should not be necessary to buoy the discharge end. Variations of design may be required for specific circumstances: 110 mm (4.5 in) corrugated plastic hose can be used at the lower mouth end to get into awkward areas of a site, but it is essential to have the air-flow lever within easy reach of the excavator so that s/he can shut off the air supply in an emergency.

The airlift must be used with great care. When excavating archaeological contexts, as opposed to removing backfill or weed accumulations, it must only be used as a means of removing spoil, normally swept gently towards its mouth using the hand, a brush or a trowel. It is best held in a comfortable position by the excavator some 20–30 cm (8–12 in) away from the surface being excavated, possibly further if there is anything extremely delicate being exposed. The valve on the air supply allows control of the strength of suction, and so allows very fine adjustment of the rate of silt removal. If it is not possible to control what is entering the tube during excavation, then either the excavation is progressing too quickly or the end of the airlift is too close to the working surface (plate 15.1).

A mesh on the suction end of an airlift or dredge should not be required to prevent objects being 'sucked up'; nor should there be any need for devices at the top to 'catch' objects that get sucked up'. Sieves may be fitted periodically at the discharge end but only as a means of monitoring the standard of excavation.

Water-dredge

The water- or induction-dredge is similar to the airlift, except that it operates more or less horizontally, and it is water rather than air that is pumped in at the mouth (figure 15.11). It has the advantage of being cheaper to set up (because suitable water pumps are less expensive to buy or hire than compressors) and can work effectively in very shallow water. The water-dredge can have a flexible tube attached to the suction end to reach difficult places and increase mobility (plate 15.2) but, as with airlifts, the valve controlling the effectiveness of the device must be within easy reach for safety reasons.

The amount of water delivered to the dredge-head is probably the most important factor related to its efficiency. As a rule of thumb, a portable fire-pump with a 75 mm (3 in) outlet diameter will provide adequate power for two 110 mm (4.5 in) diameter dredges. Anything more than 1000 litres per minute is sufficient. Smaller water pumps with a 50 mm (2 in) outlet usually have insufficient delivery to provide anything more than a mild suction but, in many circumstances, this may be all that is needed. The smaller the water pump, the cheaper they are to buy or hire, and the less space they take up; the larger the pump the better the chance of having an effective dredge.

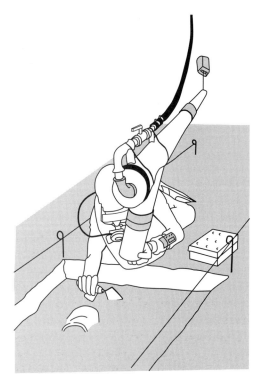

Figure 15.11 Water-dredge operation. (Based on original artwork by Ben Ferrari)

The water-dredge is used in exactly the same basic way as an airlift in that it should only be used to transport spoil fed to it by the excavator. It should not be used as a digging tool except when removing archaeologically unimportant material such as backfill from previously excavated areas, or collections of weed or other debris washed into the site during pauses in the investigation. Unlike the airlift it has no inherent buoyancy and so it is necessary to adjust it to what is most comfortable in any given situation. The dredge can be made neutrally buoyant by securely attaching one or two air-filled plastic containers (5 litre/1 gallon) to it (plate 15.3).

One of the disadvantages of dredging is the effect created by the water leaving the discharge end at speed, and the disturbance this discharge can cause on the sea-bed. This jetting can be dramatic, particularly if the pump is switched on when the diver is not ready. It is also potentially damaging to archaeological remains, and so must be neutralized. This can be achieved in a number of ways. Lengthening the discharge pipe so that its weight rests on the sea-bed can ease the problem but it increases the risk of discharge-end damage, and makes the dredge less moveable.

Discharge pipes can be positioned above the bottom by use of weights, or other forms of anchorage, and buoys. The pipe end can then be fixed at a suitable height. Anchoring the dredge-head end is also a way of reducing

the effects of jetting, but it restricts manoeuvrability. If mobility is not a problem the discharge pipe can be extended well off site (distances of over a kilometre have been achieved) provided that more water is injected into the system along its length. A simple way of achieving this is to fit the discharge end of one dredge into the suction end of the next one in the chain.

One of the simplest ways of overcoming the jetting problem is to baffle the discharge stream. This can be achieved by attaching a flat plate or board across the discharge c.0.75 mm (3 in) from the end of the dredge. Alternative baffles can be devised. For instance, standard plastic soil-pipe fittings (such as a T-piece) can be attached on the discharge end, although these should be of a larger bore than the discharge pipe.

Alternatively, even a slow curve (not a tight 90 degree bend), positioned to discharge upwards, can be attached, in conjunction with appropriate buoyancy and/or anchorage. In such ways it is possible to achieve the desired effect on almost any site, and without damage to the archaeological evidence. The ingenious diver and archaeologist can no doubt think of many alternative solutions to these and similar problems.

It is possible to purchase ready-made induction-dredge heads built largely to supply the numerous operators who recover golf-balls from rivers and lakes in the USA. These are usually made of steel. However, it is possible to make a dredge-head from components readily available from hardware stores. As long as sufficient water is pumped down a tube and across the open end of a tube let into the side of the main tube, suction will develop at the other end of the side tube. The amount of suction will depend on the velocity and volume of pumped water and the overall effectiveness of the design. For instance, the water-inlet pipe should point exactly down the centre of the main tube, and there should be minimal obstructions to the flow in the large-diameter pipes.

Choosing between airlift and dredge

Generally speaking, the airlift is more efficient than a dredge, but requires greater resources to operate and is generally less effective in very shallow water. However, when choosing, it may be relevant to consider the different surface requirements for each type.

Both dredges and airlifts can be manufactured in a range of sizes: smaller for intricate work or, particularly in the case of airlifts, larger (e.g. 150–200 mm (6 in – 8 in) diameter) for tasks like the rapid removal of backfill or seaweed. To provide power for one 110 mm airlift, a somewhat bulkier and heavier compressor is required, although it may be possible to site it on shore and pipe the air out into the site, as has been done on several occasions.

Airlifts can be easier to handle than water-dredges, but where water depth is particularly shallow, a water-dredge may be the only option, particularly as it lies almost flat. However, the airlift can work well in a depth of less than 2 m (6.5 ft) provided a very large volume of air is pumped through it. Additionally, the airlift will also work with up to one-third of its length protruding above the water surface. Both tools can be controlled easily by the hand-operated valves, but can do untold damage if used in an uncontrolled fashion. As in all archaeological operations, it is in the best interests of the surviving evidence if the work is carried out in a careful and disciplined manner.

Water-jet

Another power-tool occasionally used is the water-jet. This is simply high-pressure water released through various shaped nozzles at a volume and pressure appropriate for the task. The indiscriminate nature of a high-volume and/or high-velocity jet of water, let alone its effect on visibility, limits its application on archaeological sites. Miniature versions, run off a water-dredge, however, can be highly effective for delicate work. Alternatively, a separate water-jet can be used in conjunction with a dredge. Small water-jets with a very low power are used in a similar way to brushes and are effective on organic material. As in all cases where delicate material is being excavated, however, do not experiment with a new technique or a new tool on the object itself.

Another use for water-jets is to induce an artificial current where it is otherwise very difficult to maintain visibility around the working area, such as in lakes. The technique was developed in Switzerland for excavation of prehistoric lakeside settlements. Water is pumped from the surface and directed through rows of small nozzles directed along and just above the bed.

Advantage should be taken of the unique ability of the diver to hover above the area under investigation. There are many techniques that can be tried. One advantage of a solid site-grid (chapter 14) is that it can offer additional diver support, and one technique successfully applied on a number of sites is to clasp the horizontal bar with the feet (one over, one under) to hang inclined at 45 degrees above the site (plate 15.1). Pressure applied alternately with either foot will allow the excavator to move around the area with complete control over his/her height above the archaeological deposits. When using an airlift, height can be adjusted by control of breathing or by slightly altering the buoyancy of the airlift as described above. However, the best means of control is to use the increase in the airlift's buoyancy that occurs when the intake is partially obstructed. The excavator simply holds the airlift by the rim and extends or retracts their fingers into the water flow making the airlift rise or fall respectively. This technique can be tricky to learn but once acquired allows excavators to lower or raise themselves into and out of sensitive areas with ease.

FURTHER INFORMATION

Adams, J., 2002a, Excavation methods underwater, in C. Orser (ed.), *Encyclopaedia of Historical Archaeology*, 192–6. London.

Barker, P., 1993 (3rd edn), *Techniques of Archaeological Excavation*. London.

Buglass, J. and Rackham, J., 1991, Environmental Sampling on Wet Sites, in J. M. Coles and D. M. Goodburn (eds), *Wet Site Excavation and Survey. WARP Occasional Paper No. 5*. Exeter.

English Heritage, 2004b, *Dendrochronology: Guidelines on Producing and Interpreting Dendrochronological Dates*. London.

Harris, E. C., 1989 (2nd edn), *Principles of Archaeological Stratigraphy*. London.

Institute of Field Archaeologists, 2001a (rev. edn), *Standards and Guidance for Archaeological Excavation*. Reading.

Mook, W. G. and Waterbolk, H. T., 1985, *Radiocarbon Dating, Handbooks for Archaeologists No. 3*. Strasbourg.

Nayling, N., 1991, Tree-ring dating: sampling, analysis and results, in J. M. Coles and D. M. Goodburn (eds), *Wet Site Excavation and Survey. WARP Occasional Paper No. 5*. Exeter.

Oxley, I., 1991, Environmental Sampling Underwater, in J. M. Coles and D. M. Goodburn (eds), *Wet Site Excavation and Survey. WARP Occasional Paper No. 5*. Exeter.

Roskins, S., 2001, *Excavation*. Cambridge.

Tyers, I., 1989, Dating by tree-ring analysis, in P. Marsden (ed.), A late Saxon logboat from Clapton, London Borough of Hackney, *International Journal of Nautical Archaeology* **18**.2, 89–111.

Watts G. P. Jr, 1976, Hydraulic Probing: One solution to overburden and environment, *International Journal of Nautical Archaeology* **5**.4, 76–81.

16 Archaeological Conservation and First-Aid for Finds

Contents

- Underwater burial environments
- Materials degradation and post-excavation Deterioration
- Principal risks to finds during and after recovery
- Principles and procedures for first-aid for underwater finds
- Lifting, handling and transportation
- Approaches to packing and storage
- Sampling and analysis
- Initial cleaning
- Holding and pre-conservation treatment solutions
- Record-keeping
- X-radiography and facilities
- Health and safety
- Insurance
- Checklists

Archaeological finds are irreplaceable and contain valuable information that may contribute to knowledge and understanding of the past. Materials are likely to have survived only by reaching a physical and chemical 'equilibrium' with the surrounding context or burial environment, and waterlogged objects are particularly vulnerable to loss if not properly looked after. The removal of finds from their burial environment is likely to speed up processes of corrosion and decay, sometimes radically, potentially leading to destruction of archaeological evidence. Properly planned and applied 'first-aid' procedures (the care for finds prior to full conservation treatment) are therefore often crucial. Factors affecting decomposition or corrosion of archaeological materials in underwater burial environments vary extensively. This chapter is aimed, in particular, at those likely to work closely with archaeological finds. Thus the content focuses primarily on planning procedures, storage methods, equipment and supplies, rather than the history of materials technology and manufacture, which is complex and has been well described elsewhere.

Archaeological conservation is based on a sound understanding of materials science and the way in which materials deteriorate in the burial environment. An overriding principle is that of information retrieval – much data not recorded in the written historical record can often be revealed through the use of analytical techniques. In addition, the selection of an appropriate treatment takes into account many factors, including: the condition of an object or group of objects (assemblage), research objectives, significance, future use and the advantages and disadvantages of various treatments available. It does not consist, therefore, of the application of standard processes or 'recipes' for the indiscriminate chemical 'stabilization' of specific materials. Instead, archaeological conservation involves continuous complex decision making in consultation with other team members, and is likely to be based on the following issues:

- the effects of the burial environment and context (types of deterioration);
- general condition (such as fragility, active corrosion);
- significance, according to site research objectives (comparisons with material from similar sites);
- potential dating (such as the identification of coins, clay pipes, or ceramics);
- potential evidence of applied surfaces (signs of use, wear and tear; original surface treatments; preserved organics and methods of manufacture);

- retaining the potential for future analysis;
- ultimate use (such as display, handling or research, including potential mounting methods);
- packing and storage;
- handling in the event of publicity.

Intervention often results in irreversible changes and, therefore, professional conservation practice also incorporates ethical approaches:

- *Minimum intervention:* The aim is to use the least possible intervention required to achieve the desired result, with minimum long-term effects.
- *Reversibility of treatment:* Chemicals, materials and treatments used should be reversible in the long term.
- *Information retrieval:* The collection of all information related to associated archaeological evidence with the potential retained for future analysis, wherever possible.
- *Documentation:* The recording of all methods and procedures used (important in health and safety and in the understanding of the long-term behaviour of materials).

Conservation advice should be sought at the earliest stages of all projects, as this is likely to help reduce long-term storage, conservation and analytical costs. Sites liable to full excavation or those encompassing large structures or fragile material are likely to require the presence of an archaeological conservator on site, who should be included in financial planning at all levels and allocated a budget well in advance of any work.

All other projects, however, should ideally also have pre-arranged access to a professionally trained archaeological conservator. Staff from contracted conservation facilities may be able to provide help remotely or may be available to be 'on-call' in the event of an emergency. They are also likely to be able to help with the identification of materials, which is crucial to the appropriate storage and treatment of finds, and they should be involved in planning relating to the following issues:

- Financial planning for storage, analysis, X-radiography and full conservation treatment of finds.
- Lead-in time for the acquisition of necessary supplies and equipment in advance of project work (projected requirements can sometimes be based, to some extent, on previous site records and the nature of extant assemblages).
- Details of immediate 'first-aid' requirements (e.g. storage facilities and local amenities such as electricity, security and the quality of local water supplies).
- The design of finds-records, registration and monitoring systems, in advance, to help contribute to

interpretation and to facilitate access and early treatment.
- Approaches to handling, lifting, storage and full treatment.
- The analysis of finds and samples.
- Long-term archival deposition of finds archives, including display, all of which should be negotiated prior to any project work.

In terms of costs, large structures may require specific provisions and housing, and are likely to be considered 'high-cost', whereas the 'first-aid' and conservation treatment of many small finds assemblages might constitute a relatively small percentage outlay of overall budget costs (somewhat dependent on location) when compared with other potential costs such as dive-boat, equipment maintenance and team provisions. All members of project teams should be fully briefed in advance of work and have a full understanding of their responsibilities regarding recovery and recording methods and procedures to be used on site. It may be helpful to provide instructions in the form of induction packs and/or to provide talks at the beginning of projects or for new team members: this will help standardize recording and recovery methods. At least one member of a team should be delegated to receive finds and help with the lifting and acquisition of heavy or delicate material, in addition to those appointed to record material and oversee storage arrangements ('finds officers').

The establishment of a fully equipped conservation laboratory in association with the full excavation of a specific site may take several years to complete, so it is always advisable to seek the use of existing archaeological conservation facilities for the provision of full treatment (see the 'Conservation Register': www.conservationregister.com). Previously contracted archaeological conservation facilities, either those in the private sector or associated with museums, are particularly important in the event of unforeseen finds or potentially damaging sea conditions. They are likely also to be able to facilitate the 'rescue' of fragile and/or significant finds and structures.

UNDERWATER BURIAL ENVIRONMENTS

The way in which objects may be preserved in burial environments is extremely complex and based on chemical, physical and biological processes and local conditions. The factors involved include the following:

- the material(s) from which the objects are made;
- method of manufacture (e.g. metal alloys/applied surfaces/composite materials);
- the history of an object (wear, tear and use);

- conditions and nature of the immediate burial context;
- the juxtaposition of items and structures throughout a site;
- the potential for galvanic corrosion across a site, potentially causing the enhanced preservation of one metal to the detriment of other(s).

Underwater environments can be considered to be solutions of potentially reactive chemicals, with high levels of dissolved and partially dissolved salts (often held in solution within voids and interstices in objects and also in the outer surface layers). Seawater can be considered to be one of the most concentrated solutions and therefore one of the most potentially reactive. Depth, temperature and levels of acidity or alkalinity, oxygen, light and pollution, amongst other factors, also all have an effect. Thus the chemical and physical behaviour of each item within one area can be considered to be 'individual' according to the local burial environment, even on a very small scale. No two objects are likely, therefore, to be identical or to behave the same the way.

The effects of biological organisms, both microscopic and macroscopic, are likely also to have an impact on preservation. Microscopic organisms include bacteria, which can survive in many different types of environment. Macro-organisms include: wood-boring molluscs (such as shipworm (*Teredo navalis*) – figure 16.1), crustacea (gribble), seaweeds, barnacles and fungi, all of which will often attack wood for food. Wood-boring molluscs are particularly voracious and thus wooden components lying exposed above the sea-bed are vulnerable to destruction within a matter of years rather than centuries. Some organisms, such as barnacles, will use materials and structures as a substrate for attachment rather than as a food source, which may result in surface markings or some degree of change. Physical deposition, such as in the initial formation of a wreck-site, is often dependent on factors such as the nature of wrecking, geographical and topographical aspects of the location and the type of sea-bed. The latter may be subject to the effects of continually shifting sand and/or strong tidal flow, potentially causing scouring (erosion) or silting (deposition of sediments and sea-bed materials). Despite all these factors, wet and waterlogged deposits have often yielded artefacts in remarkably good states of preservation, particularly those made of organic materials, such as wood, leather and textiles.

MATERIALS DEGRADATION AND POST-EXCAVATION DETERIORATION

One of the most important aspects of finds conservation associated with materials recovered from underwater

Figure 16.1 *Teredo navalis* (shipworm): a large shell, evidence for historical shipworm infestation of the timber, is removed from the upper stem timber of the *Mary Rose* as part of the recording and cleaning process, prior to active conservation. (Photo: Doug McElvogue)

burial environments relates to the presence of soluble salts dissolved in surrounding solutions. Damage is likely on drying because soluble salts will re-dissolve in solution in conditions when the air humidity is high and re-crystallize when it falls, causing potentially destructive physical pressures. Thus a continuous cycle of damage may occur: even slight continuous fluctuations in relative humidity (known as RH: a measure of the amount of water vapour in the air, as a percentage, at a given temperature) may cause considerable damage, sometimes in objects made of materials that may appear to be robust. In the worst cases, such processes may lead to the complete destruction of an object. Thus, the drying of an object, particularly those from marine waters, even for seconds, may result in the weakening of structures and surfaces, even if not visible to the human eye. A vulnerable find, passed around a team during the general excitement following recovery is highly susceptible to such damage and a better way of ensuring that such objects are not put at risk entails arrangements for everyone involved to view the objects in waterlogged storage soon after recovery.

In addition, burial in underwater environments may lead to:

- leaching of components from physical matrices – leading to the weakening of structures that may still appear robust while on or in the sea-bed;
- the development of layers incorporating sediment and attached debris – often known as 'concretions' (chemically bonded to original surfaces)

or 'accretions' (overlying layers) – which may mask original detail;

- problems relating to organic materials associated with metals finds – copper corrosion products, for example, are toxic to some organisms, while lead, iron and silver corrosion products may result in the replacement and/or retention of impressions of organic materials attached to some surfaces.

Other aspects of potential post-excavation materials degradation are outlined below. Although not comprehensive, they are intended to highlight the importance of 'first-aid' procedures, based on specific material types.

Inorganic materials: These materials are derived from non-living things, including ceramics, glass and vitreous materials; stone; and metals, including metal alloys:

- Ceramics may be susceptible to salt damage, including the forcing off of outer layers and glazed surfaces, particularly in low-fired wares.
- Glass is particularly prone to deterioration in alkaline solutions.
- Objects made of metals of different types may be subject to ongoing galvanic corrosion (similar to the chemical reactions in batteries). They should never, therefore, be stored together in the same container.
- The occurrence of pure gold or silver in burial contexts is rare, except, perhaps, in the case of bullion. Such finds are usually alloyed with other metals to impart strength and alter working properties during manufacture. Some corrosion is likely, therefore, during burial, with the risk of further corrosion post-excavation. For these reasons, cleaning, or the removal of corrosion products, should be undertaken only as part of full conservation treatment.
- Iron tends to corrode outwards, with potential distortion of shape. In addition, the corrosion products on marine iron are likely to be concreted with insoluble salts and debris from the burial environment, sometimes rendering identification difficult.
- Finds made of wrought iron (iron 'worked' by hammering, extruding or other mechanical methods) are likely to corrode slowly along 'slag inclusion lines' formed originally during manufacture. If this process has not reached an end-point during burial (i.e. if there is remaining metal core), then such objects are likely to be unstable. Wrought iron was eventually replaced for many purposes, such as for making guns and anchors, by cast iron (manufactured by heating and then pouring into pre-shaped moulds).
- Cast iron corrodes by a process known as 'graphitization', usually leaving behind a metal core, which

may be retained in a potentially reactive state. Cast-iron in marine environments may be inherently chemically unstable, with associated exothermic reactions (giving out heat) leading to the forcing off of surface layers and possible rapid disintegration, with objects in some instances becoming hot, or even, on very rare occasions, exploding when exposed to air.

- Active corrosion in iron objects may be indicated by 'flash-rusting', a term applied to rapid oxidation, indicated by the development of spots of bright orange corrosion products and/or trails of corrosion in storage water.
- Copper (rarely found in a pure state) and copper alloy finds often corrode in layers, with associated potential for the preservation of original surfaces and attached organics. However, they may also become covered by masking layers, sometimes concreted to underlying surfaces and should be cleaned, therefore, only as part of full conservation treatment.
- Lead, tin and associated alloys (such as pewter) are not always as chemically stable as often thought, whether wet or dry, being particularly susceptible to corrosion in the presence of organic acids. They may also retain evidence of applied surfaces (such as plating), stamps or markings, which may be revealed in the X-radiography of thin objects.

Some finds recovered from underwater burial environments are often referred to as 'concretions' due to the development of thick surface overgrowths that mask the shape of the contents (figure 16.2). Metal finds (particularly iron) may develop overlying conglomerates, which may expand to incorporate complexes of objects of several materials together. This often renders contents unidentifiable, but considerable detail can be obtained through the use of X-radiography and appropriate conservation treatment (figures 16.3 and 16.4).

Organic materials: These materials are those derived from living things, including: wood; skins and leather; textiles; bone, horn, ivory and related materials. Burial may lead to physical weakening and chemical decay, with much of the original being supported by water. The evaporation of water post-excavation is likely to result in shrinkage, loss of shape, cracking, warping and, in some circumstances, complete disintegration (plate 16.1).

Composite artefacts: These are objects made of more than one material, often incorporating both organic and inorganic materials, which may pose specific conservation issues requiring discussion with team members.

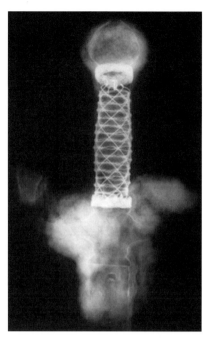

Figure 16.2 A concretion recovered from the Duart Point wreck (1653) site, Mull, Scotland. (Photo: Colin Martin)

Figure 16.3 An X-ray of the concretion shown in figure 16.2 shows that it contains an elaborate sword hilt. (Courtesy of the Trustees of the National Museums of Scotland)

Figure 16.4 The Duart Point sword hilt shown in figures 16.2 and 16.3, after conservation. (Photo: Colin Martin)

PRINCIPAL RISKS TO FINDS DURING AND AFTER RECOVERY

Removal of finds from their burial context is likely to cause extensive changes, as outlined in the list below:

- Drying may result in cracking, delamination of surfaces, irreversible shrinkage, salt crystallization and potential mould growth.
- Increased temperatures and oxygen access may lead to increased speeds of decay, biodegradation (algae and mould growth, often turning finds 'green') and corrosion, and/or may initiate new corrosion reactions in metals. Changes in temperature may cause differential expansion and contraction, leading to damage, particularly in composite objects.
- Increased light may lead to photo-oxidation, fading and increased decay rates, promoting the growth of green algae.
- Storage of different types of metals in solution together may result in galvanic corrosion (with increased corrosion reaction rates).
- Poor handling or lack of physical support may result in changes to centres of gravity, in turn leading to cracks or fractures.
- Lack of attention to labelling and monitoring of storage water levels is likely to result in the loss of

context and, therefore, any meaningful archaeological information.

These can be mitigated, to some extent, by using appropriate 'first-aid' procedures.

PRINCIPLES AND PROCEDURES FOR FIRST-AID FOR UNDERWATER FINDS

The importance of good planning, attention to detail, high-standard recording and the proper care and supervision of finds, both during and after excavation, cannot be overemphasized. Overall approaches to 'first-aid' provisions for finds (Robinson, 1998) within any project may need to be tailored according to:

- the predicted nature of the material to be recovered;
- the potential size of an assemblage; and
- planning for the recovery of large or bulky archaeological structures.

A 'first-aid' field facility should provide suitable storage for recovered finds (plate 16.2) whereas full conservation treatment should be undertaken in a fully equipped conservation laboratory, in which conditions are more easily controlled and where material may be properly examined, analysed and treated using appropriate facilities (such as with good lighting, refrigeration, microscopy and so on).

In long-term projects, all finds should be re-prioritized for analysis and conservation treatment, in the light of recently discovered material, at the end of each excavation season. Such work is best undertaken with the help of a conservator.

Immediately upon recovery, all finds should be kept:

- *waterlogged* – ideally in waters associated with the original context;
- *cold* – as cool as possible and ideally refrigerated (but not allowed to freeze);
- *in the dark* – wherever and whenever possible;
- *clearly labelled* – using at least two waterproof labels;
- *separated*, according to the materials of which they are made – different metals should never be stored together in solution;
- *in inert containers* – such as polyethylene bags and lidded boxes (never directly in contact with metal containers);
- *safely* – guns and artillery rounds, munitions and anything else potentially explosive should be handled with caution and 'made safe' in accordance with the laws and regulations of the relevant state.

In addition there are many general storage requirements:

- Conservation advice should be sought at the earliest opportunity, regarding any questions relating to storage requirements.
- All finds should be stored completely covered by water, with the minimum of air (oxygen) access. This can be achieved, to some extent, by filling lidded containers to the top or by covering water surfaces using conservation-grade inert polyethylene sheeting. Storage water should be topped up on a regular basis, to ensure that all objects are covered completely. The extremities of objects, if exposed, may act as a 'wick', leading to rapid deterioration and/or permanent damage. Regular checks on water levels (including those in bagged objects) are essential.
- Storage containers should be cleaned and rinsed out on a regular basis, to remove any potential build-up of algae and/or 'slimes' (the wearing of gloves is recommended). Proprietary household cleaners should not be used. During container cleaning, finds should be transferred to alternative wet storage, with air-exposure during handling and transfer kept to a minimum. Details of cleaning and the changing of solutions should be recorded, with dates and times.

- Waterlogged glass, small significant organic objects, damp material and all organic samples should be kept in the dark and refrigerated (at around 4 degrees Celsius/39 degrees Fahrenheit). On no account should they be allowed to freeze, as this may cause irreversible damage. (Note that the removal of large quantities of material from a refrigerator all at once may result in significant cooling within the fridge and thereby risk the possible freezing of any remaining material. Fridge thermostats should be adjusted accordingly.) They should not be allowed to dry, to any extent, during preliminary photography and drawing.
- Finds should be labelled at least twice using waterproof materials. This is essential because labelling may wear off in the short term. All labels should be checked regularly (at least monthly), with replacements added as necessary. A waterproof label such as those made of Tyvek (spun-bonded polyester) should be inserted inside bags and/or containers and all reference numbers written on bags prior to use.
- All finds should be monitored regularly, with records kept on an ongoing basis (ideally weekly), detailing condition, solution levels, topping-up and container cleaning. Conservation advice should be sought at the earliest opportunity in the event of changes in condition.
- All metal objects should be handled wearing gloves: some lead and copper compounds are toxic and acid from the skin may cause damage to finds.
- Concretions should be desalinated quickly because chlorides in seawater may accelerate corrosion. Do not dismantle concretions. They should be X-rayed and then handled by a conservator.
- Robust ceramics should be transferred into fresh water. However, earthenware should be handled carefully and monitored in case the glaze detaches itself from the body.
- Jet, shale, amber and glass should be stored wet and should be refrigerated: all are extremely prone to deterioration, with irreparable damage likely even on partial drying. Wet glass can be particularly unstable, indicated by the loss of thin scales from the surfaces and/or iridescence.
- Worked stone, especially the softer ones such as sandstone, may have weakened and therefore need careful handling. Only large quantities of bulk unworked material, such as stone ballast, should be allowed to dry out. (If in doubt, seek specialist advice.)
- Smoking, eating and drinking should not be permitted in the vicinity of artefacts or samples.
- The time elapsed between recovery and full conservation treatment should be kept as short as possible. Significant small finds, composite objects

and those incorporating glass, ivory, jet, shale, amber, leather or textile, should be referred to a conservator immediately if possible.

The process of the removal of seawater should not be confused with that of 'desalination'. The removal of seawater entails replacement using a number of baths of fresh water immediately on recovery, thereby helping to remove potential chemical and biological reactants. All finds should be constantly monitored. In most cases robust materials should be transferred into fresh water immediately on recovery, with the first two to three baths consisting, perhaps, of a 50 per cent dilution using fresh water and each bath lasting 30 minutes to 2 hours. Thereafter, changes of water should be aimed at 'desalination' (the removal of salts held deep within the interstices of objects), a process that may take many weeks or months to achieve, during which time most materials should, ideally, have been submitted for archaeological conservation. Initial baths for desalination might consist of a 50 per cent dilution using fresh water, with each bath lasting perhaps 6–8 hours, carried out over 2–4 days, with regular changes every 4–6 days thereafter.

Arrangements for the slow movement of water through large storage tanks may help to alleviate the build-up of mould growth, algae and biological 'slimes'. Very vulnerable materials, such as composite objects or those made of organic materials, may require slower levels of dilution, as they may be subject to collapse under 'osmotic pressure', although this may be somewhat rare. Once again, if in doubt, conservation advice should be sought at the earliest opportunity.

Damp artefacts (i.e. those neither waterlogged nor dry), sometimes found in deep sea-bed deposits and some of the most vulnerable to damage post-excavation, should be kept in conditions as similar as possible to the original context. They should be recovered in associated deposits and be kept damp rather than topped up with water but should not be allowed to dry out. Small quantities of water from the burial environment can be used in packing. All damp material should be 'double-bagged' in self-seal polyethylene bags and/or packed in lidded boxes with air excluded, refrigerated and sent for conservation and/or analysis as soon as possible (preferably within 24–48 hours following recovery). Materials and procedures for recovery should be checked with relevant specialists.

LIFTING, HANDLING AND TRANSPORTATION

The lifting, handling and transportation of artefacts is often complex and likely to require pre-planning and adaptation to suit the needs of specific projects. The approaches adopted may vary depending on a number of factors, including fragility, significance, location, size and mass of artefacts, project objectives, timescales, resources and arrangements for conservation. All team members should have a good understanding, in advance, of the systems and methodology to be used, and details of conservation facilities and environmental scientists to be consulted in relation to the retrieval of large or complex finds and structures.

When excavating material, all finds should be completely uncovered and exposed from the burial context prior to lifting, unless retention of the context (surrounding sediments or deposits) is important. Many waterlogged objects may appear to be robust despite being physically weak. Thus, pulling on, or attempting to lift an object where one end is still embedded, may cause it to snap. This applies equally to large objects (such as cannon) as to small finds. Finds that appear to extend downwards, and therefore, through several layers of context, should be discussed in advance. Their removal may often involve a compromise. For example, partially exposed wooden structures may be subject to environmental damage, while complete removal may leave holes in surrounding stratigraphy, with possible subsequent destabilization of the area.

- Finds should be moved slowly under water because the pressure exerted by the water may cause physical damage.
- Suitable procedures should be planned for the recovery of finds in the event that decompression stops are necessary.
- All strops on air-lifting bags should be cushioned at points of contact with finds using soft, inert materials. The numbers of points should be as extensive as possible in order to spread the load and avoid the creation of pressure points.
- Storage bins and containers of the appropriate size should be present on dive platforms, boats or barges, or the shore base.

Methods of support for finds during excavation and lifting, particularly those unable to support their own weight (figure 16.5), include the following (which may be combined):

- Flat sheets, such as inert seasoned wooden planking, steel baking-trays or polyethylene box-lids (all of which should be removed once the object is in storage).
- Self-seal plastic bags, which are useful for incorporating water and/or burial sediment from surrounding contexts (which may provide additional cushioning). The bags should not be too full of water and all air should be excluded on sealing.

Figure 16.5 Excavating a small fragile object. This leather water-bottle from the Armada wreck *La Trinidad Valencera* (1588) has been extracted by hand and placed on a supporting board with a sheet of muslin fastened to one end. When the object is in place the muslin will be secured around the remaining sides with the plastic clothes pegs at the foot of the board. (Photo: Colin Martin)

Figure 16.7 The leather water-bottle shown in figure 16.5 is brought to the surface and handed to the waiting conservator. (Photo: Colin Martin)

- Block-lifts in surrounding burial sediment/context, for fragile or large objects, may result in potentially heavy loading but may retain enhanced amounts of archaeological data.

The exposure to air of all finds raised should be kept to a minimum while handing finds from divers to personnel on barges/boats/platforms etc. Handing over (diver safety and surface conditions notwithstanding) should be undertaken as slowly and as gently as possible, given that potential loading and/or excess water run off at speed, may cause considerable damage, particularly to fragile or organic materials (figure 16.7).

Large objects: Large, heavy or complex objects, such as guns, hull components and associated structures, despite appearing robust, may be weak due to corrosion or decay and may be too heavy to support their own weight. Water may constitute much of the remaining mass, such as in waterlogged wood, which should be supported along the full length (e.g. by using planks of wood). Such finds may also have soft and fragile surfaces, retaining valuable archaeological detail (such as cut-marks) and so should be covered during lifting using a cushioning material such as bubble-wrap or polyether foam. They should be strapped using ties such as strips of bandage and splints, arranged so that they will not cut into the surface on lifting and also so as to ensure that as much of the loading as possible is carried on the supports (plate 16.3). Containers (such as boxes, barrels or amphorae) should be recovered in their entirety and lifted in an upright position (possibly using bandaging to help support overall structures), with the original contents

Figure 16.6 Raising a large organic object. This spoked wooden wheel from the Armada wreck *La Trinidad Valencera* (1588) weighs, with its associated concretions, about half a tonne. Because it is no longer structurally capable of bearing its own weight, a supporting frame has been built around and beneath it prior to lifting. It is seen here after its successful transfer to the conservation tank. (Photo: Colin Martin)

- Bubble-wrap, plastic string, or cotton ties such as archival tape.
- Purpose-built pallets or large trays with holes in the base, plus handling ropes, to support multiple boxes of bags or larger objects for transfer to boats or the shore. (figure 16.6).

retained *in situ*. Fine nylon netting, polyethylene sheeting or secured polyethylene bags can be used to cover vessel tops to help prevent the loss of contents from ancient storage containers. Removal of contents may lead to:

- collapse, due to lack of internal support;
- dilution, contamination and possible loss of contents, prior to analysis;
- poorer conditions for detailed examination and excavation in controlled conditions (content samples should normally be passed on to environmental specialists as soon as possible).

The use of air-lifting is likely for potentially heavy loads and all such lifts should be discussed and planned in advance with all team members. All such large finds should be immersed immediately in suitable containers, ideally containing water from their immediate environment. Storage containers of the correct size should be located on the dive platform, boat or barge, prior to lifting. The contents of ancient storage containers, in particular, may start to deteriorate in ordinary conditions, to the extent that the reliability of analytical results may become seriously questionable: such vessels should be sent as soon as possible following recovery to a conservator who will be able to remove the contents and send them for analysis while stabilizing the original vessel.

Fragile objects: The handling of waterlogged organics, including very fragile small finds, such as rope, textiles, bone or ivory, should be kept to a minimum. They should be recovered on rigid supports wherever possible, ideally incorporating some of the original context or sediment, using either encapsulation in self-seal polythene bags, or wrapping in thin polythene sheeting followed by gentle bandage strapping.

The packing of material for transportation should aim to:

- maintain a waterlogged environment and minimize the risk of drying, particularly where finds are not due to be unpacked for several days;
- incorporate packing materials as cushioning to protect against vibration;
- minimize the risk of undue pressure on objects or abrasion to surfaces;
- keep the amounts of water used to a minimum (to prevent objects from 'sloshing about' and to reduce overall weight);
- render objects visible (to minimize the necessity for unpacking and handling);
- exclude air wherever possible;
- incorporate clear permanent labelling with full details of senders and recipients and additional labelling such as 'fragile' or 'this way up', as necessary;

- pack samples according to the instructions of analytical specialists (plate 16.4).

Note that large waterlogged finds, even if well-wrapped in plastic, are likely to dry out even over a few weeks. This method should therefore be used only for transport, with finds being returned to wet storage immediately afterwards.

APPROACHES TO PACKING AND STORAGE

- The availability of water (and electricity) supplies should be checked well in advance of a project, including access, local daily or weekly restrictions and water-table height (which may affect the use of water stills and de-ionizers).
- All finds should be stored separately, according to the materials of which they are made. In particular, metals of different types should never be stored together in solution due to the risk of galvanic corrosion.
- Gloves should be worn when handling objects: vinyl, latex or nitrile (non-powdered versions are available, which are often useful for those with sensitive skin).
- All packing and storage materials should be made of conservation-grade inert materials and should be sourced through conservation suppliers wherever possible. They may include: lidded polyethylene (polythene) buckets, boxes and bins; self-seal polythene bags, bubble-wrap; Plastazote (expanded polythene foam, which floats in water) polyether foam (which sinks in water) and bandages (to help provide support for large fragile objects).
- Basic recording materials, such as pens, labels, ties, scissors, tools etc. should be available to all staff responsible for finds handling.
- Small finds can often be stored waterlogged in self-seal bags (double-bagged) within polyethylene containers prior to conservation.
- Large-scale plastic bags can be made up from conservation-grade plastic sheeting to the appropriate size, using a heat-sealer (budgets permitting) which melts the plastic and creates a welded seal.
- Non-bagged material should be stored in lidded, airtight polyethene boxes, filled to the top with water to exclude air and cushioned using, for example, polyether foam, bubble-wrap or Plastazote, to help to reduce the risk of abrasion against container lids, sides and bases. Delicate objects may require custom-made envelopes of bubble-wrap and an added layer of thin foam.
- Sections of plastic drainpipes, suitably cushioned, may be useful for providing support for long, thin delicate artefacts, such as pieces of wood or rope.

- Some types of object, such as quantities of wood or leather from specific contexts, may be stored together in bins, by tying off within sections of plastic 'tree' mesh tubing (available from garden centres) secured at each end and suitably labelled.
- Large objects can be stored in inert bins, drums, tanks or vats, which should be made, ideally, of polyethylene rather than polyvinyl chloride (PVC) or metal. Tall containers tend to be difficult to access, so long and low containers are preferable. Plugs or taps fitted in the base of containers can facilitate the replacement of storage water and are well worth the extra cost.
- Covering the top of storage water in tanks or bins (e.g. using bubble-wrap or inert polythene sheeting) may help to prevent the risk of freezing in winter, or evaporation in summer.
- Tanks or vats made of metal (e.g. used for the storage of ship timbers) should be lined (e.g. using polyethylene sheeting). All such lining materials are likely to require replacement from time to time due to potential weakening and/or the growth of algae or slimes.
- Where *in situ* preservation has been decided upon, such as for the storage of very large structural elements, then material might be returned to an environment simulated to be as close as possible to that from which it was excavated. Methods for storing large timbers might involve digging holes in the ground or in the sea-bed, using a variety of liners. However, storage in this way should be used only on a temporary basis, and conservation advice should be sought as necessary.
- Visitor access should be arranged so that crucial work is not interrupted, and should be limited to avoid excessive light and humidity in the storage area.
- Boxed small finds can be stored on shelving or racking, which should be made of metal rather than wood (thereby possibly helping to reduce risks of fire or insect damage).

SAMPLING AND ANALYSIS

Conservators work closely with objects and are likely to be able to help with identification and analysis. Methods of retrieval and requirements for packing and storing samples should be checked in advance with specialists. In general:

- The holding time for samples, prior to analysis, should be kept as short as possible (ideally no longer than 24–48 hours) because prolonged storage may affect the results obtained.

- Optimum environmental conditions for short-term storage should resemble the original burial conditions as closely as possible (e.g. wet/damp/dark).
- No water should be added to samples without first checking with specialists.
- The type of gloves to be worn, materials used, methods for collecting samples and storage requirements should be checked in advance with the relevant specialists.
- Smoking, eating and drinking should not be permitted in the vicinity of samples and/or artefacts due to be sampled because this may cause contamination (tobacco smoke, for example, is likely to affect radiocarbon dating results).
- Samples for despatch should be packed according to the requirements of customs authorities, airlines and postage companies and the analytical specialists awaiting receipt, particularly if samples are to be sent abroad. The importance of making such arrangements well in advance should not be underestimated, and all such material should be clearly labelled.

INITIAL CLEANING

Some initial cleaning may be necessary to remove macro-organisms (e.g. shrimp, barnacles, and seaweeds) so as to reduce the risk of biological decay during storage. If possible, all living organisms should be removed gently and returned to their original environment. Further cleaning may be necessary to remove sand and sediments and should be undertaken gently, in a controlled manner, using small water jets and soft brushes.

- Any substantive cleaning should be undertaken only as part of full conservation treatment because the rigorous washing and/or cleaning of finds may destroy evidence of remaining applied surfaces, decoration or working marks.
- The use of good lighting is essential. Illuminated magnifiers are often helpful.
- Tools used for cleaning should always be made of something softer than the material to be cleaned (e.g. wooden satay sticks for metals).
- All cleaning and sampling procedures should be recorded and passed on to the conservator(s) undertaking treatment.

HOLDING AND PRE-CONSERVATION TREATMENT SOLUTIONS

The addition of biocides and other chemicals to storage water should be used only as a last resort and on the advice of a conservator, particularly as some of them may affect other types of materials in the vicinity (such as composite objects). Proprietary household products, particularly those containing bleach, should never be used. It is often better to maintain finds in regular changes of cold water and to initiate full conservation treatment as soon as possible, ideally within days or a couple of weeks of recovery.

Chemical additives introduced to storage water – as part of 'first-aid' treatment – can be categorized generally as follows, and should be added only on the advice of an archaeological conservator:

- biocides and fungicides aimed at reducing biological decay;
- chemicals aimed at reducing potential corrosion; or
- pre-treatment solutions as part of full conservation treatment.

Biocides and fungicides may cause potential corrosion reactions or interfere with conservation treatment and long-term preservation. In addition, many are toxic and so solutions should be handled wearing gloves and other personal protective equipment, in accordance with the relevant health and safety regulations.

'Holding solutions' include the use of chemicals such as sodium sesqui-carbonate, caustic soda (sodium hydroxide), or benzotriazole (BTA), which are aimed at reducing corrosion in some metals on a temporary basis. However, they may interfere with subsequent analysis or conservation treatment.

Pre-treatment solutions are aimed at providing internal support and stabilization and are introduced to material in the wet state before controlled drying as part of conservation treatment. For example, polyethylene glycol (PEG) is often used in the treatment of waterlogged wood and glycerol is sometimes used in the treatment of various types of organic material. The efficacy of such treatments is likely to be dependent on material type, composition, construction, size and condition. They may also lead to potential damage to archaeological material if not used correctly (PEG, for example, may break down in contact with iron and iron-corrosion products, potentially resulting in organic decay).

RECORD-KEEPING

The design of finds registration and record systems is critical to the process of archaeological analysis and should be pre-planned carefully (see chapter 8). It should be designed to include information about storage type and location and any first-aid conservation carried out.

Other records relating to finds and their conservation should include:

- Exit and entry registers, recording dates material has left storage, whether temporarily or any similar information. Such records should be signed and dated by the relevant people.
- Monitoring records, which should be clear, well organized, and kept up to date with details of treatments administered or changes in the condition of finds. Simple tabulated forms may be helpful and all such details, however insignificant, should be passed on to the conservator involved, as the process of conservation documentation starts with excavation and recovery.
- Comprehensive systems for labelling containers and for recording the location of finds and samples.
- Hard-copy 'day-books' (diaries) of dive-related events.

There are additional requirements to bear in mind:

- Computers should be located away from wet working areas, and hard-copy back-ups kept in case of computer failure.
- Entries to registers should be made using clean, dry hands. Clean, dry towels and paper towelling should therefore always be made available for those involved in the handling of wet finds.
- Lists and/or notices should be made available to all those involved, giving contact details for emergency services, conservators, specialists and details of materials and equipment suppliers.

Photographs of finds are often taken for a number of different reasons, including publication, publicity or record photography for conservation (Dorrell, 1994). In terms of first-aid, shots taken immediately on raising material may help considerably in the long-term assessment of post-excavation changes: both traditional and digital photographic methods are applicable (see chapter 10). Material removed from storage containers for photography should be kept out of water for the minimum time possible and wetted regularly (e.g. by using handheld garden pump-sprays) in order to minimize potential damage. However, the need to keep objects wet during the process may result in distorted images due to the run-off of surface water. Alternative methods involve taking plan-view shots of objects in glass tanks, placed on a suitably coloured backing and filled with clean, still water. These may provide better records of shape and colours than photographs of objects removed from water. However, care should be taken to reduce possible light reflections from water surfaces (e.g. by angling cameras in the same plane with objects). In addition:

- a specific member(s) of a project team should made responsible for finds photography, with types of shot and lighting discussed and arranged in advance;
- care should be taken using hot lamps or studio lighting to avoid overheating;
- photographic 'publication grade' scales should be used in shots (rather than rulers, fingers or pencils), along with finds numbers;
- all photographs should be stored in an organized way;
- photographs of significant finds should be taken immediately following excavation;
- an overall photographic register should be kept, detailing shots taken for different purposes; and
- copyright settlements should be agreed with external organizations whenever appropriate.

X-RADIOGRAPHY AND FACILITIES

X-radiography is a non-destructive technique likely to provide considerable amounts of information and is widely used in the analysis and prioritization of archaeological material, particularly in the examination and recording of iron finds from terrestrial sites. X-rays can provide permanent records of objects immediately post-excavation and are likely to be important in the analysis and identification of marine iron finds, which are often particularly unstable and prone to extensive deterioration. The technique is also useful in the analysis of many other types of material, including organics and thin copper and lead finds. The selection of finds for X-radiography should be undertaken with the help of an archaeological conservator, ideally previously contracted to help with project work. (Note: the use of industrial X-ray equipment, such as in hospitals, is not always a suitable alternative (Sutherland, 2002).) In addition to providing other information, such as constituting useful 'maps' for dissembling concretions or in cleaning as part of conservation treatment, X-rays may reveal the following (English Heritage, 2006a):

- details of manufacturing technology and construction;
- the shape and identity of artefacts (e.g. within concretions);
- the extent of corrosion and/or remaining metal cores;
- evidence of markings, applied surfaces and preserved organic remains.

HEALTH AND SAFETY

- Full health and safety procedures should be implemented in advance of any project, in accordance with the relevant laws and regulations.
- It is recommended that all personnel should be fully informed of procedures as part of their induction. Briefings or information packs may be helpful, including information on the signing of relevant forms.
- Visitors accessing first-aid conservation facilities should be made clearly aware of relevant procedures in the event of emergencies (e.g. fire exits and evacuation).
- All buildings, equipment and equipment-related procedures should be risk assessed.
- It might be necessary to train personnel in the use of specific equipment (e.g., the use of hoists to assist manual lifting).
- Within the UK, any procedures involving chemicals should be risk assessed under COSHH (Control of Substances Hazardous to Health), requiring the compilation of COSHH forms by the user (Health and Safety Executive, 2002). The suppliers of chemicals or other such substances are required to provide data-sheets, which can then be used in the completion of such forms. All personnel involved in handling chemicals should be informed of COSHH procedures and made aware of the appropriate files, their location and associated information.
- All risk assessments and COSHH forms should be signed by those in authority.
- All personnel likely to be handling finds should have up-to-date immunizations (particularly for tetanus).

INSURANCE

All project premises should be fully insured in accordance with the relevant regulations. Full public-liability indemnity is likely to be required, particularly if visitors are allowed onto project premises. Insurance may also be required for the transportation of objects to and from premises. If in doubt, professional advice should be sought.

CHECKLISTS

The following lists, although not intended to be comprehensive, may be helpful to those involved in project work, including those involved in small projects, with minimal budgets. It is often cheaper and more effective to start the planning and acquisition of equipment, materials and supplies well in advance.

Project planning

The following issues should be addressed in advance of project work:

- Arrangements need to be made with professional archaeological conservation and specialist analytical facilities (including details regarding transportation and insurance).
- In the UK, the Receiver of Wreck (Maritime and Coastguard Agency) must be notified of any 'wreck' material, recovered or landed in the UK. This is a legal requirement.
- A location should be designated for 'first-aid' facilities and for the storage of finds.
- Finds recording and registration systems need to be established.
- Potential sources of funding should be identified in the event of unforeseen finds (particularly, provision for large objects, such as guns or ship structure).
- A receiving depository, museum or archive needs to be confirmed for material recovered.
- Provision should be made for displays relating to the project, particularly within local communities and museums.

First-aid storage facilities

'First-aid' facilities for finds should:

- comprise a lockable and secure room or building, located close to the site or the project headquarters;
- have a working area that is clean, light and airy, and a storage area with reduced light levels;
- be sufficiently large to provide space for the accessioning, storage, recording and illustration of finds;
- be separated from the main project office, equipment storage, recording and documentation areas;
- be fitted with smoke-detectors, fire-extinguishers and any other appropriate fire-protection measures, which should be serviced regularly;
- incorporate electricity and water supplies, with sinks, good lighting and ventilation.

The first-aid facilities should also have:

- good access for both visitors and finds, with access ramps for large finds;
- non-slip floors (preferably with drainage);
- tables or benches with waterproof surfaces;

- refrigerators for the storage of finds and samples (but not for food);
- ample space for storage containers, preferably with drainage systems;
- robust shelving;
- pest-monitoring systems;
- a relatively stable environment, including heating in winter;
- allocated areas for the safe storage of chemicals;
- burglar alarms (if considered necessary);
- emergency and conservation-related contact details (with a first-aid kit and accident book on site).

Additional measures might include:

- a water still or de-ionizing column;
- hoist and pulley systems for moving large artefacts;
- an incoming finds registration and evaluation area;
- allocated wet and dry work areas;
- a storage area for packing materials;
- a dry post-conservation storage area for the long-term storage of finds.

Materials and supplies

Ideally these will be sourced from conservation suppliers and should include:

- small tools (stored in a tool box and accessible to those handling finds);
- paper towels and regular supplies of clean towels for drying hands;
- polyethylene (polythene) string, book-binding ties or archival tape (available from conservation suppliers) and stretch and non-stretch bandages (potentially useful in lifting);
- Tyvek (spun-bonded polyester) waterproof labels (with or without punched holes);
- conservation-grade acid-free tissue, bubble-wrap, polyether foam, polyethylene sheeting and self-seal plastic bags, ideally with write-on label areas;
- polyethylene foam ('Plastazote'), available from conservation suppliers in varying thicknesses, in black or white (often useful for back-drops in photography and in finds packing, examination and support);
- gloves – vinyl, latex or nitrile (for compliance with health and safety regulations: e.g. neoprene or butyl rubber gloves should be worn when handling biocides and/or fungicides) or household rubber gloves (for heavy-duty work such as in changing storage water) – and check sizes and types with those expected to wear them (some personnel may require non-powdered or non-latex);

- inert polythene containers or glass tubes for sampling (check with specialists);
- 'tree tubing mesh' (available from garden centres);
- garden pump-sprays (made of polypropylene or polythene);
- Melinex (fine polyester sheeting, often available in different thicknesses and useful for a variety of purposes);
- strops for lifting, together with pads of inert foam cushioning (e.g. polyether foam);
- watering cans (plastic), hoses, hose fittings and reels for topping up storage water;
- waste bins or dustbins;
- small step-ladders;
- inert plastic storage tanks and bins (lidded, with drainage taps/plugs where possible);
- lidded polythene containers and trays for wet storage;
- clean (new) flat rigid baking trays (useful for lifting);
- plastic drainpipes, halved along their length, for supporting long objects;
- torches (with spare batteries) for viewing objects in storage containers and bins;
- glassware (beakers and measuring cylinders);
- large magnifying or examination lenses (or stereo-microscope on extendable arm);
- thermometer(s);
- cameras, photographic scales, 'back-drop materials' and associated equipment;
- glass tanks and smaller glass containers (for finds photography and they are useful in the examination of finds);
- heat-sealer (budget permitting) for making up plastic tubing or bags;
- laminator (useful for waterproofing signs for bins, storage bays, notices, etc.);
- archival stationery, including field record books, registers, exit/entry books, pens, registration cards, printouts and monitoring records for use in everyday work;
- computers, software and printers for registration records, as required.

Packing materials to avoid include:

- plastic or glass food containers, unless thoroughly washed and rinsed;
- polystyrene 'worms' (used in household packing), which may contain chemicals that can leach into objects or packaging;
- household string, which may rot in wet conditions;
- aluminium trays, which may bend when laden with water and/or cause reactions in finds;

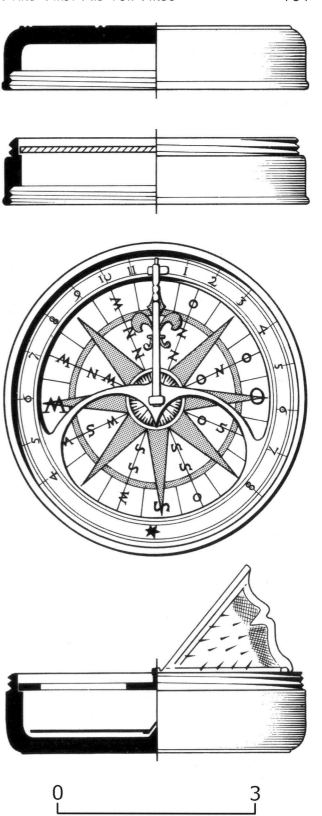

Figure 16.8 This tiny pocket sundial with integrated compass was found on the wreck of the *Kennemerland* (1664). Its extraction from concretion was a triumph of the conservator's skill and patience – even the painted cardboard compass card was intact. (Colin Martin)

- ordinary brown packing papers, paper labels, newspaper, cardboard and cardboard boxes, which are all likely to fail if damp and may be subject to insect attack;
- rubber bands and general household sticking tapes.

FURTHER INFORMATION

Dorrell, P., 1994, *Photography in Archaeology and Conservation.* Cambridge.

English Heritage, 1996, *Waterlogged Wood: Guidelines on the Recording, Sampling, Conservation and Curation of Waterlogged Wood.* London.

English Heritage, 2006a, *Guidelines on the X-radiography of Archaeological Metalwork.* Swindon.

Health and Safety Executive (HSE), 2002, *The Control of Substances Hazardous to Health Regulations* (Statutory Instrument 2002 No. 689). London.

Institute of Field Archaeologists, 2001b, *Standards and Guidance for the Collection, Documentation, Conservation and Research of Archaeological Materials.* Reading.

Neal, V. and Watkinson, D., 1998 (3rd edn), *First Aid for Finds.* London.

Robinson, W. S., 1998 (3rd edn), *First Aid for Underwater Finds.* London and Portsmouth.

Sutherland, A., 2002, Perceptions of marine artefact conservation and their relationship to destruction and theft, in N. Brodie and K. Walker Tubb (eds), *Illicit Antiquities: The Theft of Culture and Extinction of Archaeology.* London.

WEBSITES AND ORGANIZATIONS

The American Institute for Conservation of Historic and Artistic Works (AIC) (http://aic.stanford.edu).

Canadian Conservation Institute (www.cci-icc.gc.ca).

The Conservation Register (www.conservationregister.com).

Getty Conservation Institute (www.getty.edu/conservation).

The Institute of Conservation (ICON) (www.icon.org.uk).

International Centre for the Study and Restoration of Cultural Property (ICCROM) (www.iccrom.org).

International Committee on Monuments and Sites (ICOMOS) (www.international.icomos.org).

International Council of Museums (ICOM) (http://icom.museum/).

The International Institute for Conservation of Historic and Artistic Works (ICC) (www.buildingconservation.com).

Site Monitoring and Protection 17

MONITORING

Monitoring of submerged archaeological sites involves observing, surveying and sampling sites to detect signs of both short- and long-term change. A monitoring project can be of short or long duration, depending on what questions are being asked about the site or sites involved.

Sites are monitored to:

- *Find out about how sites are formed:* The processes that form submerged archaeological sites are poorly understood. Monitoring a site not only helps to answer questions about how that site has been formed and reached its current state, but also contributes to a wider understanding of the formation and subsequent behaviour of submerged sites in general.
- *Observe and understand the processes affecting the condition of sites:* Very few sites are completely stable. Most are changing in some way and the degree of change can be relatively slow, rapid, dramatic or subtle. Observing and measuring these changes helps archaeologists to understand the complex processes that aid or threaten the preservation of a site. Such understanding can enable predictions concerning the effect of future changes on a site (plate 17.1).
- *Establish whether or not protection is required:* Data collected in pursuit of the previous two points will result in a more informed decision-making and site-management process. This enables archaeologists

to identify if, when, and what type of active management might be required. It can help determine whether or not a site needs to be stabilized or artefacts recovered.

Monitoring of all submerged archaeological sites is impracticable because of the enormous number of sites and the limited resources available. Despite a number of well-publicized initiatives, such as the 'Adopt A Wreck' project run by the NAS (see NAS website), monitoring of underwater sites is likely to remain highly selective.

It is not possible to give comprehensive advice on when sites should be monitored, because requirements will depend on local and national priorities and the perceived importance of individual sites. However, monitoring should definitely be considered in the following circumstances:

- following any intrusive investigations;
- where significant instability is suspected; and
- for sites considered to be of national or international significance regardless of their perceived stability.

The type and amount of data required in a monitoring programme will depend on the reasons for monitoring the site in the first place, and also on:

- whether an individual site or a whole landscape is being monitored;
- the location;
- the level of resources available; and

- the length of time that monitoring is likely to be required for.

As a result the scale and complexity of monitoring projects will vary greatly. The following examples indicate a range of projects and the scale of work that might be involved:

- *Investigation of suspected human interference on a legally protected site:* This may simply involve a single visit to the site to check for signs of interference. Alternatively the site may be checked at regular intervals. The work involved is unlikely to go beyond visual survey unless signs of interference are found.
- *The completion of a civil engineering project, leading to fears that nearby submerged archaeological sites are being affected by erosion:* In this case the work required may involve a single visit to each site to check for changes, with visual survey recorded using video or still photography and measurements. A repeat visit may be required to check for further or longer term change. If long-term effects are likely, further visits at regular intervals may be necessary. Coring or test-excavation may be required if little is already known about the sea-bed and stratigraphy. Geophysical survey may be beneficial.
- *A funded research study of site-formation processes on an important site:* In this case the full range of scientific monitoring techniques may be deployed, including sampling and experimental work, with repeated visits on a long-term basis.

The role of biological survey should not be underestimated in site-monitoring. The survey and sampling of marine-boring organisms, bacteria and fungi can be vital in understanding the way that sites are changing. For example, some species of marine growth can provide information on the currents affecting the site and may avoid the need to employ more expensive techniques such as data-loggers. At the Duart Point site in Scotland the distribution and age of organisms such as barnacles has given important information about the loss of sea-bed to erosion (Martin, 1995a).

To ensure that data is obtained and used in an organized and effective way, it is essential to have a plan for monitoring work. Project planning has already been discussed in chapter 5. However, there are a few issues specifically relating to a monitoring project that will be considered here.

The desk-based assessment (see chapter 5) for a monitoring project should ensure that there is a clear understanding of:

- what the shipwreck material is composed of and how it is distributed;
- what the sea-bed is composed of and how it is likely to be affecting the shipwreck material;
- the biology of the site (e.g. marine growth and any marine (wood) borers present and how they are affecting the shipwreck material);
- the water movement on the site (currents and waves);
- the water itself (e.g. its salinity and pH); and
- outside factors, such as human activity, which may be affecting the site.

Visual survey: The simplest way to collect monitoring data is by simple visual survey using divers. This type of survey is sometimes called a 'general visual inspection'. It requires the divers to simply swim around a site and then record their observations either during the dive or immediately after it. It relies for its effectiveness on divers being able to recognize changes in the condition of a site between visits or over multiple visits. It is therefore carried out most effectively by divers who are already familiar with the site and who have a reasonable technical knowledge of it. It also tends to be most effective when the divers follow a fixed route around the site and, if the survey is to be repeated at intervals, the same divers are involved.

However, visual survey alone is of limited value in monitoring archaeological sites under water. Although the human eye is a very sophisticated observational tool, the information that can be derived from it depends entirely on the often-variable knowledge and powers of observation and recording of the divers involved. Its highly subjective nature means that visual survey is best supported by other techniques.

Sketch plans are probably the simplest method of providing supporting information for visual survey. Provided that the diver is a reasonably competent draughtsperson, a considerable amount of information that is difficult to describe in a written or verbal format can be recorded. The results, however, are still likely to be subjective.

Video and still photography: These are very important methods of supporting diver observations. In the right conditions, photographic images can provide far more information than can be gained from the verbal or written descriptions of divers. They have the advantage that they allow a site to be studied remotely, both by divers and non-divers, without having to revisit it. In addition, photographs can be used to identify errors in descriptions given by divers. Photographs taken at different times are also generally more easily comparable than diver descriptions. As with visual survey, the use of photography to monitor sites requires multiple site visits over an extended period.

Video and/or still photography should therefore be used in all situations where the equipment is available and the environment, such as underwater visibility, allows. For still photography, the project design should identify what needs to be photographed and from what viewpoint. All photographs should incorporate some form of easily read and consistent scale and if the photographs are repeated at a later date, they should be taken from the same viewpoint. Similarly, the project design should identify what video footage is required. Great care should be taken to ensure that the footage shot is steady and slow. If practicable, video footage should also incorporate scales.

Where conditions allow, the possibility of taking photo-mosaic and panoramic photographs should be considered. Unless they are also to be used for measured survey, photo-mosaics used for monitoring do not usually need to be precisely scaled and can be prepared using simple 'swim over' techniques.

For more detailed information about obtaining, collating and storing photographic records, see chapter 10.

Taking measurements: Measurements are a valuable method for identifying and quantifying change, provided that they are taken accurately and in a way that is repeatable. Examples of the way in which measurements can be used to monitor sites include:

- Measuring changes in the depth of burial for a site. If a site is buried or partially buried, the depth of burial can be established by probing at fixed points. Alternatively, the vertical distance between a fixed point and the sea-bed surface can be measured. This fixed point could be a nail attached to an exposed archaeological feature or a stake or spike driven securely into the sea-bed.
- Using an inclinometer to measure the angle of the plates or frames of a metal wreck to detect signs of movement.
- Counting the number of species or individual specimens and measuring the size of marine growth in sample areas.
- Using a depth-gauge to take measurements across the sea-bed. When compared with previous results, changes in the shape of the sea-bed can be detected.

If it is intended that repeat measurements should be taken over multiple visits to a site, then the position of these 'measurement points' should be marked in some way to ensure that they can be relocated and the measurement accurately repeated. This is particularly important if the work is to be carried out by different members of a group or by a different group altogether.

Sampling: This can take many forms. For example, sea-bed sediments can be sampled and identified to give information about the vulnerability of a site to erosion, or the chemical and biological environment in which artefacts are buried. Examination of timber samples can give information about wood-boring organisms and other biological threats. They can also provide important information about how long exposed wooden artefacts are likely to survive. Sampling of marine growth will allow different species to be identified precisely. This can help determine which parts of a site are in high- or low-energy environments, which in turn can highlight areas of particular vulnerability. Sampling should only be considered when the information sought cannot be obtained from other, non-intrusive, methods. The benefit that is likely to be obtained should always be weighed carefully against any damage that may result.

Non-destructive testing (NDT): A sophisticated industry has developed around the regular inspection of large metal structures in the offshore oil and gas industry. Some of the techniques used in this industry have been applied or adapted to the monitoring of metal shipwrecks and other metal artefacts by archaeologists.

Measurement of corrosion potential (CP) can be used to provide information about the corrosion history and vulnerability of metal artefacts. It has been used to study both large twentieth-century metal wrecks and concreted iron ordnance. In some cases, the technique can be used to assist conservators to identify *in situ* conservation methods, such as the attachment of sacrificial anodes. Although the application of this technique requires a certain amount of specialized knowledge, the equipment is readily available or can be constructed at modest cost and CP measurement should be within the capability of most reasonably committed projects.

Similarly, ultrasonic thickness measurement (UTM) can be used to monitor the thickness, and therefore corrosion, of metal structures such as iron or steel plates. Again the equipment is readily available.

Geophysical survey: Equipment such as side-scan sonar and multibeam swath bathymetry can be used to measure the depth and shape of the sea-bed and to detect the presence of archaeological material. It can therefore be used to monitor changes over time (e.g. the movement of sandbanks over sites and the erosion or deposition of protective layers of sand and other sediment).

Geophysical survey (see chapter 13) has traditionally required 'ground-truthing' by divers or remotely operated vehicles to check and interpret the results. However, as both the quality of data and the interpretative skills of archaeologists have improved, the need for complete ground-truthing is gradually diminishing. This trend is likely to continue. Work that might take a diving team weeks or even months can now be achieved in a few hours using geophysical equipment. As a result, this type of survey is likely to be used increasingly for monitoring in both government- and developer-funded projects. Nevertheless, it should not be regarded as a complete solution or one that works equally well on all sites. It is therefore unlikely that the need for ground-truthing will be eliminated in the foreseeable future.

The effectiveness of this technique for monitoring largely depends on the detection of changes between surveys. Therefore, more than one survey, separated by an appropriate period of time, is required (plates 13.7 and 13.8). The high cost, particularly of multibeam equipment, and the specialized knowledge required to operate it and interpret the results, means that its application to projects with modest funds is likely to be limited. Nevertheless the possibility of its use should always be considered.

Other, readily available equipment can be used for monitoring work and require less specialist skills (e.g. echo-sounders). When combined with suitable differential global positioning system (see chapter 11) equipment, echo-sounders can provide detailed information about changes in sea-bed profile around and within a site.

Hydrodynamic environment: Water temperature, pH and salinity measurements can be taken relatively easily for a modest cost. If sufficient funds are available, data-loggers can be deployed to measure current-strength, direction, suspended solids and other factors likely to influence site stability (plate 17.2). Alternatively, information on the hydrodynamic environment can be gained from the type and distribution of marine growth and often from other public or private sources such as hydrographic agencies. Liaison with other groups or organizations studying this environment can provide valuable information and technical input and is therefore always advised.

Other techniques: Experimental techniques can be very useful. Projects such as MOSS (Cederlund, 2004) have demonstrated that valuable information on the degradation of wood can be derived from experimenting with buried and exposed sacrificial test objects on site (plate 17.3). Such research can reveal much about site formation and stability. Other projects, such as recent investigations on the wreck of the late seventeenth-century warship *Hazardous* in the UK (Holland, 2005; 2006) have shown how the use of 'tracer' objects can provide information on burial processes and on the mobility of artefacts on the sea-bed. These techniques can be relatively cheap to use and many do not require specialist technical knowledge. They do, however, usually require multiple visits to the site.

Specialist knowledge may be required to understand the biological environment of a site and to devise a method for monitoring it, but advice is often available at little or no cost. It is unlikely to be beyond the resources of a reasonably committed project, particularly if it involves a multidisciplinary approach.

Many of the techniques described above rely on the comparison of data gathered during more than one visit to a site. The initial work will provide the 'baseline' data against which subsequent data sets will be compared. It is therefore essential that the way in which the baseline data is obtained is very carefully recorded and that subsequent data is obtained in the same way or in a way that is comparable. Careful thought should be given at the project-planning stage to ensure that this repeatability can be achieved.

Careful thought should also be given to ensuring that the work is repeated at an appropriate time. For example, if it is suspected that the profile of the sea-bed is affected by seasonal storms, then the site must be visited before, during and after the appropriate season. By way of contrast, if it is thought that the site is only subject to very slow long-term or negligible change, then it may only be necessary to repeat the work every year or perhaps even less frequently.

The requirement to repeat monitoring work at intervals emphasizes the need to keep the work as simple as possible, particularly if resources are limited. If the work is to be shared by a number of groups or individuals, then repeatability is even more essential. Proforma style recording is a sensible way to approach such situations.

It is not possible to give comprehensive advice on interpreting monitoring data in a book such as this because of the limitless permutations of processes affecting individual sites. However, the following are examples of what might be considered:

- *Surface distribution of material:* The movement or disappearance of artefacts on a site can indicate many processes. Missing non-ferrous metal artefacts are usually taken as good indications of human interference. However, the movement or disappearance of artefacts can be due to a wide range of factors, including erosion or natural burial and care should be taken in interpretation.
- *Changes in the shape of the sea-bed:* A sea-bed that changes shape is probably being affected by erosion and the site may, therefore, not be stable. Analysis of the monitoring data may help to determine whether this is a natural repeating cycle or a single event and whether active intervention is required to stabilize and protect the site. Changes in the shape

of the sea-bed may also explain the disappearance of artefacts.

- *Deterioration of artefacts:* The manner and rate of artefact deterioration should be considered. This can be assessed through, for example, the comparison of close-up photographs taken at regular intervals or by the close visual examination of artefacts, perhaps looking for the effects of marine wood-boring organisms. Such assessments may help to determine whether the recovery of the artefact or *in situ* protection is required. Deterioration may be due to a number of factors, including erosion, biological damage and human interference. Rapid deterioration may require active intervention to either remove the artefact or protect it *in situ*.
- *Corrosion potential:* High or fluctuating corrosion potential may mean that an artefact needs to be recovered or stabilized *in situ*.

The project plan should provide for the storage, in one place, of all the data and documents (including video and still images) collected during the monitoring project. This is called the project archive (see chapter 19).

It is incumbent upon the organizers of a monitoring project to ensure that sufficient time and resources are allocated to publication of activity and that publication is undertaken to the recognized archaeological standards of the country concerned, or to internationally accepted standards if no national standards exist (see chapter 20).

Monitoring reports and the data on which they are based are likely to become increasingly available through local and national archives. Innovative web-based publication projects using geographical information systems, such as the pan-European MACHU project (www.machu-project.eu/), may well transform the accessibility of both data and technical expertise.

PROTECTION

Protection is a physical or other *in situ* intervention that results in the slowing, halting or reversal of a process that is believed to be having a negative impact on an archaeological site. It might therefore best be described as 'stabilization'. Although attempts have been made to protect sites for many decades, until recently the study of *in situ* archaeological protection under water has received little attention and is therefore still relatively poorly understood. The legal protection of a site from the adverse consequences of human activities is considered in chapter 7 and so has not been dealt with here.

Protection should be considered for sites where a monitoring programme has shown that the condition of a site is deteriorating significantly or where serious in-

stability is otherwise apparent. Typical circumstances might include instances where:

- vulnerable archaeological material has become exposed or is deteriorating and it is undesirable or impracticable to recover it (e.g. if a fragment of wooden hull became exposed, which was unique or had important characteristics, but which was too large to recover);
- archaeological material has been deeply buried but the depth of burial is no longer great enough to prevent its condition from deteriorating (e.g. if it is no longer in an anaerobic environment);
- short-term protection is required (e.g. if a site that is normally buried is uncovered by an exceptional storm event and reburial by natural processes is uncertain).

Protection measures should only be considered where sufficient resources, both time and money, are available or are very likely to become available. Long-term monitoring may be required and a project design-led approach to protection, similar to that required in monitoring work, should be adopted. As with monitoring, best practice would involve thorough recording and the publication of work carried out.

A full discussion of all of the techniques that have been used for the *in situ* protection of shipwreck sites is beyond the scope of this book. Nevertheless, a few examples can be given to show the range of options that can be considered.

Re-burial: If a site has been destabilized because sand or other overburden has been removed from it, then simple re-burial can be considered. This is, however, likely to be successful only in the short term, unless the process that resulted in the exposure is no longer active. In the absence of any means of stopping the destructive process, re-burial with a more resistant material could be considered (e.g. gravel instead of sand). The physical and chemical impact of this different material on the site would have to be considered, together with the risk of intrusive wreck material being added to the site. The latter may be particularly significant if dredged sand or gravel is used to cover the site and it is known to come from an area of high maritime activity, such as the approach channel to a port.

The cost of re-burial can be significant, particularly for a large site, unless a partnership can be forged with an existing commercial operation.

Sandbags: Sandbags can be used to cover a whole site or, selectively, to reinforce or cover individual or groups of artefacts. They are resistant to currents and, to a

Figure 17.1 A covering of sandbags placed over a fragile area on the Duart Point wreck. (Photo: Colin Martin)

moderate extent, the force of waves. Sandbags can be successfully used to fill in scours or excavation trenches and to 'weigh down' artefacts, such as small fragments of wooden hull that may be prone to disturbance. They are best positioned so that each bag is touching or preferably overlapping, so that the gaps between them fill with sand or other mobile sea-bed material (figure 17.1). They are less effective when used individually or in small groups and should not be regarded as a permanent or 'fit and forget' means of protection because the fabric, whether man-made or natural, will decay. Unless the sandbags are subsequently buried under overburden, they will also require regular inspection.

Sandbags should be laid so that they promote the smooth flow of the current across the site. This will reduce the risk of the sandbags themselves causing erosion by creating current eddies in the immediate vicinity or elsewhere on the site. It will also increase their useful lifespan. If sandbags are to be laid across a whole site, they should be closely packed and have a profile that is as low and smooth as possible. Examples of sites where sandbags have been laid to prevent erosion include the fifteenth-century Studland Bay site off the Dorset coast of England and the Duart Point site in Scotland. At Duart Point, sandbagging has been used, after partial excavation, to encourage reinstatement of the original sea-bed surface (Martin, 1995b).

Deployment of sandbags can be a time-consuming process and planning should ensure that it is carried out as quickly as possible. Slow or partial deployment could actually promote erosion rather than reduce it. It is also a common mistake to underestimate the number of sandbags that will be required.

Geotextiles: Membranes of artificial textiles can also be used to cover a site and promote stability. A wide variety of materials are available and the choice ranges from rolls of weed-inhibiting textiles, available from gardening suppliers, to more complicated textiles used by the civil engineering industry for stabilizing vulnerable surfaces, both under water and on shore. These textiles work by providing a smooth, continuous barrier across the sea-bed, which promotes a stable, often anaerobic, environment beneath it. Textiles manufactured specifically for a sea-bed environment are often also designed to promote the re-deposition of sediment where they are laid. For example, they may be manufactured with a layer of 'fronds' designed to slow the current that comes into contact with them, causing sand or silt in suspension to fall to the sea-bed.

Geotextile sheets are often large and cumbersome and can be very difficult to lay. Securing a geotextile cover to the sea-bed takes care and thought and it is likely to require regular monitoring. This is particularly significant in busy areas for shipping or recreational use, as the geotextile is likely to represent a significant hazard if it becomes loose. Geotextile sheets can be expensive, particularly for large sites. Nevertheless they can be very effective and their use is therefore likely to increase. Although Terram is usually cited as the example of a geotextile, there are many different types available. The possibility of setting up an experiment to determine which is likely to be most suitable, either close to a site or in a similar environment, should therefore be considered.

Anodes: Work at sites such as Duart Point has shown that anodes can be used to help stabilize metal artefacts, such as iron guns, on the sea-bed (MacLeod, 1995). An added benefit is that if it is anticipated that the artefact concerned will eventually be recovered, then the use of anodes may allow part of the usual conservation treatment to be undertaken while the artefact is still on the sea-bed (plate 17.4). However, the use of anodes requires some specialist knowledge, involves a long-term fieldwork commitment and can be fairly expensive. It is therefore only likely to be attractive if the artefact concerned is considered to be exceptionally important, or if it is planned to recover it subsequently.

In certain circumstances *in situ* protection may be impracticable or ultimately prove unsuccessful. In such circumstances, the recovery of vulnerable artefacts may be the only practicable means of ensuring their survival, even if recovery runs counter to the prevailing heritage-management policy. The possibility of this situation arising should therefore be considered in the project plan and contingency arrangements allowed for. This should always be done in consultation with the relevant curator and conservator. If an owner of the material concerned has been identified, then that person or body should, of course, be involved at an early stage.

FURTHER INFORMATION

Archaeological State Museum of Mecklenburg-Vorpommern, 2004, *Management plan of shipwreck site Darss Cog, MOSS Project*.

Camidge, K., 2005, *HMS Colossus: Stabilisation Trial Final Report*. Unpublished report for English Heritage. www.cismas.org.uk/docs/colossus_stab_trial_final.pdf

Cederlund, C. O. (ed.), 2004, *MOSS Final Report*, web-published final report (www.nba.fi/internat/MoSS/download/final_report.pdf).

Det Norske Veritas Industry A/S, 2006, *Recommended Practice RP B401: Cathodic Protection Design*. Laksevag, Norway.

Gregory, D. J., 1999, Monitoring the effect of sacrificial anodes on the large iron artefacts on the Duart Point wreck, 1997, *International Journal of Nautical Archaeology* **28**.2, 164–73.

Gregory, D. J., 2000, *In situ* corrosion on the submarine *Resurgam*: A preliminary assessment of her state of preservation, *Conservation and Management of Archaeological Sites*, **4**, 93–100.

Grenier, R., Nutley, D. and Cochran, I. (eds), 2006, *Underwater Cultural Heritage at Risk: Managing Natural and Human Impacts*. ICOMOS.

Holland, S. E., 2005 and 2006, Following the yellow brick road. *Nautical Archaeology Society Newsletter* **2005**.4 and **2006**.1: 7–9.

MacLeod, I. D., 1995, *In situ* corrosion studies on the Duart Point wreck, 1994, *International Journal of Nautical Archaeology* **24**.1, 53–9.

Martin, C. J. M., 1995a, The Cromwellian shipwreck off Duart Point, Mull: an interim report, *International Journal of Nautical Archaeology* **24**.1, 15–32.

Martin C. J. M., 1995b, Assessment, stabilisation, and management of an environmentally threatened seventeenth-century shipwreck off Duart Point, Mull, in A. Berry and I. Brown (eds), *Managing Ancient Monuments: An Integrated Approach*, 181–9. Clywd Archaeology Service, Mold.

Palma, P., 2005, Monitoring of Shipwreck Sites, *International Journal of Nautical Archaeology* **34**.2, 323–31.

18 Archaeological Illustration

Contents
- Basic drawing equipment
- Drawing archaeological material
- Recording 'by eye'
- Recording decoration and surface Detail
- Recording constructional and other detail
- Post-fieldwork photography and laser scanning
- Presenting a range of complex information

The old adage that 'a picture is worth a thousand words' is particularly appropriate to the field of archaeological recording. Drawings and illustrations of various types provide a convenient way of conveying a great deal of information very quickly. For detailed information about archaeological illustration, the reader is referred to existing literature (see the Further Information section at the end of the chapter). This chapter offers basic advice about equipment and techniques for small-finds recording and the range of different methods for presenting information in archaeological drawings. Information on the recording of larger finds, particularly appropriate to underwater archaeology, is provided in appendices 1 and 2.

BASIC DRAWING EQUIPMENT

There are several volumes available that deal comprehensively with the options in terms of drawing equipment (e.g. Green, 2004 and Griffiths et al., 1990): therefore, only a brief guide is offered here. A few basic items are all that is required to start with. An archaeological drawing 'toolkit' can be accumulated over time and this is sensible because the range of equipment required is dependent on the type of drawing undertaken. A list of initial purchases might include paper/drafting film, pencils, drafting pens, erasers, measuring equipment, etc.

The smoother the drawing surface the smoother the lines that can be drawn. Drafting film (permatrace) has a very smooth surface but it is expensive. Paper is available in a wide range of thicknesses. In preference, choose a medium weight, double matt smooth paper. Thin, 'bristol', board can also be used. If in any doubt, the staff of a shop specializing in equipment for graphic art and technical drawing should be able to offer plenty of advice. CS10 paper is favoured by many illustrators because it is ideal for artwork and has a very smooth surface that can be carefully scraped to remove mistakes.

It is always worth asking about the long-term stability of the drawing materials on offer (ink, paper, drafting film) and the suitability of the ink and pens for the drawing surface. Ink that fades quickly and paper that readily tends to yellow and distort should be avoided. Tracing paper is particularly unstable.

Having pencils in a range of hardnesses should allow for work on drafting film as well as paper (4H to HB should cover most situations). Note that when working on drawing film, a 4H can behave more like a 2H. Mistakes will occur, so accept the inevitable and buy a good eraser as well.

Drafting pens, also known as technical pens, are available in a range of sizes. The 0.35 mm nib is a frequently used size for many outlines, although 0.5 or 0.7 mm may be required for larger drawings that will be greatly reduced for publication whereas a 0.25 mm nib may be used to outline a small drawing that will not be reduced.

In general, detail can be drawn with a 0.35 mm or 0.25 mm nib, with a 0.18 mm nib used for fine detail. Remember, however, that if the drawing is going to be reduced by more than 50 per cent, the finer lines are likely to be lost.

Specialised ink erasers are available. Choose the type suitable for the ink and drawing surface being used. It is common practice to use a scalpel with a rounded edge (no. 10) to scrap off mistakes on drafting film. Special drafting powder is then used to buff the surface prior to inking.

Measuring equipment is important because some means of taking accurate measurements is required. Dividers and rulers are also basic requirements and are available in a wide variety of forms. Vernier callipers are very useful (essential, in fact, for drawing ceramic vessels) and should be available whenever possible.

Most drawings will require some form of labelling or annotation. For registration and record purposes, handwritten text will often suffice. For publication this is rarely acceptable, unless expertly executed. Lettering stencils are widely available and, although they take practice to use well, they are cheap. Label printers are also readily available: they allow text and captions to be typed onto clear adhesive labels that can be stuck onto the drawing or plan. A word-processor attached to a high-definition printer (such as a laser printer) enables the rapid and convenient production of lettering to an acceptable standard. Lettering and captions can be printed out as required and attached to the drawing for reproduction. Alternatively, a drawing can be scanned (to high resolution) and an image-editing software package (such as Adobe Photoshop) can be used to add lettering. It is best practice to save the final drawing as a TIFF file, and when scanning a drawing to do so as greyscale at 600 dpi minimum.

Other basic equipment that might be considered could include a scalpel, set squares, an engineer's square and graph paper.

A designated area will be required for drawing activity and it should enable people to work comfortably for extended periods. Essential points to consider include the working surface and lighting.

It is vital to have a suitable surface on which to support the drawing. A purpose-built, adjustable drawing table is ideal but expensive. Drawing boards can be bought or made. If making one, it is very useful to make sure that one corner is machined straight so that it can be used as a 'T-square'. It can be very useful to tape a sheet of graph paper onto the drawing board. If drafting paper is then used, the lines on the graph paper will show through and can be used to align drawings, labels and datum lines.

Drawing is based on observation and a strong, directable light source will help enormously in picking out

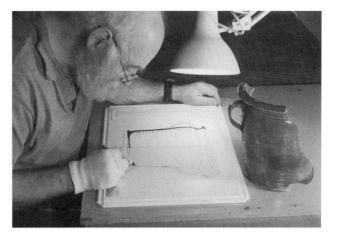

Figure 18.1 Once the outline and other details of an object have been recorded in pencil they are traced in ink onto a clean sheet of tracing paper or film to produce the final drawing. Note the well-positioned lamp and the use of a cotton glove to avoid smears on the drawing surface, which can reduce its ability to accept clean ink lines. (Photo: Edward Martin)

detail that might otherwise be missed. An angle-poise lamp can be ideal (figure 18.1). Conventionally, objects are drawn as if lit from the top-left corner. It can be useful, therefore, to place a lamp on the top-left corner of the drawing board and also to position the board so that natural light is coming from that direction. Examine the object from a number of angles in different lighting conditions; new features might well become apparent.

With experience comes a more extensive drawing toolkit. More pens, an adjustable drawing table or a reducing machine can all add significantly to the ease and convenience with which high-quality work can be produced. Accuracy and clarity, however, can be achieved with basic equipment of the kind listed above.

Whether drawings are made on film or paper, they should be stored in a safe, dry place. A plan-chest is ideal because drawings can be laid flat and inspected with ease. Variants in which the drawings are suspended from racks are also available. If there are to be a lot of drawings, then it can help if each sheet of paper or film has its own number marked in a place that is easily visible when the sheets are in storage. All artefacts drawn on that sheet would have that sheet number attached to their record card so that the relevant drawing can be found with ease. This helps to prevent loss of information (see chapter 19).

DRAWING ARCHAEOLOGICAL MATERIAL

The point of archaeological illustrations is to convey information, not to break up long passages of text. Accuracy and clarity are of the utmost importance.

Illustrations need to convey technical information about each object – exact dimensions, types of material, method of construction, traces of use and many other features – accurately enough for a researcher to recognize parallels, similarities or differences with material from elsewhere.

To satisfy these criteria, archaeological illustrations follow set standards and conventions. This subject area is wide, and the reader is referred to publications listed in the Further Information section at the end of the chapter. On a large project where a lot of drawing work is required, it is well worth appointing a supervisor to be in charge of ensuring consistency in terms of conventions used. Artefact photography has been covered in chapter 10. This section will deal with the basic concepts and techniques involved in recording archaeological material by drawing.

Archaeological drawing can be described as a mechanical process. Artistic flair may be an advantage and expertly executed archaeological record drawings can often be extremely attractive works in their own right. Having said that, the ability to make careful observations and transpose a measurement from an object onto paper are the main skills required to produce acceptable results.

Choosing between drawing and photography: Drawings have many advantages over photographs. Clarity of, and emphasis on, particular detail can be achieved more readily with a drawing. While photography can be used to record a number of objects relatively quickly, photographs cannot be used to present sectional information of complete or near-complete objects in the way that drawings can; nor do they tend to offer quite the same opportunities for comparison of form. Illustrations showing hypothetical reconstruction are also much more readily achieved with drawings than photographs. Most people would agree that a photograph alone is an insufficient illustrative record of an archaeological object. Quite often, during the process of making a measured drawing of archaeological material, more information comes to light, which helps in refining the final record. Notes of observations made while drawing therefore need to be kept and added to the written record.

Recording by drawing can be divided into a number of areas:

- recording shape and dimensions;
- recording decoration and surface detail;
- recording detail related to composition and manufacture.

All the above are essentially objective processes based on careful study and measurement. However, an element of interpretation is involved in many drawings – for example, the orientation of the object on the page, the emphasis given to each feature, the selection of views recorded. This is unavoidable and an illustrator must try to ensure that such decisions do not become the source of distortion and bias. Explicitly interpretive drawings are dealt with separately below.

There are a few general conventions that should be noted. As mentioned already, archaeological material is most usually drawn as if lit from the top left-hand corner of the page. Record drawings do not generally involve perspective; elevations are drawn from as many views as required to convey the available information. More views may be recorded for the archive than are eventually published. Sections through the object are very useful as an adjunct to other views and should always be included when drawing ceramic vessels.

All record drawings should be clearly and permanently marked with the object's record number, the draughtsperson's name and a linear, metric scale.

Originals and reductions: Most artefacts are drawn at full-size in pencil, and completed inked drawings reduced for publication as necessary. Some smaller objects are drawn at more than life-size (often using tracing from carefully scaled and prepared photographs) while others (such as amphoras) are drawn at half or quarter scale. The final inked version of any drawing should be the version that is reduced for publication when necessary.

It is clear from the outset that reduction must be considered when planning the drawing. A linear scale is used so that it will remain true, even when a drawing is reduced or enlarged for publication. Line-thicknesses and levels of detail are also important factors to consider. Lines less than 0.25 mm thick will often be lost with a 50 per cent reduction, and intricate detail will blur if not carefully drawn. It is worth remembering that plans and section drawings will have to be drawn for reduction as well as objects. In these drawings, particular attention must be paid to ensuring that any lettering or labelling remains clearly legible after reduction. Adding the lettering to a drawing/plan after it has been scanned and reduced can help avoid this problem.

Consideration of the effects of reduction can sometimes leave the full-size drawing looking a little plain but this is preferable to merged and blurred detail when reproduced at published size. In extreme cases, it may be necessary to produce a highly detailed record drawing and another, less-detailed but still accurate, drawing for reduction and publication.

There are some advantages to reduction.

- Reducing a drawing by 50 per cent can remove many slight blemishes from view, and smooth out slightly rough lines.

- Drawings for reduction can be checked quickly for choice of degree of reduction, and its effect, by use of a photocopier.
- Effective illustration for reduction is a skill but like everything else it can be learned and practised.

Recording shape and dimension: An initial step in making a drawn record of an object is to establish its outline and general dimensions. The techniques used will vary from object to object but the example offered below, of a technique useful for drawing the outline of a ceramic vessel, will serve to introduce methods of working which are applicable in a wide range of situations. Other objects may need to be drawn from a greater number of angles to convey the required information.

Pottery is usually drawn full-size for eventual reduction. This allows the maximum number of measurements to be made. Amphoras form an exception, being drawn at a reduced size (1:2 or 1:4), or being traced from reduced photographs. There are a number of different ways of drawing pottery and methods will vary with the completeness of the vessel.

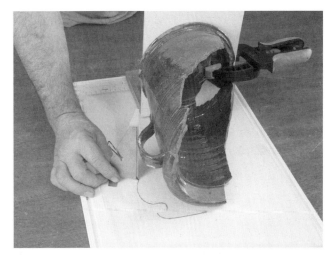

Figure 18.2 A simple method for recording the shape of an object is to place it securely on its side and trace round it with a set-square which incorporates a pencil lead at its apex. This is a commercially produced version, but they are easily home-made. The base of the pot is secured to a vertical plate by means of a clamp with soft rubber pads. Care must be taken not to damage objects when drawing them. (Photo: Edward Martin)

Example: drawing complete and near-complete ceramic vessels

Recording the outline and dimensions of a complete or near-complete vessel can be achieved using a simple method involving a right-angle block and an engineer's square (or any similar device), as follows.

Place the vessel on its side on a sheet of drawing paper with its rim or base flat against the right-angle block as if it were standing upright on a table. It is likely that the vessel will have to be supported by plasticine or modelling clay (never *Blu Tack* which will strip off surface layers and leave behind oil marks) and should not be allowed to move. A line drawn along the base of the block will represent the line of the base or rim of the vessel.

Place the engineer's square against the edge of the vessel. Where the square touches the paper is directly beneath that point on the vessel and this can be marked, on the drawing paper, with a pencil (figure 18.2). By moving the square around the edge of the vessel and marking the points with a pencil on the paper, the outline of the vessel can be plotted. The more marks that are made, the easier it will be to join them up into an accurate outline. Special attention should be given to areas where the outline changes dramatically (such as rims) and to where handles or spouts are attached. The outline of a wide range of objects can be recorded in this way.

An alternative method of drawing the outline of a near-complete vessel or large object is to position it firmly on a level surface and then establish a vertical datum. Offset measurements can then be taken from the datum to the object to describe the shape, as shown in figure 18.3.

Example: drawing incomplete ceramic vessels

Drawing the outline of incomplete vessels can be much more complicated. Sometimes only fragments of a vessel will be available, but it is still possible to achieve a useful record of the shape of the object by the careful application of simple techniques.

If less than half of the rim or base of the vessel is available, it may not be possible to use the techniques mentioned above to draw the outline of both sides of a vessel. Other techniques can be used which make this possible. Clearly, it is important to find out how wide the vessel was at the base and at the rim to allow its shape to be reconstructed. This can be done by using radius templates (figure 18.4) or by placing a rim or base fragment on a radius chart (a sheet onto which semi-circles of known radius have been drawn – see figure 18.5), as follows.

Place the rim upside down on the radius chart, in its proper plane. Its true position is then checked by looking along the plane of the paper, and adjusting the angle at which the sherd is held until the arc of the rim or base lies flat on the paper. Rounded or abraded rims can be difficult to orientate correctly, as can very small fragments (figure 18.6). Move the base or rim across the

Figure 18.3 The shape of an object can be recorded by establishing a vertical datum (here a plastic set-square on a wooden base) and taking offset measurements. (Photo: Edward Martin)

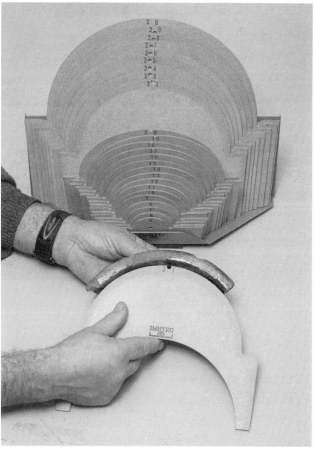

Figure 18.4 The diameter of sherds can be determined by the use of radius templates. (Photo: Edward Martin)

radius scale maintaining the same orientation until the outer curve coincides exactly with one of the curves of the radius chart when viewed from above. If the rim fragment fits the curve of, for example, 10 cm radius, it can be deduced that the vessel was 20 cm wide at the rim, despite the fact that only one piece of the rim remains. The dimensions of the vessel can now be indicated on the drawing.

Once the dimensions of the rim or base are established, the outline of the sides of the vessel can be drawn. This can be done by using a block as described above, but great care must be taken to make sure that the piece of pottery is orientated correctly. Position the pot rim or base against the vertical wooden block, ensuring the rim is in its correct plane (as when using the radius chart). When isolated sherds are being drawn, they should be orientated to provide the maximum complete profile, although the angle at which this profile is orientated may have to be estimated to a degree. Plasticine or modelling clay can be used to stop the pot or sherd from moving. Use an engineer's set square as described previously.

Figure 18.5 A chart on which circles of measured radii are drawn allows rim or base potsherds to be aligned with successive curves until an exact fit is found. Segments that show the percentages of the full circle enable the amount of the fragment being measured to be assessed. (Photo: Edward Martin)

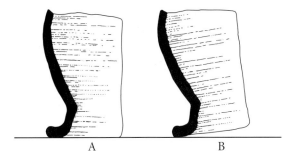

Figure 18.6 When attempting to orientate a sherd of pottery correctly on a radius chart, the presence of throwing marks (horizontal lines created during manufacture) can be very useful, especially when rims are rounded or abraded: A) the marks can be seen to be horizontal, indicating that the sherd is in its correct orientation; B) the rim is not correctly aligned and would give an incorrect reading on the chart. (Based on original artwork by Ben Ferrari)

Figure 18.7 A simple pottery drawing. The convention is to show the outside of the vessel on the right and the inside in section on the left, with a central line separating the two (imagine that the quarter of the pot on the near left has been cut away). The outside surface detail appears on the right, with the hard lines around the rim circumference and base shown complete and the less dominant throwing lines on the body incomplete and fading to emphasize their softer nature. On the left-hand side we see the section of the vessel, emphasized in black, and an indication of throwing lines on the inside wall. This pot has a handle on each side. The right one is shown externally, with a cross-section, while the left-hand one is sectioned vertically, and hatched to indicate that it has been added to the body of the vessel. These simple conventions, which should be followed wherever possible, can be adapted to accommodate the peculiarities of individual pots. (Colin Martin)

Check, and repeat if possible for the other side. The measurement obtained using the radius chart provides the position of the rim for the opposite side of the vessel. Pots and other objects are very rarely symmetrical and it is not appropriate to make a measured drawing of one side and then use a tracing of this to form the outline of the opposite side. However, where only one side of the vessel is available, the same piece will have to be used to reconstruct both sides of the outline. It might be possible to extend the outline by studying the shape of the fragments from the body of the vessel. The more pieces that are available, the more easily this can be done. Conjectural sections of any outline should be indicated using a dotted line.

If no rim or base fragments are available, then it may not be possible to reconstruct the shape of the vessel being studied. Some indication of shape can be provided by drawing cross-sections of the body sherds, but an unacceptable amount of guesswork may be involved to reconstruct a complete outline. This does not mean, however, that such pottery fragments are useless; they may have distinctive decoration or construction marks that should be recorded and not ignored.

Once an outline has been drawn, it is usual to draw a line through the mid-point of the pot, dividing it in half. The convention is that the left side is used to show internal detail, while the right shows external features (figure 18.7).

Using callipers, measure the thickness of the wall (figure 18.8) and draw a cross-section through the vessel. This is shown on the left-hand side of the drawing. Careful attention should be paid to internal details at this stage. In the case of complete pots these may not be easy to observe but the study of broken examples of similar pots may help in assessing what might be there, and in developing a strategy for recording it. On the right-hand side of the pot, indicate external features such as decoration, or technical traits. Outlines of handles and spouts often cause problems and various conventions exist for showing them in section as well as in profile (figure 18.9). However, handles usually appear on the right of the drawing and spouts on the left. Check the resulting dimensions of the object carefully by direct measurement and with callipers.

RECORDING 'BY EYE'

Objects can also have their outline recorded 'by eye'. The object is carefully positioned on the paper and supported by modelling clay. It is important that it does not move at all throughout this process. The draughtsperson then positions him/herself over the edge of both the object and

Figure 18.8 Thickness-gauge callipers can be useful in determining a pot's section. (Photo: Edward Martin)

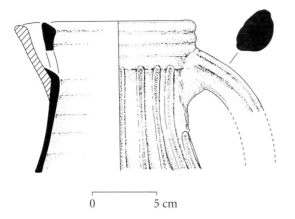

0   5 cm

Figure 18.9 Following convention, the spout of this jug has been shown on the left and details of the way in which it was attached to the vessel can be seen in the drawing. Additional views of the spout could be added if required. (Based on original artwork by Ben Ferrari)

Figure 18.10 Profile gauges can provide a quick way of checking the shape of an object but they can easily damage delicate surfaces and should only be used on robust items. Their use is not generally recommended in recording small finds (for which better techniques are available), but they can be useful for recording features such as the mouldings on guns. (Photo: Colin Martin)

the pencil tip, and draws around the object keeping directly above the pencil point and object edge. Accurate outlines can be drawn in this way, particularly for fairly flat objects with well-defined edges, but it requires considerable practice and the results should be very carefully checked by measurement. A slight misalignment of the pencil, object edge or eye can produce significant errors.

If the illustrator has access to a copy-camera with precise reduction and enlargement facilities, flat objects or materials may be photographed on the copy-camera screen and printed to the desired size. The printed image can then be placed beneath draughting film to be used as a basis for the line drawing. If this method is used, it is essential that a series of individual check measurements is made to ensure that the alteration in scale is accurate. A variety of pieces of equipment are available to aid the recording of outlines (e.g. flexi-curves and profile gauges) but they do not record the finer details

of shape if used carelessly. With practice, however, they are useful tools (figure 18.10).

RECORDING DECORATION AND SURFACE DETAIL

Once the outline and dimensions of an object have been accurately recorded and checked, surface detail can be added. Again this should be done through a process of careful observation and measurement. Tracings from carefully scaled photographs can be used to illustrate complex decoration and the outlines of objects can be recorded in the same way. Careful attention should be paid to reproducing decoration and deliberately applied surface detail accurately (figure 18.11). Others may want to use drawings as a means of comparing with archaeological material elsewhere and, therefore, small details of style and shape may be important.

A number of methods for representing three-dimensional shape and decoration exist. Stippling (a series of dots) and linear shading are used widely and conventions exist in terms of the way that various features are represented. Looking at a wide range of published work will enable familiarization with these conventions and the way that

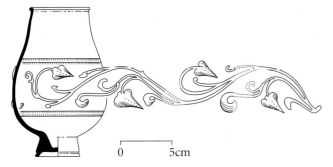

Figure 18.11 When complex decoration is being recorded it can be 'rolled out' as shown here. If the design is repetitive it might not be necessary to show it all in this way. (Based on original artwork by Ben Ferrari)

RECORDING CONSTRUCTIONAL AND OTHER DETAIL

In addition to decorative surface detail, it is important to record other surface features. For example, is there evidence of the way in which the object was made or repaired? On pottery this might take the form of its fabric, the internal shape of the vessel or the coil-construction indicated by the undulating surface (figure 18.14). Ironwork may show evidence of rivets, welding or hammering. Tool-marks might be apparent on stonework. All such features are relevant in terms of recording all the available information.

Care should be taken to distinguish between different materials in composite objects. For example, an iron knife with a wooden handle might be drawn using stippling for the iron and linear shading to convey the grain of the wooden handle. A wide range of methods exists and the most appropriate option for each illustration should be chosen with care (figure 18.15). Clarity is essential and it is usually useful to provide a key to explain the conventions used.

they can be both informative and attractive. When effectively applied, stippling can greatly enhance a drawing, suggesting shape and texture (figures 18.12 and 18.13). It can, however, also be misleading if carelessly done and can obscure as much detail as it shows.

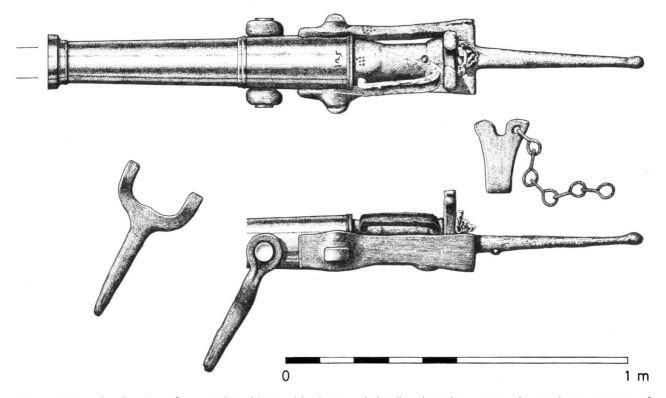

Figure 18.12 The drawing of a complex object, with views and details selected to convey the maximum amount of information. Different techniques have been used to convey the texture and character of the various materials of which this swivel gun from *La Trinidad Valencera* (1588) is made. The dull reflective surface of the bronze barrel is indicated by stippling, while fine lines of various thickness depict the granular structure of the wrought-iron breech, swivel crutch, and locking wedge. The leather pad behind the wedge is represented by black and white blocking, with a little dotting to convey the curves of the folds. (Colin Martin)

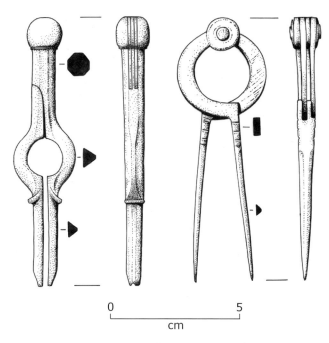

0   5
cm

Figure 18.13 Two pairs of brass dividers from the *Dartmouth* (1690), of different designs. Front and side views of each instrument are shown, with sections where appropriate. Stippling is used to convey details and the matted surface of the metal. (Colin Martin)

Lastly, it is important to avoid misleading additions or omissions. A total absence of stippling on the surface of a pot drawn with beautifully smooth sides might suggest a fine fabric and possibly a wheel-turned production. If, however, the vessel is in fact made of rough fabric and is asymmetrical then this should be indicated in the drawn record. Similarly, avoid putting hard lines where none exist on the object. The process of drawing involves a certain amount of decision making but if an edge is rounded it should be drawn as such. If a junction or feature is not clear and sharp, dotted and broken lines can be used to describe it in preference to a solid line, which suggests a definite boundary.

POST-FIELDWORK PHOTOGRAPHY AND LASER SCANNING

Alternative or additional techniques for post-fieldwork recording of artefacts include photography (see chapter 10) and laser scanning. Laser scanning uses a sophisticated combination of mirrors, cameras and lasers to record the surface detail of objects, structures and landscapes. A triangulation laser-scanner uses mirrors to move the laser over the surface of an object. As the surface deforms the laser stripe, the camera records the shape of the laser line. This allows the surface of small objects to be recorded and stored digitally on a computer (figure 18.16). Larger

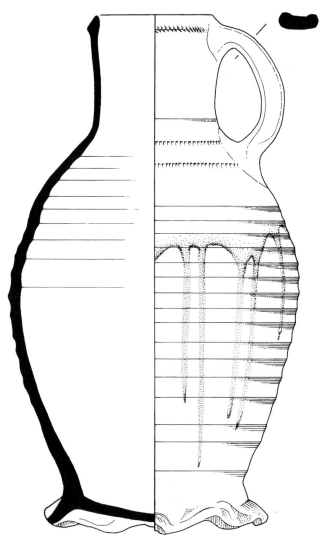

Figure 18.14 A more complex pot. Here the throwing lines on the outside are more pronounced, and the top of the vessel is glazed. Some of the glaze has dribbled down the side, and this has been indicated by variable dotting. Cut and stamped lines run around the upper part, and the foot has been thumbed into a 'pie crust' form after throwing. Simple techniques have been used to convey these features within the general conventions of pottery drawing. (Colin Martin)

structures, such as entire vessels, can be scanned using time-of-flight laser scanners, which simply bounce laser pulses off the object millions of times to build up a map of the structure (figure 18.17).

PRESENTING A RANGE OF COMPLEX INFORMATION

Isometric and axonometric representations: A drawing method commonly used in presenting structural information is a projection (possibly isometric or axonometric).

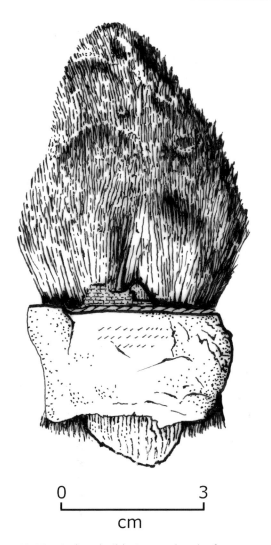

0   3

cm

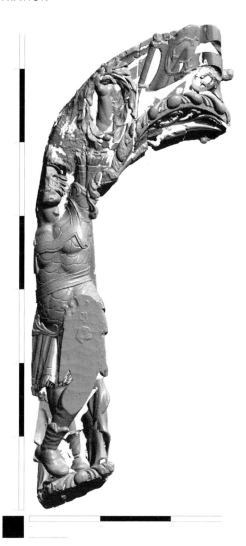

Figure 18.15 A boatbuilder's tar brush from western Scotland. This artefact is made of different organic materials, and these are indicated by various drawing techniques. The remains of the handle, seen protruding at the bottom, are of wood, and this is depicted by lines suggestive of grain. The band above is of leather, indicated by restrained stipple and black cracks. Above that are fragments of cord and fabric, which are given a simple realistic treatment. The hair of the brush has been worked up with a series of fine black lines, clustering in areas of shadow. Fine separation of the 'hairs' has been achieved by careful scraping with a scalpel. (Colin Martin)

Figure 18.16 A render of the 3-D laser-scan data of the figurehead from HMS *Colossus* (1798). The high accuracy of the 3-D laser-scanner is suitable for detailed recording of maritime artefacts and this can be seen on the recording of subtle tool-marks on the flat knee section. (Copyright Archaeoptics Ltd; reproduced courtesy of the Mary Rose Trust)

These are simply ways of converting the information contained in single-plane views derived from the site recording into composite three-sided dimensional views. An isometric drawing (using a 60 degree and 30 degree set square) is correct along any axis whereas an axonometric (greater than 30 or 60 degrees) representation is correct along one or more.

The representations are usually viewed from a 'common corner', a point at union with the three viewed sides of the object. All the principal lines of the projection are parallel to each other within the same plane. The disappearing effect of perspective is not drawn, and there is no foreshortening needed. All isometric projections of timber structures (e.g. for details of joints) should always be checked against the originals if they survive. For standing structures, front, side and back elevations are desirable at a scale of 1:10 to provide the necessary information. Plans (for quays, river wharves, etc.) should be at superstructural, base-plate and foundation levels, and sections across such structures should show the relationship of the timbers to adjacent deposits.

Computer-aided design (CAD):   Computer draughting has the ability to produce and manipulate two-dimensional

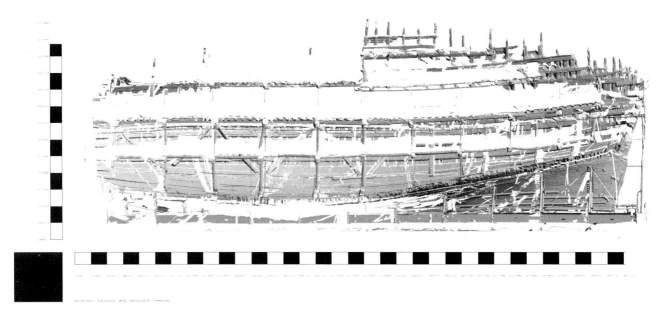

Figure 18.17   An orthographic view of the shaded 3-D laser-scan data of the *Mary Rose* (1545) hull. As the scanner operates in a 'line-of-sight' manner, the timbers near the top have been obscured from the scanning viewpoint near the ground. (Copyright Archaeoptics Ltd; reproduced courtesy of the Mary Rose Trust).

and three-dimensional images from survey information. The use of CAD does not restrict one to working from the keyboard or to inputting data 'from scratch'. A scale drawing made on the site may be entered into the computer by placing the drawing on a digitizing pad (a sensitized electronic drawing surface) and tracing carefully with a pen-like instrument, or by importing the image as a Bitmap and tracing it digitally within the software package. Highly accurate two-dimensional drawings have been produced in this manner from original drawings. The relative height information can then be added to produce three-dimensional images.

The effects of sag in a building or hogging influences on a ship are readily apparent in three-dimensional perspective views whereas they are difficult to see in two dimensions. The ability to rotate a structure or to change perspective reveals features that would otherwise only be obtainable through tedious re-drawing. This feature is equally dramatic in the three-dimensional display of the archaeological stratigraphic sequence. The surface contours, strata, and artefacts can all be displayed in their original spatial relationships. This level of representation allows the archaeologist to analyse the site in ways not previously available. Beyond the potential insights revealed by CAD drawings, there is also a significant reduction in the amount of effort required to reproduce graphics. CAD software can replace the task of photographic reproduction, allowing the production of drawings of any size at the press of a key.

Interpretive drawings: Interpretive drawings, reconstructions, views of objects as they might have been used and objects placed together in assumed association have their place in the depiction of archaeological evidence. However, they should be rigidly separated from record drawings, which are representations of the objects as they are, not how it is thought they should be (i.e. observations not interpretations). Any element of a record drawing involving hypothetical reconstruction (e.g. the shape of missing sections of a pot) should be clearly indicated. Of course, some reconstructions can be made with more confidence than others, and such drawings can be fundamental to archaeological work in terms of communicating ideas and exploring interpretations. It cannot be overstated, however, that these types of illustration are not an adequate sole record of an object.

FURTHER INFORMATION

Eiteljork II, H., Fernie, K., Huggett, J., and Robinson, D., 2003, *CAD: A Guide to Good Practice*. Oxford.

Green, J., 2004 (2nd edn), *Maritime Archaeology: A Technical Handbook*. London.

Griffiths, N., Jenner, A. and Wilson, C., 1990, *Drawing Archaeological Finds: A Handbook*. London.

Institute of Field Archaeologists, 2007, *Draft Standard and Guidance for Nautical Archaeological Recording and Reconstruction*. Reading (www.archaeologists.net).

Steiner, M., 2005, *Approaches to Archaeological Illustration: A Handbook*. CBA Practical Handbook 18. York.

Post-Fieldwork Analysis and Archiving

19

Contents

- Handling material and keeping records
- Post-fieldwork treatment of survey work
- Specialist analysis
- Interpretation and gathering supporting evidence from other sources
- Producing an archaeological archive

During fieldwork, evidence will have been collected in a variety of forms and a wide range of techniques will have to be applied in order to extract the maximum information from it. In all probability, this post-fieldwork phase will take a great deal more time (possibly three to four times as long), and require more effort, than the fieldwork itself. Any post-excavation work is time-consuming, and can be expensive. Even without excavation, the amount of evidence collected can be enormous and it is very important to include in the initial project planning (chapter 5) the appropriate time and resources needed for post-excavation work.

These are the main principles of post-fieldwork activity:

- Post-fieldwork analysis is not a chore left for non-fieldworkers to do. It is the reason for doing the fieldwork in the first place, and is best done by those directly involved.
- A high standard of fieldwork is meaningless if the information collected is not recorded, analysed thoroughly, and made readily available to other workers as conveniently, promptly, and fully as practicable.
- All post-fieldwork activity must continue to relate to the recording system. Drawings, plans and conservation records must all be cross-referenced to artefact/finds numbers and the project register (chapter 8). The collection of information continues throughout this phase as material is studied and recorded. The process can generate considerable quantities of notes, records and drawings. Therefore, maintaining an efficient system of documentation is as fundamental to this part of the archaeological process as it is to fieldwork.

- The aim of post-fieldwork processing of archaeological evidence is the establishment of the 'site archive' (see below). From the archive, information can be synthesized to create reports and publications.
- Mistakes and inadequacies in the fieldwork are likely to appear as the evidence is analysed. It is important that these failings are not glossed over or hidden. Everyone makes mistakes and everybody can learn from them. Other researchers should be given all the information they need to assess conclusions objectively. This should include access to the archive.
- The difference between analysis and interpretation is fundamental to post-fieldwork processing and should be taken into account at every stage. Analysis might be described as making and recording

measured, quantified and objective observations. Interpretation can be defined as using a set of observations to support a conclusion; it can very rarely prove a point beyond any doubt.

- The way evidence is collected and analysed, the strategy employed, the reasons for it, along with the success or failure of the techniques used, are of interest and should be recorded. Such information may help others plan their work and will enable more effective assessment of the validity of any conclusions drawn.

HANDLING MATERIAL AND KEEPING RECORDS

If material is brought to the surface during a project, it should already have been recorded to some extent *in situ* (chapter 8). Its position relative to other material, its context and its position relative to site control points will have been noted and drawn on a site-plan. Photographs may also have been taken before, during and after lifting (chapter 10). Limited visibility or a rescue scenario are not acceptable excuses for poor standards or lack of recording. The information is there, and it is the responsibility of the archaeologists to recognize and record it.

Once the material is on the surface, there is an opportunity to undertake more detailed recording than is usually possible under water. It is important to consider that material that has not been conserved can be unstable and fragile. Recording and analysis can involve a lot of handling and movement of an object from place to place. During any pre-conservation recording and study, it is an imperative to keep handling to an absolute minimum and avoid exposing the material to dramatic changes of environment.

The importance of ensuring that any material remains are securely identified, with the number assigned during the initial recording process, has been emphasized already (chapter 8). During the post-fieldwork processing phase, objects and samples may be moved from place to place and examined by many different people, some of whom will be less careful than others in terms of putting things back where they found them. It is therefore important to make sure that everyone handling material is aware of the need to avoid detaching or defacing labels. A specifically appointed finds-assistant can be very helpful in ensuring that the material and the written record remain connected. It may be useful, if material is likely to be moved to a number of different places, to devise a system for keeping track of changes of location. In this way it will be possible to track past locations and current whereabouts.

Figure 19.1 A plan chest is an ideal way to store plans and drawings. They can be laid flat and inspected with ease. Variants in which drawings are suspended from racks are also available. (Photo: Mary Rose Trust)

POST-FIELDWORK TREATMENT OF SURVEY WORK

Maps, plans, and sections: Survey work on and off site is likely to have produced a large amount of information that must be processed and prepared either for publication or the archive. The methods used for obtaining and plotting survey results are covered in chapter 14. This discussion focuses on ordering the records and preparing them for future use, analysis and archiving. The following types of survey drawing are typically found in an archive or report (figure 19.1):

- location maps;
- site-plans showing archaeological, topographic and environmental features; and
- sections and profiles.

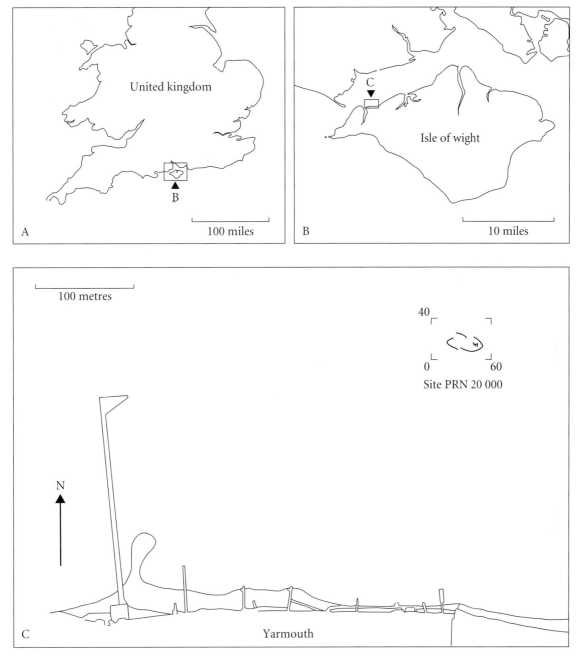

Figure 19.2 An example of a site-location map for the protected wreck-site at Yarmouth Roads. (Drawing by Kit Watson)

As a general rule, such drawings should be organized according to scale – from large-scale, site-location map to close-up plans of specific parts of the site or sections of a deposit. They should be presented and captioned in such a way that cross-referencing between the site master-plan and details is easy. Off-site information, such as the results of survey work around the main site, may also be included.

It is advisable to keep all drawings free from unnecessary information. Location maps are more effective when uncluttered with roads, buildings or anything else not directly relevant to the subject (figure 19.2). If drawings become cluttered or confused, it might be appropriate to create overlays (drawings showing the same main features but providing different details). This can be particularly useful in terms of presenting results (e.g. of magnetometer or probing survey, or the location and extent of any trenches excavated). Changes in the nature of sediment across a site can also conveniently be shown in this way. The important point to remember about such overlays is that common control points (chapter 14) linking them to the master-plan or section must be clearly marked.

Other maps or drawings can be added. These might include maps showing the nature of the local geology, maps showing the site's location relative to other similar sites in the same country or abroad and distribution maps or plans to aid discussion of the site's general significance.

Finished drawings of field-recording: Producing finished illustrations from section drawings and site surveys made in the field is not a process of enhancement or an opportunity for embellishment. However, conventions and styles of presentation may be found which greatly improve the clarity of drawings. This is entirely legitimate and should be encouraged if it does not obscure or distort the original information. Decorative borders do little to make up for badly recorded evidence. Similarly, detailed renderings of cannon and anchors which bear no resemblance to those actually on the site are best avoided.

When inking-in section drawings, remember that boundaries between contexts are rarely precise. Therefore solid lines should be avoided in favour of broken lines or stippling. Conventions can be used for clarity when showing different sediment types. It is common practice to offer a record drawing of the section alongside an interpretative drawing, where solid boundary lines are shown, to aid clarity in discussion (figure 19.3).

Each plan, section or overlay should include sufficient information for it to be usable by others. For example:

- Site name and code, together with the record number of the drawing and the date drawn.
- Subject (plan or section of what).
- A clearly marked linear scale (in proportion to the drawing, so that it is visible but not intrusive).
- Position (site-grid coordinate on plans; reference to plan on which position of section is indicated). The grid coordinates used should be consistent throughout the project (see chapter 11).
- Orientation (true or magnetic north on plans and maps; the direction exposed sections face on section drawings).
- Key to symbols used.

Although it has rarely been indicated on archaeological drawings in the past, archaeologists should be more honest about the value of their survey, and so it is

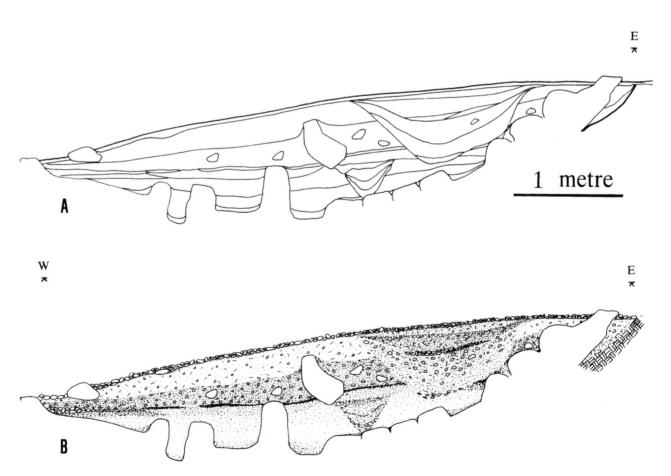

Figure 19.3 Interpreted (A) and 'naturalistic' (B) representations of the same section prepared for publication. (Drawing by Kit Watson)

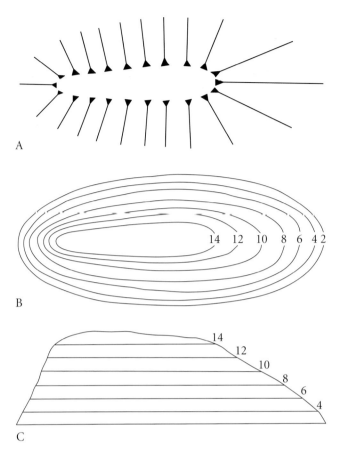

Figure 19.4 By using A) hachures or B) contours, the topographical relief of mound C) can be represented

recommended that a quantified estimate of the survey accuracy be included.

Presenting topography: Site-plans should include an indication of the local topography and environment. Some thought should be given to how such features should be represented. Indicating the direction and degree of slope is very useful and can be done very simply. At the most basic level, direction of slope can be indicated on a plan with arrows or, following more detailed survey work, deduced from contour lines (figure 19.4).

Successful representation of the underwater terrain requires effort both in the field and at the post-excavation stage. If the area on and around the surveyed area is relatively flat, an efficient method of representation may be to simply put spot heights at suitable places on the site plan. More complex topography is often shown in contour plans. There are other ways of representing the sea-bed graphically, including wire diagrams and surface rendering. These are normally only produced by computer printers or plotters following analysis of the data by suitable programs.

Changes on the sea-, river- or lake-bed may be in texture or material rather than height. Differences in sediment, the size and shape of large stones or boulders, bedrock or vegetation type may all be indicated, and this should follow the UK Hydrographic Office's *Symbols and Abbreviations ued on Admiralty Charts, Chart 5011* (www.ukho.gov.uk). An informative topographic site-plan may have both contours and symbols to help bring out the detail in a readily understandable way.

SPECIALIST ANALYSIS

Almost all projects will benefit from the analytical input of specialists who will be able to contribute to both the identification and interpretation of the evidence. Specialists are particularly valuable when the analytical techniques employed require special training and/or equipment and when there is a need for somebody with more experience of dealing with a particular class of material. The level of specialist input may vary from a major study of material over a considerable time to comments made over the telephone, by e-mail or letter.

Different specialists will have different requirements in terms of the form in which material should be presented to them, but some general guidance can be offered in terms of maintaining a productive working relationship:

- Obtain a firm commitment from the specialist that s/he is willing to undertake the work and on what terms. Do not assume that s/he will be willing to look at the material immediately or that s/he will do the work for free.
- Specialists have skills and knowledge that have often taken a long time to accumulate. They should be accorded the same consideration and respect as other team members and their expertise acknowledged.
- Establish the timescale. How long will the specialist require to study and report on the material? Agree on a timetable. It is easier to hold people to a deadline if they are being paid for their time. Someone working for free may be less inclined to respond positively to unreasonable pressure.
- A specialist will often be able to make extensive and perceptive observations on material presented for examination but may not be able to provide all the answers hoped for. S/he is likely to take a very rigorous approach to what can or cannot be said about the material with certainty. This close attention to detail may seem frustrating but in fact it is a reflection of the problems involved in dealing objectively with archaeological evidence.
- Information such as the context and association of the material being analysed should be given to the specialist.

- Additional information may become available from the processing of samples. The need for a continuing dialogue between all specialists, including archaeologists and scientists, throughout this phase is self-evident. The evaluation of the archaeological evidence will probably raise new questions or modify those originally formulated. The scientific studies may also raise problems or provide additional information with a direct and significant bearing on archaeological analysis and later interpretation.
- Finds-specialists will often be able to make observations on accurate drawings and written records, but on occasion they will want to see the original material. This can raise security problems, both in terms of safe transport and storage at a new location. Keeping a very careful record of what has gone where is crucial. It may be less trouble to bring the specialist to the material.
- It must be borne in mind that specialists may damage, alter or destroy material as part of their analyses. Material that must not be changed should be clearly identified (e.g. organic remains for dating purposes should not be treated with biocides and objects for display should not be sectioned).
- Finally, the work of the specialist should be acknowledged in any report or publication and the issue of copyright clarified in the terms and conditions under which the work is carried out (see chapter 20).

The need to establish and maintain an organized recording system that allows easy cross-referencing – for example, between finds, drawings, written records and plans – has already been discussed (chapter 8). Evidence collected in post-fieldwork recording should be entered into the same system. Indeed, it is during the post-fieldwork analysis stage that the flexibility, reliability and efficiency of the system will really be tested.

INTERPRETATION AND GATHERING SUPPORTING EVIDENCE FROM OTHER SOURCES

After the evidence has been analysed, conclusions about the site and interpretations of the material and its significance can be offered. Some people argue that interpretation should be avoided until the excavation or investigation is completed; that the objective collection of evidence is the most important part of fieldwork; and that preconceptions should not be allowed to influence what is recorded. However, human nature being what it is, most people find it impossible not to interpret and formulate ideas as investigations progress, and many would consider it very useful to formulate explanations for the evidence as it appears. Most important, however, is the need to prevent preconceptions and prejudices from affecting judgements made when collecting information, and to maintain the distinction between observation and interpretation through to final publication.

At this stage it is often useful to cast the net very widely in the search for information that will assist in this phase of the investigation. The evidence from the site itself may well have produced more questions than answers. One of the first places to look for supporting, or even contradictory, information and ideas is the developing corpus of archaeological literature. A poor researcher is one who does not make considerable efforts to become familiarized with the results of similar projects elsewhere. What evidence was recovered? What conclusions were drawn? What preconceptions were derived and what mistakes were made? Do these agree or conflict with the observations from the site currently under investigation?

It is not always easy to track down every publication concerning comparable sites and material. A reasonable way to start is to attempt to locate an article that gathers together the available evidence. Following up references within this article may lead to more detailed discussions of particular aspects of the subject concerned. Synthesized works are not generally satisfactory on their own. As a general rule, it is important to get as close to the primary source or original publication as possible. Each step away from the original publication may mean that some information has been filtered out by subsequent writers or presented in a way that was not intended by the original researcher (see chapter 9).

Open and regular communication with other researchers and frequent attention to journals are good ways of tracking down useful articles. It can take some effort to stay in touch with what other workers are doing and writing but it is also a fundamental part of archaeological research and essential for effective interpretation and publication (see chapter 20).

PRODUCING AN ARCHAEOLOGICAL ARCHIVE

The complete collection of all records and finds from a site is called the archive. An archaeological archive in a publicly accessible place is a valuable tool for researchers, as it allows reassessment of the evidence in the light of new techniques that become available, and new information gained from other sites. The fuller the archive, the more effective any reassessment will be. It is not good archaeological practice to exclude material from the archive because it is considered unimportant. It is impossible to guess what questions or analytical techniques will be applied to material in the future. Therefore, as far as possible, everything should be included in the archive. It would be arrogant to think that all present-day archaeological and forensic techniques will be acceptable in the future. If the decision is taken to disturb a site for either rescue or research motives, future generations are denied the opportunity to study the evidence *in situ*. At the very least, therefore, the full results of the excavation must be left for posterity in the site archive.

The concept of the 'site-archive' has been described as containing all the evidence gathered during fieldwork, which must be quantified, ordered, indexed and internally consistent (English Heritage, 1991). The first objective of the site archive is to maintain the integrity of the primary field-record, which may contain:

- original record forms;
- reports on site surveys, such as text and illustrations of specialist pre-excavation survey reports (e.g. remote sensing, sediment sampling, diver search);
- all drawings produced on site (the drawings not being inked-in or amended in any way after the fieldwork is completed);
- all photographs produced on site, above or below water;
- all artefacts and ecofacts recovered from the site (e.g. human bone, animal bone and samples resulting from basic processing);
- copies of correspondence relating to fieldwork;
- interim report(s) (any interim reports, whether published or produced for restricted circulation).

To make the final archive readily accessible and usable by other researchers, it should also contain:

- a brief account of the events and personnel of the project, with a summary of the results;
- a copy of the research design and excavation strategies (with an explanation of any changes during the course of the work) which should also include an assessment of the success of the research;
- a description of the understanding of the formation of the site, the character of the objects and structure;
- an explanation of the archive structure and contents, including a breakdown of documents and records present;
- the metadata relating to any databases or spreadsheets;
- a copy of the published report.

Every effort should be made to ensure that computerized records are in a format that can be easily transferred

to new future technologies (i.e. 'future-proofing' – see chapter 8). The same applies to photography. This will help maintain accessibility in the future and help to minimize the maintenance costs of the archive. For an archive to be of use, it must not only be physically preserved, but must also be organized in a logical manner, to enable other researchers to easily interrogate the resources (Brown and Perrin, 2000). For a simple description of the component parts of an archaeological archive, see *MAP2* (English Heritage, 1991).

The site archive, together with the finds, has to be prepared for presentation at a level suitable for consultation. This should aim to make the evidence comprehensible to those who have never seen the site but wish to, for example:

- reassess the results;
- prepare programmes of work on adjacent or contemporary sites; and
- produce works on topics that include results from your investigation.

It is important that a suitable repository is arranged for the archive as soon as practicable. As a general rule the arrangements for deposit of the archive and its accessibility should be made at the project-planning stage (chapter 5), before any work or research is carried out on the site. If the results of a project cannot be made available within a reasonable period of time, the director has to consider whether the excavation or investigation has been justified. Information that has not been disseminated is no more useful than if it had remained uncollected. In

Figure 19.5 One of the Mary Rose Trust organic stores. Following conservation, all objects are stored on acid-free tissue in environmentally controlled conditions with daily temperature and humidity checks. (Photo: Mary Rose Trust)

Figure 19.6 All of the *Mary Rose* finds are recorded on a card-file system, which pre-dates the use of computers in archaeology. There are in excess of 30,000 cards. Everything from initial identification, through every process (photography, radiography, conservation, illustration, etc.), is recorded on the cards, which are kept in a controlled and secure environment and updated regularly. While a computerized system now exists, the card-file index is still active. (Photo: Mary Rose Trust)

such a case, the site will have been as effectively lost as if it had been destroyed by dredging or treasure-hunting.

Since the aim of archaeology is to gather information for the use of future generations, it stands to reason that the information has to survive intact. Complete archives can be bulky and contain a wide range of materials, including paper, video film, conserved finds and samples. An environment suitable for one type of material may not ensure the survival of another, and, if the archive is split between locations for this reason, then very clear records of what is stored, and where, should be kept at each location. Particular attention should be paid to the safety of computerized records stored on disk as part of the archive, as well as their accessibility, owing to changing technology. Dust, humidity and direct heat will cause severe problems for such records. As stated earlier, arrangement of suitable storage space for this corpus of information should be addressed at the project-planning stage. Museums are sometimes willing to store site archives, but often only when they have had a direct involvement in the project.

Attention should also be paid to the materials used to create and store the archive. Some tissue papers are acidic and will eventually damage artefacts packed in them. Some slide-holders can also cause deterioration in transparencies over time, although 'archive quality' variants are available. Videotape and digital media are currently thought to survive reasonably well, though it is not a good idea to skimp on quality of things such as videotape or CDs. The provision of suitable packing and storage materials for the archive should be considered when planning the costs of the project because 'archive quality' materials are not cheap.

FURTHER INFORMATION

Brown, D. H., 2007, *Archaeological Archives: A Guide to Best Practice in Creation, Compilation, Transfer and Curation.* Archaeological Archives Forum (www.archaeologists.net/modules/icontent/inPages/docs/pubs/Archives_Best_Practice.pdf).

Brown, A. and Perrin, K., 2000, *A Model for the Description of Archaeological Archives.* English Heritage (www.eng-h.gov.uk/archives/archdesc.pdf).

English Heritage, 1991, *Management of Archaeological Projects (MAP2).* London (www.eng-h.gov.uk/guidance/map2/index.htm).

English Heritage, 2006c, *MoRPHE Technical Guide: Digital Archiving and Digital Dissemination.* London.

Institute of Field Archaeologists, forthcoming, *Standard and Guidance for the Creation, Compilation, Transfer and Deposition of Archaeological Archives.*

Perrin, K., 2002, *Archaeological Archives: Documentation, Access and Deposition: A Way Forward.* English Heritage, London (www.english-heritage.org.uk/upload/pdf/archives.PDF).

Richards, J. and Robinson, D. (eds), 2000 (2nd edn), *Digital Archives from Excavation and Fieldwork: A Guide to Good Practice.* Archaeological Data Service, Oxford (http://ads.ahds.ac.uk/project/goodguides/excavation/).

Presenting, Publicizing and Publishing Archaeological Work

Contents

This chapter examines the importance of disseminating archaeological work in both the public arena and academic circles. It provides practical advice about identifying and satisfying potential audiences and highlights some important points concerning when and how to publicize a project and its results.

THE IMPORTANCE OF PUBLICIZING (WHERE AND WHEN)

Recent years have seen a dramatic expansion in media coverage of archaeology and history. This is a reflection of the public appetite to learn more about our past; interest in the human story has never been greater. Maritime archaeological projects, in particular, have the ability to spark the imagination of individuals, from children to retired scholars and everyone in between.

Public interest in marine cultural heritage is vitally important. Through a public appreciation of maritime archaeology and an understanding of its potential, the perceived value and importance of the resource will grow. This will have a positive long-term impact on the discipline of archaeology, promoting respect for the finite and non-renewable resource and ensuring its inclusion in future policy and planning frameworks. Similarly, it will have an effect on the degree and availability of funding for continuing research and investigation. Public interest in marine cultural heritage should therefore be positively encouraged.

In addition to engaging public interest, it is necessary and desirable to publicize any project among fellow researchers, funding bodies, sponsors and appropriate heritage agencies. Presenting results to fellow researchers is a vital aspect of dissemination for any archaeological project. Fellow archaeologists may offer constructive comments or draw attention to references or examples of which the project team were unaware. Even if this is not the case, peer-review is an essential part of thorough archaeological research. It may also be necessary to prepare a presentation for a heritage agency, funding body or company representative, with a view to attracting support, funds or assistance in kind for a project. Note that, prior to approaching any funding body, it is wise to carefully consider the type and scope of projects that they can and will be likely to fund.

With such diverse audiences, the way a project is presented and publicized will vary considerably. The target audience and reasons for publicizing a project will also affect *when* a project is publicized. It can be beneficial to publicize a project right at the very beginning. This may attract offers of help (personnel, funds or equipment) and could engage local interest from the start. However, unless the project has a high profile, attracting media interest at an early stage may prove difficult. Alternative means of awareness raising may therefore be necessary.

Preliminary results of project work can be publicized while research is continuing. This can raise the project profile among other researchers and attract offers of support. Interim publications are mentioned below but there are many other methods for communicating results and information that do not oblige researchers to commit themselves to definitive statements before work is finished. Use of the media, internet and electronic publication to promote results can be an efficient and cost-effective way to reach a large international audience. Towards the end of a project, when fieldwork and post-fieldwork processing have been completed, the final project report should be compiled and disseminated (see below).

IDENTIFYING AND SATISFYING AN AUDIENCE

When publicizing an archaeological project, it is important to differentiate between the various target audiences mentioned above and to tailor presentations and materials to suit each of them. The nature of maritime trade and communities means that maritime archaeological sites often evoke strong feelings of identity and association. A shipbuilding town may have a close association with a vessel wrecked thousands of miles away, for example, or a fishing community may proudly identify with archaeological evidence of a long local fishing tradition. This means that the local community may have more than a passing interest in a local project and every effort should be made to encourage this enthusiasm and interest.

Informal talks and seminars provide many opportunities for publicizing an archaeological project. These are regularly organized by local diving, history and archaeology clubs. Even apparently unrelated clubs and societies (sports and social clubs, Women's Institutes, University of the Third Age, etc.) are often happy to host a visiting speaker talking on the subject of maritime archaeology (figure 20.1). It is well worth asking an audience at the start of a talk what their particular interests are. This enables the presenter to tailor the talk to the audience and further engage interest and help.

Presenting project work to such groups can be rewarding in more ways than one. Members of the audience may be able to contribute to the project with knowledge of historical detail, other sites in the area or other significant information. Another desirable by-product of such talks is that they may equip the audience with the understanding and inclination to approach such sites in a responsible, archaeological manner. Furthermore, individuals or groups may feel inspired to become involved in archaeology as a result of hearing such a talk. They may even offer help for future fieldwork or post-fieldwork processing (see chapters 18 and 19).

Fellow archaeologists (those already converted) will more readily be reached via a presentation and/or display stand at an appropriate conference. Conferences come in a wide range of forms, from major international events

Figure 20.1   A public talk is arranged during a NAS project at Stourhead, Wiltshire, UK. It is vital to disseminate information about a project and its findings. (Photo: Vicki Amos)

to local seminars. Speaking at a conference provides researchers with the opportunity to present their work to a group of people with similar interests. The importance of exposing work to peer review has already been mentioned and conferences/seminars provide an unparalleled opportunity to present material directly to peers. Similarly, a conference/seminar provides a rapid overview of who is doing what in the field and offers a chance to meet the people concerned. Discussions face-to-face can often achieve much more than written communications. International conferences are particularly valuable in this respect.

Some conferences publish proceedings (a collection of papers presented at the conference). A paper presented at a conference is not expected to be a definitive and final statement about a project. Therefore, not having completed research to final publication standard should not discourage anyone from offering a paper for publication in this form. Some of the most interesting papers are often those that take a discursive approach rather than those consisting of a bland catalogue of facts.

Communicating the results of research involves addressing all levels of society, including the archaeologists of the future. To present information to a younger audience in an effective manner may require additional effort to ensure presentation materials are appropriate. The whole style and content of a display or talk may have to be modified and yet still convey the essential ideas of archaeology and the project results. It is obviously of benefit to generate an awareness of the remains of the past in schoolchildren. Discovering about the past can be both fun and educational. It may offer a future career or just an area of interest for a young audience but it will certainly be good for the discipline (plate 20.1).

One of the most effective means of reaching a wide audience, comprising many of the different elements of society mentioned above, is through the internet. A well-designed and well-thought-out website can attract attention from around the globe. With increasing access to the internet, this is a rapidly growing method of publicizing work (see below).

METHODS OF PRESENTATION

A project can be publicized through a variety of media: written reports, websites, leaflets, displays, public talks. Whichever combination of media are chosen for publicizing a project, it is vital to have a clear understanding of the target audience and what the presentation is trying to achieve. This section gives some practical advice about making the most of publicity opportunities in a variety of settings.

One of the most common ways to publicise a project is direct face-to-face communication. As previously discussed, the actual content and style of any talk will depend on the subject matter, personal presentation style and the audience. However, there are some general guidelines that can help in delivering an effective and entertaining talk.

- *Plan and structure the talk* – writing an outline or plan will help put the material in logical order. Think carefully about what to say. Standing before an audience and reading a presentation needs careful consideration because it can prevent a more natural delivery of the subject matter. 'Prompt cards' summarizing the main points can be effective. One approach, especially if time is tight, is to read from a script that has been carefully planned, targeted at the specific audience, designed to be listened to rather than read (e.g. shorter, punchier sentences) and is marked with the places where the slide should be changed (to save turning around to see what is on the screen).
- *Use effective presentation materials* – visual aids should be chosen to complement the talk. Take the time to explain what the audience can see on screen. Use text sparingly, and try to avoid reading it out and repeating what is already displayed on screen.
- *Know the venue and equipment* – arrive early enough to get familiar with the venue (seating, lighting, acoustics and equipment). It is preferable to have a practice run with the equipment and to try to anticipate any potential problems that may arise

and ways to overcome them. This will help steady the nerves and make the whole experience more enjoyable.

- *Practise* – good preparation is vital for an effective talk. Thorough knowledge of the subject matter and presentation material will enable last-minute changes or unexpected queries to be dealt with calmly and effectively. Practice will also help overcome nerves and allow the timing of a talk to be perfected.
- *Hone the delivery* – when speaking publicly, it is important to speak more slowly and loudly than normal. It is also helpful to look up and make eye contact with the audience. Avoid talking while looking at the screen (even when wearing a microphone).

Slides: While advancing technology means that the use of slides is declining, it is worth remembering that many smaller groups and venues are often not able to provide the facilities required for multimedia presentations. What is more, slides make great visual aids – they are easy to use and colourful. The wide availability of slide projectors also makes them a good choice. It may take a while to have slides produced for a talk, so plan in advance.

Always load a slide carousel before going to a talk. (If you have practised the talk, it should be ready to go.) Double-check that all slides are the right way up and the right way round. This is particularly important for slides with text and important diagnostic features (e.g. a portside rudder). However, if a back-to-front slide does inadvertently slip through the net, it isn't necessary to point it out to the audience and risk interrupting the flow of a talk (you may just get away with it!).

Slide projectors are usually operated by remote control, which allows the speaker to position him/herself in a spot where it is easy to check which slide is showing on screen (although it is best to mark up slide-changes in your notes or script, and not to have to check the screen). Always remember to face the audience; do not turn around and talk to the screen. When presenting detailed information, a pointer may be required to highlight particular features on the slides.

Multimedia presentations (PowerPoint and video):
All of the above tips for slide presentations also apply to the use of multimedia presentations. The most common program used is Microsoft 'PowerPoint', which allows the creation of an innovative talk using a range of digital media. The use of video clips, layered images and a range of annotation possibilities combine to give a stunning production. The ease with which such presentations can be created allows talks to be quickly adapted for a range of audiences.

Bear in mind when designing a presentation that it is not necessary to use all the available 'flashy' media options. The visual aids should not overwhelm the facts and ideas that are being presented. Ensure images are clear, and as large as possible, and avoid excessive use of text – an audience wants to listen to the speaker and see the pictures. Also avoid gimmicks such as the range of slide transitions on offer – when repeated they can become distracting. Take care to test each presentation in advance. When using projection equipment provided by a venue it is wise to check which version of the software their system is running and the graphics capability of the computer. This will help prevent problems such as having video clips that are too large to play or text that disappears off the screen.

Online presentation: One of the fastest and most effective ways to make information available to a wide international audience is through the internet. The massive growth in affordable electronic media has led to significant expansion in the number and variety of websites. Establishing a website can promote a project far and wide and the content can be regularly updated with new discoveries or information. A wide range of graphics can be used to great effect.

When planning a website there are some basic points that will help create a dynamic but informative product:

- *Structure* – plan the website on paper in advance and organize material into logical groupings to help determine the most suitable structure. This is essential even when a web designer is being employed/consulted.
- *Basic details* – these should include the purpose of the site, the author/body, and formal details such as a charity number and contact details.
- *Navigation* – a user might not enter the site from the home page, so make sure there are good links (including one to the home page) on every page of the site. Work on the 'three click' rule (namely, a user should be able to get to any page of the site within three clicks of the mouse).
- *Conventions* – be aware of current guidelines on the design of web pages, use of text, colours, graphics, animations and sounds. Many books and online guides exist on website design. As this field is constantly developing, it is essential to ensure that recently published sources are consulted.
- *Accessibility* – in the UK, the Disability Discrimination Act (1995) requires by law that website design addresses accessibility issues for the disabled user. This means using things like clear text of a reasonable size, appropriate colours and avoiding gimmicks or over-cluttered backgrounds. If such issues are addressed at the design stage of a website, they are much more economically viable than attempting to incorporate them at a later stage.

Displays: Organizing a display of project work is another effective way of reaching a wide audience. Such displays can range from large-budget exhibitions produced by a professional designer, to relatively cheap displays prepared entirely by team members. Venues for displays include museums and libraries, the foyer of a sponsor's premises or even the local church hall. Planning timescales can vary from weeks to years. A wide range of material can be used effectively. Static displays can involve photographic and drawn material, text and interpretive illustrations. Film footage can either add interest as part of a display or form an easily distributed and attractive presentation in its own right. The same rules about accessibility apply. It is better to have a few bold pictures with captions readable from two feet away than include everything at an unreadable size.

If archaeological material is going to be displayed, careful attention must be paid to questions such as security, the extent to which the environment can be controlled and whether the material concerned is sufficiently robust to be transported and exhibited (see chapter 16). Unconserved material is rarely suitable for display due to its potential instability and the need for a carefully controlled environment. Much conserved material will also require a controlled display environment. This may be available in a museum setting but can be a problem in a mobile exhibition. Some displays allow visitors to handle archaeological material and there is no doubt that this can be very rewarding, especially for those with impaired sight. However, it is clearly important to select the material to be subjected to this treatment very carefully. Much archaeological material will not be appropriate, and certainly unrecorded or un-conserved material should never be put at risk in this way.

Displays of information gathered from project work are often very popular with people local to the area of the site. Such displays are not only good in terms of informing local people, they can often foster an interest in the well-being of the site and encourage local vigilance; hence the potential returns for a little effort can be considerable. Displays of project work can often be a very useful focal point for fund-raising activities. Current sponsors' names can be prominently displayed and potential sponsors may be attracted by a dynamic and attractive presentation.

Press and media

The press and media can provide many opportunities for communication with a large and varied audience. Local coverage is often picked up by the regional and national media (figure 20.2). However, although press coverage can be very useful in publicity terms, it can sometimes be very unpredictable. The treatment of a story that a journalist

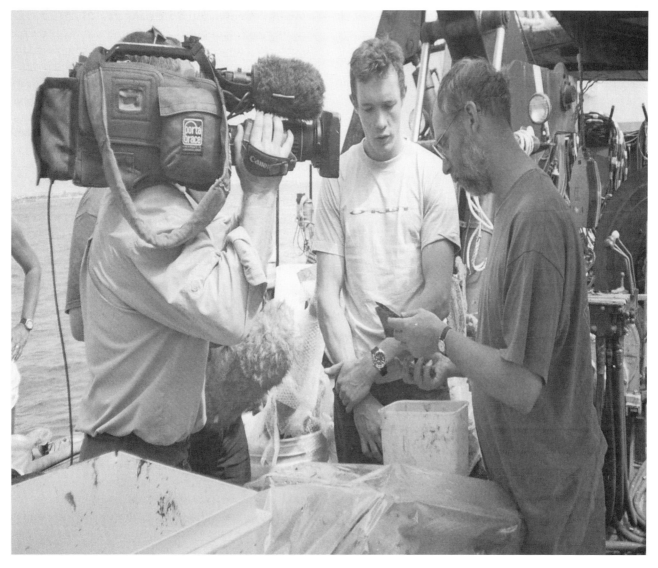

Figure 20.2 Increasing public awareness: underwater archaeology is inaccessible to many people. To engage with the public, all team members must be prepared to exploit any available opportunities for publicity, as here, where a member of the team is interviewed for TV while working on the *Mary Rose* (1545) site. This may be disruptive for the project but it is vital for the discipline as a whole. (Photo: Mary Rose Trust)

feels will be suitable for publication may differ greatly from your own. It is worth considering the following points:

- Will involving the press help the project?
- Journalists will often simplify a story (e.g. 'could be' can become 'is').
- In Britain, news about archaeology is still usually presented in one of a limited number of ways such as ritual (druids), King Arthur, human sacrifice/cannibalism, the oldest, the biggest, the best (more important than . . .) and, of course, 'treasure'. To avoid these, give the journalist another angle.
- 'Off the record' is taken to mean the information can be used but is not attributable.

- Do not be rushed into answering journalists' questions. Phone or e-mail them back with a considered response.
- Try to get as much control as possible (but remember, even the journalist does not have final control over what is printed). Review the journalist's record of the interview. Check quotations. Discuss the journalist's view of the story and how it will be presented – is it a feature article or a news item? Check who else they are talking to.

Drafting a press-release: Local television, radio and newspapers are always looking for good stories. They will recognize that underwater archaeological projects can provide interesting news if they are presented to them properly. A well-written press-release will alert the media that the project exists, highlight significant events and provide them with sufficient facts to produce a piece easily and quickly. Bear in mind the following when drafting a press-release:

- Be aware of what makes news. Interesting elements of news include discoveries, pictures (especially with people and colour in them), objects, conflicts or crises and links with local history. The local media are likely to be supportive of long-running stories with the potential for updates.
- Newspaper articles are written with all the important points in the first paragraph. These are then repeated and expanded throughout the rest of the article. Reflect this style in a press-release.
- Keep the press-release short and simple. Avoid technical language and overlong supporting detail. Be ruthless about short sentences and paragraphs.
- Stick to one topic (save other topics for future press-releases).
- Assume no prior knowledge of the subject. It is, of course, a challenge to explain everything whilst remaining concise.
- The press-release can be supported by background notes. Strike a balance between keeping the journalists informed of the background and keeping it short enough so it will be read. Alternatively offer contact details for anyone wanting further information.
- Establish contact with the news editor and follow up press-releases with phone calls.

WRITING REPORTS AND PUBLICATIONS

The quality and quantity of scholarly publication, more than anything else, will influence the future direction and value of work in maritime archaeology. The fact that other workers can use results in their research with confidence will serve to determine the credibility of the project, the team and even the sub-discipline of maritime archaeology. Scholarly publication will demonstrate to the relevant authorities that the work was genuinely archaeological and carried out to the required standards.

It is important to appreciate the difference between popular and scholarly publication. Glossy 'coffee table' publications, such as *The National Geographic*, which tell the story of a project, and include a few decorative photographs of artefacts and the odd technical-looking diagram, are not adequate publications in terms of making archaeological evidence available to other researchers. They certainly do not fulfil the fieldworker's obligation to disseminate information. Such volumes have a vital role

to play in communicating archaeological discoveries and ideas to the general public and should not be dismissed out of hand. However, they should also be recognized for what they are and no more. Projects that present such books/periodicals as a final publication rather than a preliminary communication lay themselves open to criticism.

This is not to imply that reports have to be dry and dull. Although glossy full-page photographs of 'the team', boats or equipment are unlikely to feature prominently, good writing style and presentation are very important. This said, the emphasis in a scholarly publication is on presenting fully the recorded evidence in a clear and objective manner.

A scholarly report will rigidly differentiate between analysis and inference. It will state what was found and describe the evidence as objectively as possible, clearly highlighting any subjective elements. Conclusions drawn from the evidence will be presented as a separate section. Clarity and accuracy are more important than a good story and, although fieldwork methodology is important, it should be presented as an additional way of assessing the validity of the results obtained and conclusions drawn. Details of the logistics of a project will usually only merit a brief summary, unless they have a direct bearing on the nature of the results achieved.

A scholarly publication aims at presenting evidence in a way that allows other researchers to make their own judgements and draw their own conclusions about a site. This means it is important to avoid being over-selective about what is included. Evidence presented should not be limited to that which supports the conclusions drawn in the report. Honesty about conflicting evidence is one mark of a good researcher. Evaluating and resolving such conflicts without distorting the facts or resorting to unsupported speculation is a fundamental part of archaeological work. Practical considerations mean that an element of synthesis is inevitable, but the fuller the report, the more useful it will be. Evidence not included can be referenced in the archive. Microfiche and/or databases provide an opportunity for making large amounts of data readily available.

With increasing amounts of digital data being gathered, processed and analysed, the internet is being used for a variety of forms of electronic publication (plate 20.2). Project details that have not been included in the final report could be made available on-line or through a well-established digital data storage source.

Many people find writing reports, especially for publication, a very challenging experience. The problem is often knowing where to start. Fortunately, word-processing packages make the whole process of report writing a lot less intimidating than it used to be.

The first step is to decide what type of report is required – interim statement or full report. Some projects are substantially unpublished, except perhaps for brief notices, until the appearance of the final report; other projects issue detailed seasonal interim reports and specialist reports in advance of a final report. An increase in the range of fieldwork that has taken place in recent years is reflected in the variety of resulting publications. These include leaflets that summarize progress during work, websites detailing latest discoveries, popular reports aimed at both the public and professionals, and academic publications either as reports in journals or complete monographs (subject-specific, free-standing publications).

Interim reports comprise a brief account of the fieldwork and an outline of some of the major results. They can serve to inform other workers and gather further information from them. A suitable periodical for an interim report would be the *International Journal of Nautical Archaeology* (*IJNA*) published bi-annually for the Nautical Archaeology Society. Someone working on a similar project elsewhere may read it and be able to offer assistance or advice. If the final report is likely to take a considerable time to produce, great effort should be made to produce one or more interim statements.

The key to any report is organization. Draft the structure before starting to write, paying attention to the need for specialist contributions and include deadlines for each stage. The layout chosen will not necessarily be the same as that used in other reports, but common section topics are related to strategy, methods, results, analysis and interpretation. English Heritage's *Management of Archaeological Projects MAP2* (English Heritage, 1991) provides guidance on the publication of archaeological projects.

The following list is an example of a layout for a site report:

1 Brief summary (for wider circulation than the full paper).
2 Introduction: the aims of the work and general description.
3 The site and its environment: description with plans and maps.
4 Past work in the area and/or historical research relating to the specific site.
5 The research design and the fieldwork strategy, including a discussion of their effectiveness.
6 The structural features: the hull structure (in the case of a wreck) or buildings/occupational features. Correlations of stratigraphy with features and objects.
7 The objects, environmental and scientific evidence. This should include description, numerical information, diagrams, tables, plates and scale drawings and discussion of function, parallels, significance and implications.
8 Discussion of site formation and chronology.

9 General discussion: assessment of the site in its wider context.
10 Acknowledgements.
11 Appendices: supportive specialist reports on specific topics, which are usually too detailed for inclusion in the main body of text or include extensive numerical data. These must be related to the main text and to each other.
12 Bibliography.

Before writing, ensure familiarity with the publisher's requirements. Papers submitted to academic journals may have to adhere to a set format or length, so check with the editor, or look for the 'notes for contributors' in previous issues. Complete the manuscript, and offer it to a journal that covers the appropriate region, period or speciality. If the intention is to publish in a more substantial work, such as the Council for British Archaeology Research Reports, you may have to submit an outline of your publication for acceptance by the publishers. If concerned about grammar and syntax, ask someone more skilled in this field to comment and advise. It may be possible to obtain grants towards publication and a proposed publisher may be able to advise on this, should any be required.

One responsibility an author has is to ensure that the report title accurately reflects the type of site and range of evidence discussed in the article. A cardinal rule is to assume that the reader has no prior knowledge of the site or project. A title such as 'Port Augusta 1983–5' may mean little to archaeologists unfamiliar with the country, period, or subject of the project. Geographical references should be explicit for example 'off Penzance (Cornwall, UK)' rather than 'off Penzance'. A date for the site, however approximate and tentative, will help readers identify sites relevant to their period of interest. Try to indicate the type of investigation being summarized or described, and whether the report is a preliminary account, one of a series, or final report.

Many people will have worked very hard to bring the project to completion. A public acknowledgement may be the only reward that they receive. It is common practice to acknowledge not only individuals and institutions who have provided assistance in fieldwork and report writing, but also those who have allowed the project to take place. All unpublished information, such as comparable finds from other sites, should be sourced and due acknowledgement made (as a footnote or reference if it is published, or as a 'personal communication' if it is not).

Establish at an early stage the form in which the specialist input is going to be presented and credited. Is a separate written report going to be submitted by the specialist for inclusion in the report and archive, or is information from the specialist report going to be integrated into the main text? The specialist concerned may have strong views on the subject so do not take anything for granted. Certainly, any input of this kind must be clearly and fully acknowledged. If information provided by a specialist is to be integrated into the main text, then the source of the information must be made clear. Equally, great care must be taken in editing major contributions for publication. A slight change in wording may alter the meaning of a sentence dramatically, especially where a complex technical discussion is concerned. Therefore, consult as fully as possible with the author concerned and be as accommodating as possible without allowing unnecessary wordiness.

All reports should be fully referenced, with a bibliography at the end. Most archaeological journals, including the *International Journal of Nautical Archaeology*, now use the 'Harvard' system for referencing. The reference will appear in the text as the author's family name followed by the year of publication and, where relevant, the page number, all in parentheses (brackets).

Restrict a bibliography to material cited in the report. Bibliographies should be laid out in alphabetical order of author's family name. If there is more than one work by any author, these works should be listed chronologically. Works of multiple authorship should include all the authors' names. The essential information is the author's name and initials, the title of the work and the place of publication. Some editors like to include the publisher, and/or the International Standard Book Number (ISBN). The key rule is consistency when laying out references and a bibliography.

Here are some examples of entries in a bibliography:

Baker, P. E. and Green, J. N., 1976, Recording techniques used during the excavation of the Batavia, *International Journal of Nautical Archaeology* 5.2, 143–58.
Pearson, V. (ed.), 2001, *Teaching the Past: A Practical Guide for Archaeologists*. Council for British Archaeology, York.
Riley, J. A., 1981, The pottery from the cisterns, in J. H. Humphrey (ed.), *Excavations at Carthage, Vol. 6*, 85–124. Ann Arbor, University of Michigan.

Within the text, these would be referred to as (Pearson, 2001: 56) – where 56 refers to the page number – or (Baker and Green, 1976).

Some of the factors that must be considered when planning illustrative material in a report are as follows:

- What is the illustration going to contribute to the report? Does it emphasize a point or provide material for comparison with other sites? The illustrations that are most informative and useful for making comparisons may not be the most attractive.
- Just as text can be either descriptive or interpretive, so can line drawings or electronically generated images. Descriptive plans depict artefactual and non-artefactual features in their observed archaeological associations. An interpretive plan attempts reconstruction by phase, or event.
- Include as many interpretive illustrations as are required to communicate an idea. Do not include them at the expense of record drawings and always identify them as interpretive in the caption and text.
- If drawing by hand, what size must the final drawing be, and what size must it therefore be drawn at to allow for reduction? It is useful to keep a common reduction factor in mind for as many drawings as possible (1:4 is common for pottery, for example).
- If the figure is a composite one that has involved the pasting together of multiple pieces of paper or drafting film, it is important to try to keep it to a transportable size. Alternatively, it may be better to scan each drawing and 'paste' them together digitally.
- If the figure is a digital one, make sure it is saved in an acceptable file format and at an appropriate resolution (see chapter 10) to allow for good quality reproduction. If in doubt, consult the editor/publisher.
- Has the appropriate scale been included or accurate dimensions indicated?
- Have all the illustrations been numbered, and references to them inserted at appropriate places in the text?

A SIGNIFICANT ACHIEVEMENT AND CONTRIBUTION

The dissemination of the results of archaeological work is important for individual sites and the profession as a whole. Anyone undertaking archaeological work, especially intrusive work, has an obligation to follow this work through to publication. Research without communication at every level is of very limited use. While the task of full publication of a site may seem daunting, the results can be highly rewarding. On a personal level, it means involvement in the investigation and dissemination of a piece of history that is of relevance to the human past. On a wider scale, the results of a project may increase understanding of a period in history, or even change established interpretations.

The production of a published report should instil a tremendous sense of achievement. While the commitment and dedication required to make this happen is considerable, the ultimate reward of adding to the accumulated knowledge of the past should be a more than adequate inspiration. The work has been done, now get out there and let people know about it!

FURTHER INFORMATION

Anon (ed.), 2007, *Dive Straight In! A Dip-In Resource for Engaging the Public in Marine Issues.* CoastNet, UK.

Arts and Humanities data service (http://ahds.ac.uk).

Bolton, P. (ed.), 1991 (3rd rev. edn), *Signposts for Archaeological Publication: A Guide to Good Practice In the Presentation and Printing of Archaeological Periodicals and Monographs.* Council for British Archaeology.

Council for British Archaeology Notes for Authors (www.britarch.ac.uk/pubs/authors.html).

English Heritage, 1991, *Management of Archaeological Projects (MAP2).* London (www.eng-h.gov.uk/guidance/map2/index.htm).

English Heritage, 2006c, *MoRPHE Technical Guide: Digital Archiving and Digital Dissemination.* London.

PRESENTING AND TEACHING ARCHAEOLOGY

Adkins, L. and Adkins, R. A., 1990, *Talking Archaeology: A Handbook for Lecturers and Organizers. Practical Handbooks in Archaeology, No. 9.* Council for British Archaeology, London.

Hampshire and Wight Trust for Maritime Archaeology, 2007, *Dive into History: Maritime Archaeology Activity Book.* Southampton.

Henson, D. (ed.), 2000, *Guide to Archaeology in Higher Education.* Council for British Archaeology.

Jameson, J. H. Jr and Scott-Ireton, D. A. (eds), 2007, *Out of the Blue: Public Interpretation of Maritime Cultural Resources.* New York.

Pearson, V. (ed.), 2001, *Teaching the Past: A Practical Guide for Archaeologists.* Council for British Archaeology, York.

1 Appendix: Anchor Recording

Although anchors are considered symbols of the maritime world, it is surprising how little work has been undertaken to date in terms of collecting and organizing the wealth of information that exists in the form of anchors found on wreck-sites, in museums and on public and private property all over the world. The NAS is helping to address this issue with the Big Anchor Project, which aims to develop a global tool for the identification of anchors by helping individuals to gather information in a consistent format. It will result in a freely accessible, online database of anchors, which will provide a valuable tool for anybody undertaking research or with a general interest in the subject.

This appendix provides information about stone and stock anchors and how to record them. For further information and to download anchor recording forms, see the Big Anchor Project website (www.biganchorproject. com).

STONE ANCHORS

Stone anchors were, almost certainly, the earliest type of anchor to be used, and some types are still in use today. For this reason the dating of stone anchors is immensely difficult. Any dated examples are, potentially, of considerable importance so it is vital that any dating information is recorded. Most stone anchors are, in fact, composite anchors, in which the stone provided weight to take the anchor to the seabed, while the holding power was provided by means of separate arms, usually made of wood. There may have been additional stones hung to help hold the anchor on the ground (analogous to the present-day use of a short length of chain).

Pierced stones have been used for a variety of other purposes, including weights for fishing nets/lines and for buoys and mooring, and some more unusual ones, such as driving fish into nets by beating them on the water. At present, insufficient evidence exists to define clearly these various uses for pierced stones, so all examples should be recorded.

There is plenty of evidence that a variety of objects have been re-used as stone anchors and fishing weights. This must be recognized, as it could lead to problems when it comes to ascribing a date to an anchor, or even recognizing an object as an anchor. Quern and mill-stones could be recycled as anchors (Naish, 1985), so the discovery of such a stone on the sea-bed should not automatically lead to the assumption that it was being carried as cargo.

Stone anchors have been found with a variety of inscriptions, ranging from simple symbols to an elaborate carving of an octopus (Frost, 1963). Carvings can sometimes provide evidence of date. A Christian cross, for example, suggests a date after the introduction of Christianity to the area.

So far, five basic types of stone anchor have been recorded throughout the world. In many cases, anchors are made of either a naturally shaped stone, or a worked block.

No hole: These are stone anchors without a hole drilled through them. They may have a groove cut around them (waisted). Stones of this type were sometimes used to weigh down a wooden anchor. Small versions were used as fishing weights and larger versions as buoy weights. Worked rectangular and rounded versions are known, as well as those based on an un-worked stone

One hole: These are stone anchors with a single hole through them. They were also sometimes used to weigh down a wooden anchor. Smaller versions (maximum length 30 cm (1 ft), weight less than 10 kg (22 lb)) were used as net-weights and very small examples (maximum length 12 cm ($4^1/_2$ in), weight less than 500 g (1 lb)) as line-weights (dimensions taken from British examples). They can be found in two main forms:

- *With a central hole* (plate A1.1). These are sometimes called 'ring anchors'. Quern and millstones can be recycled as anchors of this type. In eighteenth-century Yorkshire, worn out millstones were used as buoy anchors. Variations are known with smaller holes at the edge, either to attach the stone to a wooden anchor, or for a smaller rope to help in handling the anchor.
- *With a hole at one end.* These are the earliest known forms of stone anchor. Examples have been found in the eastern Mediterranean dated to the second millennium BC. These have continued in use until the present day, and are one of the most common type of fishing weight. Variations are known with an additional hole drilled into the top of the stone, linking with the main hole. This is believed to have been for securing additional ropes to aid recovery and prevent loss.

Two holes: This type of anchor has two holes, one at each end of the stone (plate A1.2). These anchors could be used in two ways, both of which are known from recent ethnographic examples. In the first example, one hole took a rope, the other an arm (usually of wood). In the second, only recorded on fishing gear, both holes took ropes. Wear patterns on the stone should make it possible to distinguish which method was used.

Classical: This anchor was commonly used in the Mediterranean by both Greek and Roman shipping, though plenty of examples are known from the Atlantic, Red Sea and Indian Ocean. It is often referred to as a 'classical' or 'Roman' anchor, although evidence suggests that it continued in use for many centuries after the end of the Roman Empire. The holes form a triangle. The upper hole, which took the rope, can either be in the same plane as the lower holes or it can run across the stone. There are usually two lower holes (though three are known), always in the same plane. Rough examples, on flat slabs of stone, are probably 'sand anchors' designed to lie flat on a sandy sea-bed to achieve maximum grip.

Indian Ocean: This type is currently known from the Indian Ocean and Red Sea. It consists of a long stone block with the holes for the arms cut at 90 degrees to one another. It is known to have been used during the medieval period, though when it was developed is unknown.

In addition to illustrations (in all cases the stone should be drawn and/or photographed), the following information should be recorded for all stone anchors or weights:

- *Overall dimensions* – In the case of un-worked examples, maximum dimensions should be recorded. For worked examples, the length, breadth and depth should be recorded.
- *Associations* – The association of the anchor with any other material, wreck, structure, etc. must be recorded. The dating of stone anchors is very difficult because they have been used for millennia. Any dating evidence needs to be recorded in detail. In addition, stone weights have been used for a variety of purposes and precise association may throw light on these.
- *Holes* – Each hole should be measured and its shape recorded. If possible, the internal form should be recorded: does it have straight sides or do they taper? If it tapers from one side toward the other, this indicates that the hole was worked from one side. If it tapers towards the middle, this would indicate that the hole was worked from both sides. The holes may show signs of wear from where ropes were tied and these should also be recorded.
- *Other parts of the anchor* – There may be traces of the arms surviving. These were usually made of wood, though metal examples are known. They should be recorded and treated as any other objects made of these materials.
- *Geology* – If possible, the rock from which the anchor is made should be identified. This can help identify trade routes and manufacturing centres.
- *Inscriptions* – Any inscriptions or marks should be described, drawn and photographed.
- *Weight* – If possible, the weight of the anchor should be recorded. This may help distinguish between boat anchors and gear weights.

To download a 'stone anchor recording form', visit the Big Anchor Project website.

STOCK ANCHORS

The Big Anchor Project has developed a 'stock anchor recording form', which prompts for consistent information under the following headings:

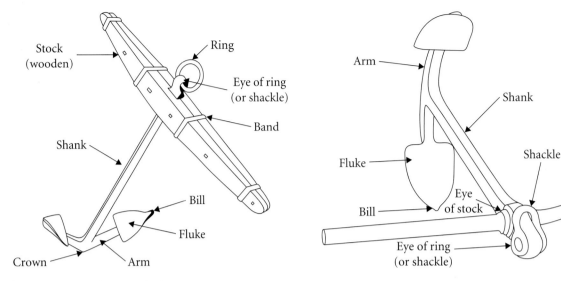

Figure A1.1   Anchor terminology

- context (category, site, location, ship name/type/ size, function, type);
- date and origin (including degree of certainty);
- features (fluke shape, stock type, shank form, crown, arms, weight);
- dimensions (for shank, arms, shackle, fluke, ring).

The form and guidance notes to aid completion are available from the Big Anchor Project website. Figure A1.1 gives the anchor terminology.

It is recommended that the data are complemented with a selection of photographs (with scale). Ideally, five photographs are suggested, the most important one being an overall view of the anchor. The five suggested view are as follows:

1 Overall view, showing the entire anchor from the front (both arms visible on each side).
2 View of the interior face (palm) of one fluke, showing its shape.
3 Close-up view of one arm, showing the shape of the arm.
4 Close-up view of the crown, showing the welding of the two arms.
5 Close-up view of the upper area of the shank, showing the ring and other features.

FURTHER INFORMATION

Frost, H., 1963, *Under the Mediterranean*. London.

Harland, J., 1984, *Seamanship in the Age of Sail: An Account of the Shiphandling of the Sailing Man-of-War 1600–1860, based on Contemporary Sources*. London.

Naish, J., 1985, *Seamarks: Their History and Development*. London.

Nelson Curryer, B., 1999, *Anchors: An Illustrated History*. London.

Upham, N. E., 2001 (2nd rev. edn), *Anchors*. Princes Risborough.

Appendix: Guns 2

Contents

Many wreck-sites are located due to the recognition of guns on the sea-bed. The increasing use of remote-sensing methods (see chapter 13), and divers striving to reach greater depths, will probably result in even more gun sites being found in the future.

The durability of the raw materials used to manufacture guns and their size (often considerable) contributes to the relatively good survival of guns on the sea-bed. The usually distinctive shape of guns increases the likelihood that they will be recognized and they may be the only visible indicators of a wreck-site. Guns can provide an initial means of discovering information without invasive excavation (e.g. about the nature of a site and type/size of a wrecked vessel). Irrespective of size, guns are portable and have a perceived market value and are therefore potentially at risk. Removal from their original position without a predisturbance survey can result not only in loss of information about the nature of the site, but also potentially of the site itself.

In the past, guns have been raised indiscriminately, their protective concretion smashed off and their contents extracted without adequate recording and analysis. Although apparently robust, they do require conservation (see chapter 16), which can be costly. Consequently, many can be found (partially or inadequately conserved) disintegrating in gardens, public house car parks or dive clubs, irreparably separated from their archaeological context.

This appendix is included to illustrate the information that can be gained from ordnance and ways in which all types of ordnance can be studied, recorded and illustrated.

THE IMPORTANCE OF SEA-BED RECORDING

As with any other artefact type, the position and relationship of other objects in association with the gun will yield important information. If the guns were present on the ship to carry out their primary function (i.e. not as ballast), then each one may have an associated carriage and collection of gun furniture (such as rammers, spongers, shot), depending on the period. Ropes for breeching the gun and as recoil-restraints may be present, together with block and tackle. The organic portions of these rarely survive except in waterlogged conditions. They are, therefore, extremely important.

The wrecking process and the greater weight of the gun can result in the whole assemblage coming to rest upside down (with the carriage uppermost) and this can be difficult to recognize. Understanding the development of the carriage is as important as studying the gun, so any carriages or fragments should be carefully recorded. Fixing bolts or straps may be fragile and degraded, but their positions and associations will also be important in understanding the structure of the carriage.

In most instances, operational guns were placed in specific places on vessels. In simple terms the lighter guns were usually higher up, with the larger guns carried as low as possible in the ship. In many cases the number of guns, their sizes and location (at least with respect to which deck) is historically known for specific classes of vessel. The fact that many guns either bear dates or can be dated and identified with respect to country of origin, means that they are extremely important and can be used to date – and potentially to name – particular vessels. Their distributional relationship can provide information about how much of a vessel may be represented and the processes of site formation.

The size, form, number and orientation of guns should be recorded in relation to each other and any other identifiable features. Many site-plans exist which consist entirely of distribution surveys of guns, anchors and predominant geomorphological features. To achieve this, each gun has to be identifiable (with an attached durable label carrying a unique number).

Once each gun has been numbered, the end of each gun should be measured to each end of every other gun. This provides the correct orientation of the guns relative to each other. In addition, a simple note of relative depth at each end of each gun should be recorded. If possible, the bore (front) and breech (back) end of the gun should be indicated, along with information about whether a gun is lying upside down. This can be an important factor in modelling site formation. For example, a distribution of two parallel rows of guns of similar sizes, with their breeches facing each other and bores facing outwards, may indicate a ship which came to rest on an even keel. One linear distribution of guns of either the same size or different sizes, with the bores all facing the same direction, can be indicative of a ship lying on its side. One would expect, in this instance, for a number of the guns to be upside down. On a site with a dispersed pattern of guns, the relative positions of the various guns may be important in understanding the formation of the site. Guns (particularly the more easily accessible upper and main-deck ordnance) were often jettisoned to lessen the draught of a vessel in distress and may leave a 'debris trail' connecting parts of a site. It should be noted, however, that the buoyancy afforded by a wooden carriage can result in guns moving considerable distances under water.

It can be difficult to identify particular features of guns still on the sea-bed as they are often covered with corrosion products. However, a great deal of information can be obtained – without the removal of surface corrosion products – using simple observation and recording techniques.

In addition to the usual information required for any artefact type (e.g. site code, unique artefact number, context, position and associations), the following need to be considered when recording guns on the sea-bed:

- orientation description (aspect, angle of slope, upturned);
- material (cast iron, wrought iron, bronze);
- length overall, and length of barrel;
- maximum diameter (external);
- minimum diameter (external);
- bore diameter (if visible);
- presence of external features (trunnions, lifting dolphins, cascabel, lifting-rings).
- distance between trunnion faces; diameter of trunnion face.
- inscriptions or markings;
- evidence of carriage;
- evidence for breeching-ropes.

Sample gun-recording forms are available on the NAS website (www.nauticalarchaeologysociety.org). It is advisable that guns are sketched/drawn and photographed *in situ*, paying particular attention to any distinctive features.

Information can be recovered from isolated, individual fragments of guns, such as lifting dolphins, trunnions and cascabel buttons (see figures A2.1, A2.2 and A2.3). Powder-chambers and lifting-rings are often found separately from larger wrought-iron gun concretions. Weak points on iron swivel guns include the junction between the barrel and the chamber holder and the chamber holder and the tiller. Spare chambers for these are relatively solid and often survive well. These objects may remain on sites where guns have been salvaged, but they can still provide important evidence if they are recorded in a systematic way.

IDENTIFICATION OF MATERIAL

The majority of guns are made of iron or an alloy of copper with tin, lead and zinc in varying amounts. In antiquity, such alloys were referred to as 'brass', although the current definition of brass is copper alloyed with zinc alone. The main constituents, however, are copper and tin, and, as such, they are referred to as 'bronze' guns. Composite guns do exist, principally of bronze and iron. Guns of copper and other metals wrapped around with cord, plaster, leather or wood have also been found. Many questions surround the place of such guns in the evolution of gunfounding.

A general trend from wrought iron, to cast bronze and finally to cast iron can be suggested, and this sequence is useful for attributing a broad date to a gun. In the UK, there is little evidence for the casting of large iron guns before the beginning of the sixteenth century and these tend not to feature ornate decoration. Such embellishments are more likely to appear on bronze ordnance because they are difficult to cast successfully in iron.

In an underwater context, iron is easily distinguishable from bronze due to the formation, over the surface of the object, of a thick layer of concretion resulting from the corrosion of the parent metal (see chapter 16). Any damage to the concretion layer reveals a black layer that

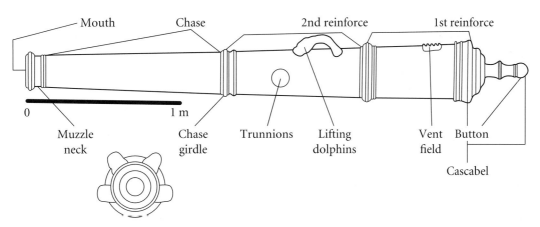

Figure A2.1 Cast ordnance: terminology – a corresponding recording form is available on the NAS website. (Based on original artwork by Ben Ferrari)

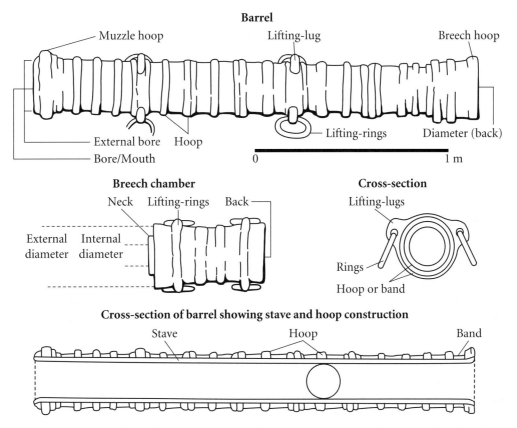

Figure A2.2 Wrought-iron breech-loading tube gun: terminology – a corresponding recording form is available on the NAS website. (Based on original artwork by Ben Ferrari)

will rapidly oxidize to produce the characteristically red-brown colour of rusty iron.

Bronze guns do not develop concretion, but they can support a surface layer of marine growth. These guns can also be stained or covered with thin corrosion products. If the latter are promptly treated by conservation specialists, a well-preserved, relatively easily identifiable object can be recovered.

The nature of the concretion that develops around iron objects will be influenced by the composition of the object and its local environment. A study of the concretion, how it was formed and how it isolated the object from the burial environment, can give valuable information to both the conservator and the archaeologist. The indiscriminate removal of concretion will re-activate corrosion processes and so should not be undertaken without

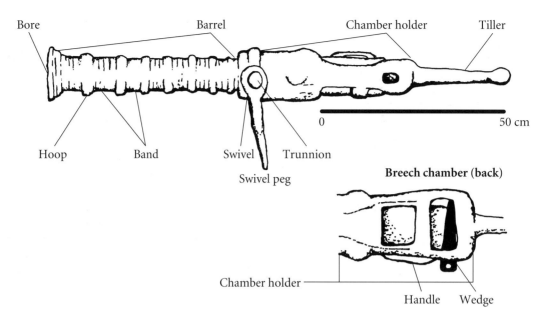

Figure A2.3 Breech-loading swivel-gun: terminology – a corresponding recording form is available on the NAS website. (Based on original artwork by Ben Ferrari)

specialist consultation. Any de-concretion process should be recorded along with any features and artefacts subsequently identified. The concretion covering wrought-iron guns is often very irregular, completely obscuring the object within it. The radiography of all concreted objects is advisable before any attempt is made to remove the concretion (chapter 16).

It is often difficult to differentiate between wrought iron and cast iron, especially under water, as diagnostic features are frequently obscured by concretion. A wrought-iron object may be indicated by laminated or layered corrosion products, perhaps visible in a break in the overlying concretion, often mistaken for wood. The final phases of the corrosion of cast iron is typified by the production of a graphitized object, steel-grey in colour with a greatly reduced weight. Cast-iron objects in this condition are easily damaged, even by touching. The last stage of corrosion may be the reduction of the object to a steel-grey fluid.

CLASSIFICATION BY METHODS OF LOADING

Guns can be classified into two types depending on the method adopted for loading the gun:

Breech-loading guns: These guns are loaded from the breech rather than the bore. Early examples typically have a separate chamber to contain the gunpowder and a tube forming the barrel. Barrels can be cast bronze, cast iron, or wrought iron. The chamber may be of a different material to the barrel and is either inserted into an integral holder within the gun itself, or it may form the rear part of the gun by slotting into the back of the barrel. This type of gun could be rapidly re-loaded without the necessity of moving the gun back out of the port.

Muzzle-loading guns: These guns are sealed at one end. The powder, projectile and wadding are all placed into the gun at the muzzle and rammed down. This group also includes some of the earliest guns made. Early bronze handguns (termed hand cannons) were loaded in this way. On a ship, this type of gun must be brought inboard to re-load, or must be re-loaded outboard. Muzzle-loaders have a wide range of sizes and include mortars (which are very short and have a large internal bore).

Muzzle-loaders that have a defined area (smaller in diameter than the bore) into which the powder is placed are termed 'chambered muzzle-loaders'.

CLASSIFICATION BY SHAPE

The shape of the gun is determined largely by the characteristics of the raw material and the technology available to work it. Cast bronze ordnance was used from the fourteenth century AD. Early iron ordnance was built using wrought iron, heated and hammered into shape, requiring simply the skills and facilities of the local blacksmith. The difficulties of casting iron (due to the high temperature required to keep the metal fluid) meant that the casting of iron guns was less common until the seventeenth century.

Tube guns (figure A2.2): Typically, tube guns are constructed of wrought iron, although cast bronze examples also exist. This technique of manufacture consisted of placing staves of iron longitudinally over a 'former' to create an open-ended tube, which was then overlain and strengthened with thick hoops and wide bands. A separate unit, called a chamber, had to be made to contain the explosive charge. Chambers may be constructed of one piece, heated and beaten into shape. Alternatively, they may be made of staves reinforced with a thick plug at one end and a projecting neck, which fits tightly into the open tube of the gun, at the other end.

Swivel-guns (figures A2.3 and A2.4): The term swivel-gun is used to describe a gun that was mounted on a 'Y-shaped' stirrup (often called a yoke and peg), allowing it to be moved or 'swivelled' horizontally. There would also have been a certain amount of vertical freedom of movement. The peg locates into a hole in a horizontal beam within the ship itself. Swivel-guns were easily transportable and designed for rapid fire at close range. They could be deployed by one person (figure A2.4).

Swivel-guns were usually made of wrought iron, cast bronze or iron, or a combination of iron and bronze. Typically, they have separate breech-chambers, forged or cast with carrying handles, and the internal bore is less than 10 cm (4 in).

Cast guns (figure A2.5): The earliest method used to cast bronze ordnance (Ffoulkes, 1937; Kennard, 1986) necessitated the making of individual 'patterns', which contained a mould both for the bore and for the gun itself. These were destroyed to remove the finished gun, thus each early cast gun is unique. Although attempts were made at standardization (using length, calibre, weight of gun, and weight of projectile) this was clearly difficult to achieve until guns were bored from the solid. A gun that did not conform was often termed a 'bastard' and this appears on many guns.

By the sixteenth century, English cast-bronze ordnance was generally termed, and ranked (in descending order of calibre) as follows: cannon royal, cannon, demi-cannon, culverin, demi-culverin, saker, minion, falcon (Blackmore, 1976; Hogg, 1970).

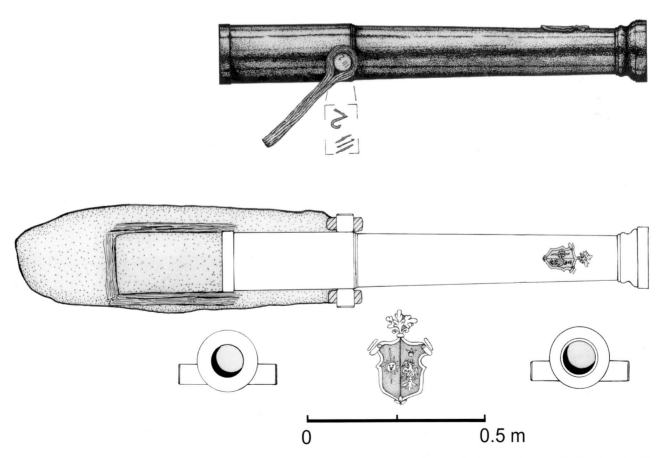

Figure A2.4 A small swivel-gun recovered from a Spanish Armada wreck off Streedagh Strand, Ireland, showing detail and the large concretion still adhered to it. (Doug McElvogue)

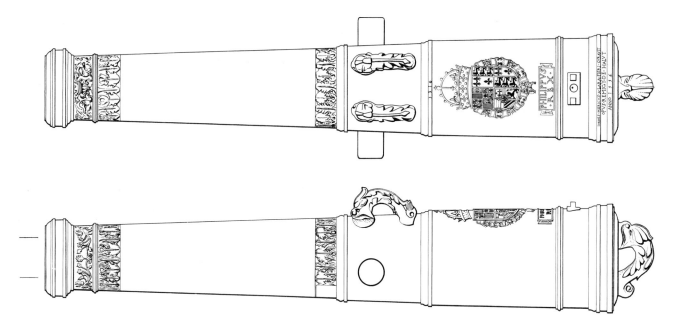

Figure A2.5 Plan and side view of a decorated cast-bronze gun from the Spanish Armada (1588) wreck *La Trinidad Valencera*. (Colin Martin)

The latter half of the seventeenth century saw the use of the weight of the shot to describe the gun (e.g. the '12-pounder'). The increased use of the shell during the nineteenth century further complicated this system, as guns primarily firing shells were described by their bore size in inches. The gradual increase in the number of iron guns over bronze, which began in the seventeenth century continued, primarily driven by the cheapness and availability of iron. Surface decoration is restricted on iron guns, and lifting dolphins are rare. By 1800 a breeching-loop situated by the cascabel button, or a loop in place of the cascabel button, was common on both bronze and iron ordnance.

INSCRIPTIONS AND DECORATION

A great deal of information can be found on the gun itself. In the case of bronze guns, the entire surface can be decorated, usually in relief (raised). This is especially true of the larger sizes of guns, which commanded attention both as powerful weapons and as art objects. Written information can be cut into the metal, be raised, or stamped. There appears to be no absolute rule regarding the positioning of inscriptions and decoration. Much of this information will only become available after conservation treatment and it is not justifiable to remove concretion under water to locate such markings. Listed below are the most common features with their usual location (conventionally guns are described from the cascabel towards the muzzle).

- *Monograms* – These comprise initials, often surrounded by a garter, usually either the reigning monarch or the chief official in charge of gun production. Monograms are usually restricted to the upper surface of the gun, either on the first reinforce, the second reinforce or the chase.
- *Heraldic/other devices* – These are a pictorial representation, often a coat of arms, and are restricted to the upper surface of the gun, either on the reinforce or the chase. They are usually in relief (figures A2.5–6).

Inscriptions comprise any other letters or numbers that appear on the gun. They give various details of interest, including:

- *Weight* – The weight of the gun is usually incised, either stamped or hand-cut, and the units used will relate to the nationality of the gun. The historical study of weights and measures is very complex, and care must be taken in interpreting the units used. A note should appear with the gun stating whether the weight quoted is as it appears on the gun, or is deduced by some other means (figure A2.7). Common locations include the cascabel, on the base ring, by the vent, on either the first or second reinforce or the chase, and on either or both trunnions, usually the face or topside.
- *Name of foundry or founder* – These can appear either as initials, full names, maker's 'marks' or as part of a general extended description. Principal locations

Figure A2.7 The weight number on the breeching ring reinforce on the *Stirling Castle*'s demi-cannon, as shown in figure A2.13. (Doug McElvogue)

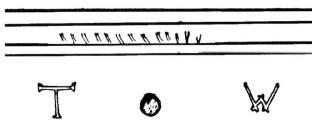

Figure A2.8 Founder's mark (Thomas Western), touchhole and details of markings on the first reinforce on the *Stirling Castle*'s (1703) demi-cannon, as shown in figure A2.13. (Doug McElvogue)

Figure A2.9 Detail of the broad arrow on the *Stirling Castle*'s (1703) demi-cannon, as shown in figure A2.13. (Doug McElvogue)

79A1232

20 cm

Figure A2.6 Decoration: a Tudor rose heraldic emblem from one of the bronze guns from the *Mary Rose* (1545). (Mary Rose Trust)

(usually on the uppermost surface of the gun) include the base ring, by the vent, on the first or second reinforce and chase, or on the end of one or both trunnions. They can be engraved or in relief.

- *Foundry number, gun number (serial number)* – Principal locations include underneath the cascabel or breech area, on the breech ring, on or under the first reinforce, on or under either or both trunnions. Occasionally they are found on the muzzle face. Guns do not always have either batch or individual numbers and can have one without the other. If one appears by itself, it is often the gun number.

- *Date* – Many guns carry the date of casting incised on the gun. The date can appear in either roman or arabic numerals and can refer to the calendar date, or to the year of the reign of the monarch, or even date (in years) following a significant event such as a revolution. The date can be incised on the base ring, the first or second reinforce, the face of either or both trunnions, or the chase.

Other markings: In addition to general inscriptions, often describing either the monarch or the founder (figure A2.8), other marks occur. One example is the British Government's Broad Arrow. These are often chiselled into the upper surface of the gun on the chase or the reinforce (figure A2.9). Large vertical lines, possibly located on the trunnions or the base ring, are often sighting aids. Quarter-sight scales may also be inscribed into the base ring. Other more commonly found marks include the bore diameter and the shot weight. These are often found incised into the bore of the gun at the muzzle. Breech chambers can carry marks that match them with their particular guns, or denote government issue.

Abbreviations: Abbreviations and initials relating to the gunfounder or foundry must be learned, or suggested by deductive techniques using the other information available.

PROJECTILES, CHARGES AND TAMPIONS

Projectiles are many and varied, as almost anything can be fired from a gun. A gun containing coffee beans has been recovered from the sea, and the heads of Turkish prisoners were fired from the guns of St Angelo during the Great Siege of Malta in 1565.

As with any other source of archaeological information, the contents of guns (including projectiles, wadding and gunpowder) should be recorded and investigated to an acceptable standard (figure A2.10) if any attempt is made to remove them (never under water). The relative positions of all components and their condition should be recorded. Contact should be made with analysts interested in the composition of gunpowder to determine optimum sampling techniques for the powder remains. There are several methods employed to ascertain whether or not a gun is still loaded. Often loaded guns are recognizable because they have a soft wooden bung (tampion) close to the mouth of the gun. Sometimes a tampion is located further down the bore (due to water pressure). A simple

method is to strap a tape-measure to a long rod and carefully measure how far down the gun it goes and compare this with the length taken to the touch-hole on top of the gun. A powerful torch is also useful.

It should be stressed that this removal exercise is best left to an experienced conservator because the various components of the charge and shot can suffer significant damage during the operation. Often the tampion and wadding are destroyed without even realizing they are present. In addition, attempts made without the benefit of conservation expertise and facilities may leave deposits in the bore that may compromise subsequent stabilization treatments. Care should be taken to look for any evidence of powder cartridge. Although combustion is not likely, gloves, safety goggles and protective clothing should be worn and care should be taken not to inhale any of the components.

Spherical shot are traditional projectiles. They can be made of stone, cast iron, cast bronze, cast lead, lead-covered cast iron or stone, or lead cast over iron dice. With regard to stone shot, the type of stone is important. Limestone is durable and easily worked, and so is an obvious choice. Granite is very heavy, and often used where it is abundant. Naturally, the type of stone used will depend on availability, either local or through trade. The nature of the working, whether finished or unfinished and whether or not tool marks are distinguishable, as well as

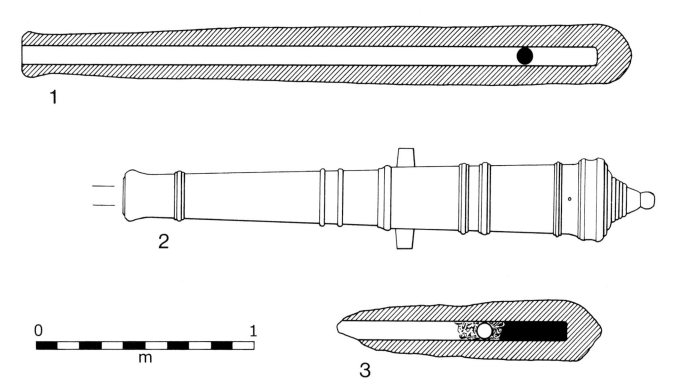

Figure A2.10 Plain iron guns from the Armada wreck *El Gran Grifon* (1588): 1) shows the sectioned shape of one gun within an abraded concretion deposit, with its iron roundshot still in place; 2) is the profile of a complete gun (a side or frontal view would be needed to show the set of the trunnions); while 3) is an abraded fragment within which the roundshot, wadding and powder charge are preserved. (Colin Martin)

the weight, diameter (measured with callipers) and circumference should be recorded.

With cast-iron shot, many carry the stamp of the reigning monarch or possibly the maker and shot weight. These should be recorded, as well as the diameter and condition of the iron. The weight of the shot should be recorded but care must be taken in the use of these data because corrosion processes in the sea can change the parent metal into lighter compounds. On-site location and association information (together with the condition of the shot) is important in assessing the effect of the burial environment. Individual shot from one assemblage can also show differential corrosion effects, so if this detailed information is required, shot should be recorded *in situ* and numbered before lifting.

In some cases, pairs of spherical shot were joined together, either with a bar or a chain. Other examples were formed of segments, which formed a sphere that separated on leaving the gun. Detailed drawings of different types of shot can be found within published texts (e.g. Munday 1987; Peterson 1969; Blackmore 1976).

Encased shot generally includes anything that can be grouped together within a container, such as grape shot. Anything can be placed within a canister or bag and fired from a gun. Particularly favoured items include pebbles, nails, iron dice and flint fragments.

Incendiary shot is a shot that has been wrapped in a cloth or rope impregnated with an inflammable mixture, usually around spikes projecting from the shot.

Shells (exploding shot) are cases into which either explosive powder or a mixture of powder and a further projectile were placed. They have a fuse-hole and often also have surface features such as lifting-lugs. They can be spherical, cylindrical and cylindro-conoidal in form. Traditional cases include cast iron or cast bronze.

Tampions are shaped blocks of wood placed into the bore of guns to protect them from damp. From muzzle to breech, the sequence can be: tampion, projectile, wadding, powder. Tampions, often made of softwood, are larger in diameter than the bore of the gun but with a suitable chamfer to make them a tight fit.

RECORDING AND ILLUSTRATING GUNS

Additional equipment may be required when drawing ordnance because of their potentially large size. These might include large callipers, scale rulers, profile gauges and tapes (which may be more suitable than rulers). For tracing and taking rubbings of ciphers and inscriptions, thin polythene can be used with fine permanent marker pens. Alternatively, soft paper (such as thin layout paper) for use with graphite, wax crayons or soft pencils might be considered.

Drafting film is preferable to paper in many situations because it is stable in wet environments and will reproduce more clearly. Being transparent, it can be taped onto a drawing board covered with graph paper to provide a grid on which to base the drawing.

When drawing reductions of guns, carriages, timbers, etc., it is better for the draughtsperson to use a hand-held board so that they can walk around the object, taking measurements, adding detail and notes to the drawing. These working drawings in pencil often get messy but final ink drawings for record and reproduction can be traced off them.

Ordnance should be drawn as a side elevation together with an end view of the muzzle to show the diameter of the muzzle and length, diameter and position of the trunnions. If the piece is highly decorated, a plan view may also be necessary to show the positioning of ciphers, inscriptions and mouldings (figure A2.11). With a wrought-iron gun, a plan view is often required to record the detail of the lifting-lugs and hoop configuration.

Large pieces of ordnance can be drawn to a scale of 1:5, 1:10 or 1:20, with any decoration or inscriptions drawn at 1:1, with the aid of a rubbing (figure A2.12) or tracing, or at a scale of 1:5 (figure A2.13). Care should be taken with inscriptions, as spelling can be unusual and even letters reversed. If a letter cannot be deciphered, it should not be drawn in as a 'best guess'. It is better to leave a space in the drawing and include an interpretation in note form.

When recording guns on the surface, the piece should be placed on a smooth, level floor with the muzzle supported so that the longitudinal axis of the piece is parallel to the floor. The central point of the muzzle should be at the same height as the centre of the cascabel. The floor can then be used as a datum line from which to measure the positions of the reinforcing rings, trunnions etc. If the floor is not level and smooth, or if it is not possible to support the piece so that the floor can be used as a datum line, the piece may be drawn using its longitudinal axis as the datum line. Points on its profile may be plotted with reference to the distance from the muzzle (e.g. diameter of gun at 0.5 m from muzzle; start of trunnion at 0.73 m from muzzle; end of trunnion at 0.85 m from muzzle). Two large set-squares (marked in centimetres on their vertical edges and supported so that they will stand vertically) can be used to measure the varying diameters of the piece if sufficiently large external callipers are not available. The horizontal distance between the two set-squares, when butted up to either side of the piece, gives its diameter at that point.

If it proves impossible to record a piece of ordnance in the conventional way, then it is possible to use a tailor's tape to record the circumference of the piece at the individual points. It is then an easy calculation (diameter

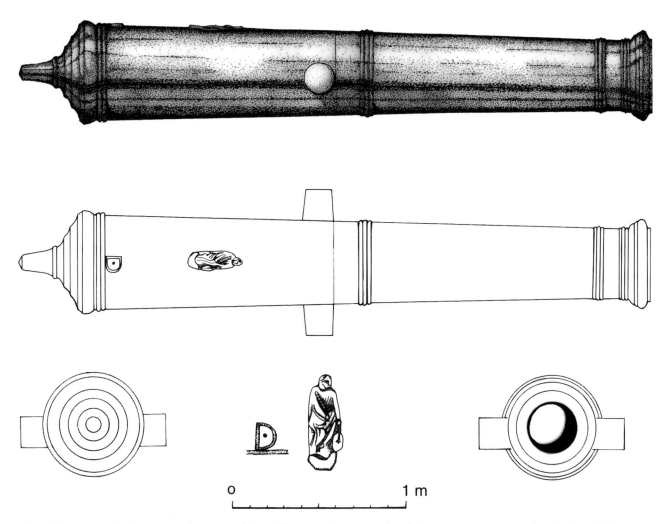

Figure A2.11   A drawing properly set out for publication of a piece of ordnance – a cannon perrier (designed for stone shot) from a Spanish Armada wreck off Streedagh Strand, Ireland. (Doug McElvogue)

= circumference divided by 3.14) to find the diameter and then the radius (diameter divided by 2). These can then be marked onto the drawing at 90 degrees to the centre line.

Diameters should be taken wherever the gun significantly changes in profile, such as at the reinforcing rings, trunnions, dolphins and at a number of positions over the cascabel. The diameters can then be joined up to give the external lines of the gun. Internal callipers can be used to measure the bore. A torch will suggest whether or not the bore is straight-sided, tapered, chambered or rifled. It should also be noted if the piece is still loaded (not an uncommon phenomenon for ordnance from wreck-sites) and, if so, what type of shot (see above).

It is preferable to add as much of the detail as possible in the initial drawing phase. However, if time is short, detailed photographs can be helpful when drawing features such as the lifting dolphins. Such details can be outlined and then added at a later stage using information taken from the photographs. Care must be taken to use photographs taken from the correct angle.

The length of a piece of ordnance is from the back of the base ring to the end of the muzzle, not including the cascabel. The length overall includes the cascabel. The main dimension and features required of the piece are given in table A2.1 for the piece of ordnance shown in figure A2.13. The minimum dimensions and features required are marked with an asterisk.

A marked metric linear scale should be included on any drawing. The original drawing should also carry the unique number of the artefact. Although it may be useful to publish length and bore in imperial (or other) units, those used on archaeological drawings should be metric. If any conversion has taken place this should be clearly stated (quoting the initial figures and the conversion factor used).

If a gun is recovered on its carriage, it will probably be convenient to draw the full assemblage first. This should be fully surveyed under water to record information such as the angle of the piece in relation to the bed of the carriage, the presence of any wedges or quoins, breeching-rope, breeching-tackle and the position of all concretions. This information will be vital for research and any reconstruction drawings. It is preferable to do post-excavation drawings at a scale of 1:5 because there may be a lot of detail to record on the carriage as well as the gun.

The surviving components of the carriage may include cheeks, bed, axles and wheels (or trucks), transoms, cap-squares, jointing tennons and dowels, wrought-iron bolts and rings. The wood should be carefully cleaned to show constructional features (e.g. bolt holes, treenails, rebates, remains of metal straps, hinges or brackets) and timber features such as grain, sapwood and rot. Different colours, hatching or stippling may be used to denote features such as iron concretion and nail holes. Each face of the components should be drawn so that detail such as variations in thickness can be recorded. Some elements may be inaccessible if the carriage is still partly assembled. In this case comprehensive measurements should be taken before drawing all parts at 1:5 scale.

Figure A2.12 Rubbing taken from the top of the barrel of the swivel-gun shown in figure A2.4. (Doug McElvogue)

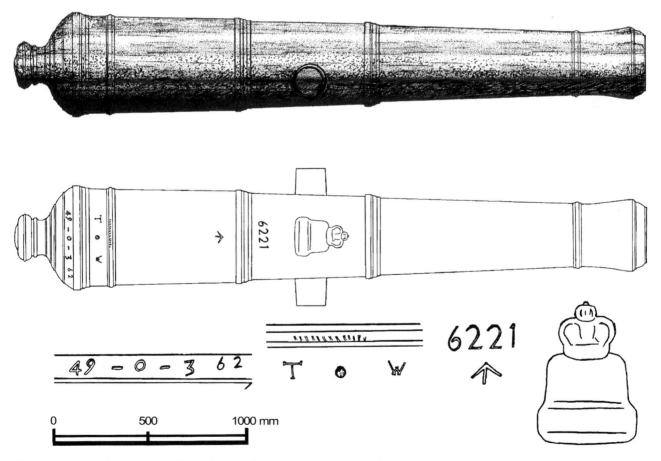

Figure A2.13 A demi-cannon from the wreck of the *Stirling Castle* (1703). (Doug McElvogue)

Table A2.1 The main dimension and features required for recording the gun shown in figure A2.13

Type	Demi-cannon (turned and annealed)
Length*	2.955 m (9½ ft)
Length overall*	3.255 m
Bore*	160 mm (6 in)
Maximum diameter at breech mouldings*	0.320 m
Maximum diameter at muzzle mouldings*	0.495 m
Trunnion diameter and length*	169 mm/169 mm
Touch-hole diameter	20 mm
Weight	2470 kg
Stamped weight	49 – 0 – 3 (49 cwt, 3 lb)
Inscription	6221, broad arrow, T W
Decoration	Charles II cypher (bare) with crown

* The minimum dimensions and features required are marked with an asterisk

Tracings are a very useful record to refer to during the conservation treatment, or when the piece of ordnance or carriage may be otherwise inaccessible. They are also the primary source for the 1:5 drawings. Final drawings for reproduction should be inked, and in black and white. A key may be necessary to denote features such as iron fittings, concretion, etc. Areas of erosion should also be indicated. The carriage will probably be incomplete so a conjectural reconstruction is useful, together with an 'exploded view' to explain the assembly of the components.

It is preferable initially to record wrought-iron ordnance and sledges found together as a single assemblage. It is best if the wheels or trucks are separated prior to lifting. Tracings will not be possible, so 1:5 scale drawings should be made. Try to have the sledge placed on a flat, smooth floor that can be used as a datum line. If possible, support the sledge so that the barrel of the piece is parallel to the floor. Draw the longitudinal axis of the gun as a horizontal pencil line on the drawing board and, using the graph paper backing as a guide, mark in the position and dimensions of the hoops on the barrel. It is better to start with a plan view of the gun and sledge, leaving space for an elevation below it. Graph paper backing makes it easy to transfer measurements and positions of features from the plan view to the elevation.

The draughtsperson should look out for details of the breech chamber, lifting-rings and markings, along with evidence of a gun-sight and wedges. An X-radiograph (if it can be arranged) is very useful in clarifying construction details.

After the barrel drawing is completed, the sledge can be drawn around it. Care must be taken to record features such as metal fastenings and breeching-rope, particularly the impressions of any rope surviving in concretions. Bolt holes on the underside of the sledge may indicate where the axle was attached. An elevation view of both sides may be necessary to show different features. An end elevation to show the bore-size is also required. Any surviving axle, wheels or tiller should be drawn separately before combining with the piece and sledge for a reconstruction drawing. When the barrel and sledge are separated for conservation treatment, each should be drawn to record any additional information. If the sledge is found without the barrel, it is still better to start with a plan view, which will also provide information about the shape of the missing barrel.

Ordnance accessories and other smaller ordnance items, known as 'furniture', should be drawn according to the guidelines given for artefact drawing in chapter 18. Handguns and other objects with a complex shape should be drawn at a scale of 1:1 if size permits. Rammers, scoops and other ordnance furniture may be recovered with long handles. In this case a 1:1 view of the head will provide detailed information and an extra view of the complete object at the same scale as the associated gun will also be useful.

FURTHER INFORMATION

Birch, S. and McElvogue, D. M., 1999, *La Lavia, La Juliana* and the *Santa Maria de Vison*: three Spanish Armada transports lost off Streedagh Strand, Co. Sligo: an interim report, *International Journal of Nautical Archaeology* **28**.3, 265–76.

Blackmore, H. L., 1976, *The Armouries of the Tower of London: Vol. 1, The Ordnance.* HMSO, London.

Carpenter, A. C., 1993, *Cannon: The Conservation, Reconstruction and Presentation of Historic Artillery.* Tiverton.

Ffoulkes, C. 1937, *The Gunfounders of England.* Cambridge.

Hogg, D. F. G., 1970, *Artillery: Its origin, Heyday and Decline.* London.

Kaestlin, J. P., 1963, *Catalogue of the Museum of Artillery in the Rotunda at Woolwich. Part 1: Ordnance.* London.

Kennard, A. N., 1986, *Gunfounding and Gunfounders.* London.

McElvogue, D. M., 1999, A breech loading swivel gun from the *Curacao*, 1729, wreck-site, *International Journal of Nautical Archaeology* **28**.3, 289–91.

Munday, J., 1987, *Naval Cannon.* Shire Album 186. Princes Risborough, Bucks.

Peterson, H. L., 1969, *Round shot and Rammers: An Introduction to Muzzle-Loading Land Artillery in the United States.* New York.

Smith, R. and Brown, R. R. (eds), 1988, Guns from the sea. *International Journal of Nautical Archaeology* **17**.1, special issue.

Appendix: NAS Training Programme 3

Contents

- An introduction to foreshore and underwater archaeology
- NAS Part I: Certificate in Foreshore and Underwater Archaeology
- NAS Part II: Intermediate Certificate in Foreshore and Underwater Archaeology
- NAS Part III: Advanced Certificate in Foreshore and Underwater Archaeology
- NAS Part IV: Diploma in Foreshore and Underwater Archaeology

---

The NAS Training Programme helps people learn more about archaeology. It offers unique opportunities to gain and practise techniques and skills and to take part in archaeological projects, both above and below the water (plate A3.1). The Programme has been running formally since 1986. Though devised in the UK, it is now being used by heritage and archaeological organizations in over 15 countries to raise awareness of threats to archaeology and to provide skills training for participation in projects. NAS courses have taken place in more than 25 countries. The NAS Training Programme is recognized by many diving organizations and heritage organizations around the world as a standard for archaeological skills training for recreational divers and non-divers alike. The NAS Training Programme is therefore an ideal way to gain experience in this fascinating subject under the guidance and assistance of NAS's team of experienced tutors/instructors (plate A3.2).

The Programme is modular, beginning with the introductory course and becoming progressively more advanced through Parts I to IV (outlined below). This book is entirely complementary to the NAS Training Programme and an essential component of its recommended reading list. The latest information about NAS training courses in the UK can be found on the NAS website (www.nauticalarchaeologysociety.org), which also lists contacts for NAS accredited training around the world.

For each course, please consult the NAS website or contact the NAS office for information on the minimum diving qualification required to take part in the underwater exercises. For non-divers, there are equivalent practical exercises on land.

AN INTRODUCTION TO FORESHORE AND UNDERWATER ARCHAEOLOGY

This is the entry-level course to the NAS Training Programme and it is aimed at introducing maritime archaeology to divers and non-divers. It provides a broad-based view of the subject by covering a wide range of topics.

Course objectives ensure that the participant will:

- be introduced to the basic principles and scope of nautical archaeology;
- gain an appreciation of the importance of underwater cultural heritage and the need for it to be recorded, protected and preserved;
- learn how archaeologists date sites; and
- learn how to undertake a basic pre-disturbance site survey.

The typical format involves a 1 day course conducted by approved tutors/instructors. Certain parts of the course can be adapted to suit the special requirements of the individual or groups taking part. The Introduction course is not examined and participants are provided with a certificate of attendance. The Introduction course is a prerequisite for attending the Part I Certificate in Foreshore and Underwater Archaeology course.

NAS PART I: CERTIFICATE IN FORESHORE AND UNDERWATER ARCHAEOLOGY

The NAS Part I builds on the basic knowledge and practical skills covered during the Introduction course. The practical elements of the course are usually held in sheltered open water or on a foreshore site for non-divers. A typical NAS Part I course will comprise the following:

Day 1
- Project case-study (e.g. 'the *Mary Rose*')
- Project logistics and safety
- Introduction to wreck-recording techniques
- Practical session using a wreck-recording form
- De-briefing session.

Day 2
- Introduction to 3-D survey programs
- Survey practical
- Data-processing using the 3-D survey software program
- Introduction to finds-handling and the problems associated with waterlogged material
- Introduction to NAS Projects

NAS PART II: INTERMEDIATE CERTIFICATE IN FORESHORE AND UNDERWATER ARCHAEOLOGY

There are two aspects to the NAS Part II Certificate:

Submission of a short survey-project report: The NAS Part II offers the opportunity to put into practice some of the things learnt on the Introduction and Part I courses by carrying out an independent survey and producing a short survey report. The choice of survey site is entirely up to the individual or group (plate A3.3). The report will be assessed against a number of criteria. 'Guidance Notes' and further help with the project can be obtained from the NAS office or via the NAS website. The report can be submitted by an individual or a group, provided each group member has contributed to the data gathering and report writing (see NAS website for further information).

Attendance at archaeological events: The second aspect of the NAS Part II Certificate promotes a broad approach and a commitment to the subject by encouraging people to attend 2 days (or equivalent) at archaeological conferences and/or events, including museum tours or diver heritage trails. The 2 day requirement can be accumulated by attending individual lectures, short seminars or longer events. The events do not have to be exclusively maritime related but must be archaeologically relevant. Participants are expected to have their attendance confirmed by the event organizer or a NAS representative. A list of events that would be particularly appropriate for this Part II component can be found on the NAS website, if in doubt, please contact the NAS office.

NAS PART III: ADVANCED CERTIFICATE IN FORESHORE AND UNDERWATER ARCHAEOLOGY

The NAS Part III component of the Programme provides the major formal teaching element. Participants will either attend a full-time field school, often based around the work on an actual archaeological site, or attend a series of modules in specific techniques. This can involve lectures, on-site demonstrations and practical exercises in many of the techniques of underwater archaeology (e.g. section drawing, control-point (datum) positioning, recording *in situ*, sampling deposits, excavation strategies and methods). A wide variety of courses enable students to obtain practical experience of, for example, handling archaeological materials, first-aid conservation techniques, finds drawing and photography.

The main objective of the NAS Part III course is to produce a competent archaeological field-worker who has the necessary technical knowledge to be an asset to any project.

The NAS Part III syllabus consists of eight core subject groups:

1. Research and information technology
2. Archaeological science
3. Survey
4. Excavation
5. Recording
6. Conservation
7. Ancient technology
8. Managing archaeology

Participants must accumulate 100 points in six out of the eight core subject groups. Points are awarded on the basis of a full point for each practical contact hour and half a point for a theoretical contact hour (this is to encourage courses with a greater practical content). The following criteria also apply:

- a maximum of 28 points can be obtained in any one core subject;
- a minimum of 8 points must be obtained for each of the six chosen subject groups; and
- experience cannot be considered retrospectively.

NAS PART IV: DIPLOMA IN FORESHORE AND UNDERWATER ARCHAEOLOGY

The final element and the highest grade of the NAS Programme is the Part IV Diploma. The primary objective is to provide a certification level for a field-worker capable of supervising a work programme on an archae-ological site, in conjunction with the archaeological director. This component does not involve formal teaching, although guidance and advice will be provided at all stages by the NAS Training staff. To qualify for the final certificate, the candidate must have fulfilled the following requirements:

- have worked on at least three different archaeological sites for a minimum of 12 full weeks since the completion of his/her NAS Part II certificate; and
- have completed a dissertation or extended portfolio of work on an approved project or topic, including a full report prepared to publication standard.

FURTHER INFORMATION

The Nautical Archaeology Society, Fort Cumberland, Fort Cumberland Road, Portsmouth PO4 9LD, UK. Tel/fax: +44(0)23 9281 8419; www.nauticalarchaeologysociety.org; e-mail: nas@nauticalarchaeologysociety.org.

Glossary

absolute dating the determination of age with reference to a specific timescale (e.g. dendrochronology and radiocarbon dating).

accuracy the precision and reliability of a measurement.

airlift tool used for underwater excavation, driven by compressed air so as to create suction at the head of the tool (*see also* dredge).

anthropology the study of humanity, its physical characteristics and its culture. It can generally be broken down into three sub-disciplines: biological (physical); cultural (social); and archaeological.

archaeological record set of data collected by archaeologists, which will preserve information about sites after they have been destroyed by excavation.

archaeology the study of the human past through its material remains.

archaeomagnetic dating sometimes referred to as paleomagnetic dating, this is based on changes in the earth's magnetic field over time. A record of the remnant magnetism in materials such as fired-clay allows the time when the material was 'last fired' to be determined.

archive the complete collection of all records and finds from a site. Also refers to the location in which these records are kept.

artefact any portable object, used, modified or made by humans.

assemblage a group of objects occurring together at a particular time and place.

association the co-occurrence of an artefact with other archaeological remains.

attribute a basic characteristic of an artefact, such as colour, decoration, form, raw material, style.

AUV autonomous underwater vehicle, a self-contained cableless/tetherless underwater robot (*see also* ROV).

ballistics the study of the functioning of projectiles and firearms.

baseline a line, usually graduated, fixed between two datum points/control points.

bathymetric survey a survey measuring depth, the results showing the topography of the sea-bed.

benthic of or relating to, or happening on, the bottom under a body of water.

bilge the compartment at the bottom of the hull of a vessel, where water collects, and from where it may be pumped out of the vessel.

Bronze Age a period in the Old World when bronze became the primary material for tools and weapons. In Europe this conventionally spans from 2000 to about 700 BC.

calibration the adjustment of data by means of systematic change to exclude extraneous or error-producing information.

carvel-built describes a way of building a vessel in which the hull planks are laid flush, edge-to-edge.

classification the ordering of items into groups or other associations on the basis of shared attributes.

clinker-built describes a way of building a vessel in which each hull plank overlaps and is fastened to the one below.

composite artefacts artefacts that consist of more than one material type.

concretion dense masses of hard material that form from corroding iron. They may be found on the surface of any material, or as a layer stretching across a whole site. They are composed mainly of the oxides of iron (Robinson, 1998).

context an artefact's context usually consists of its matrix (i.e. the materials within its immediate surrounding), its position within the matrix, both vertically and horizontally, and its association with other artefacts.

control points points around a site whose location has been fixed. These can then be used to survey other artefacts/features.

coprolites fossilized faeces, which contain food residues that can be used to reconstruct diet and subsistence activities.

core samples an intact cylinder of soil or sediment collected with a coring device and used to provide an undisturbed sample. Cores are used to evaluate the geological context of archaeological material and its surroundings.

crannog an artificial island in a lake or marsh forming the foundation for a small settlement. They are found in northern Britain and Ireland, dating from the Bronze Age (*c*.2000 BC), and in some instances were in use until the seventeenth century.

cultural heritage the legacy of physical artefacts, structures, and intangible attributes of a society which are inherited from previous generations.

datum offset survey method in which the perpendicular distance (offset) from an artefact to a baseline is recorded, along with the distance of the offset along the baseline. This is a simple and effective way of surveying a site.

datum point a reference point, either permanent or temporary, from which survey measurements are taken (*see also* control point).

dendrochronology dating by the study of tree-ring patterns, which reflect fluctuations in climate and rainfall each year (*see also* absolute dating).

DGPS differential global positioning system. DGPS uses a network of fixed ground-based reference stations to broadcast the difference between positions indicated by the satellite systems and known fixed positions.

Direct Survey Method (DSM) a three-dimensional survey method (DSM) program used extensively in maritime archaeology, also known as Web for Windows (*see also* Site Surveyor/Recorder).

distribution map a diagram showing the locations of particular artefacts around a site.

DPV diver propulsion vehicle – an underwater scooter.

drawing frame *see* planning/drawing frame.

dredge also known as a water dredge or induction dredge, this is a tool used for underwater excavation driven by water pressure so as to create suction at the head of the tool (*see also* airlift).

echo-sounding an acoustic underwater survey technique that can be used for topographic mapping, where a pulse is transmitted and the return signal (or echo) is collected and interpreted.

ecofacts non-humanmade items, such as animal and plant remains, associated with an archaeological site. Such remains can provide clues to the past environment.

EDM electronic distance measurement. EDM units use infrared or laser light to measure distances.

environmental archaeology a field of interdisciplinary research – archaeology and natural science – directed at the reconstruction of human use of plants and animals.

ethnography the study of contemporary cultures through first-hand observation. A sub-discipline of cultural anthropology.

ethnology the derivation of general principles about human society by the study of contemporary cultures. A sub-discipline of cultural anthropology.

excavation the systematic uncovering of archaeological remains by the removal of materials covering and accompanying them – a destructive and irreversible process.

experimental archaeology the study of past behavioural processes through experimentation and reconstruction under carefully controlled conditions.

feature non-portable evidence of human activity found on an archaeological site (e.g. a wall, a ditch, a floor, a hearth).

foreshore *see* inter-tidal zone.

geophysics scientific study of features below ground (or under water) using a range of instruments.

GIS geographic information system. A system for capturing, storing, analysing and managing data and associated attributes that are spatially referenced to the earth.

GPS global positioning system. A satellite-based system for determining three-dimensional positions on land or sea. The level of accuracy is dependant on the system used. Navigational grade *c.*10 m (33 ft); mapping grade *c.*1 m (39 in), Survey grade *c.*0.10 m (4 in).

ground-truthing the process by which information received from remote-sensing equipment is compared with what is there in reality so as to verify the data.

half-life the time in which half the quantity of the radioactive isotope in a sample decays (*see also* radioactive decay).

iconography the study of 'artistic' representations (e.g. paintings, sculptures, etc.).

in situ in place.

inter-tidal zone area of foreshore that is exposed at low tide and submerged at high tide.

isostatic uplift/downshift changes in the level of land caused by the spread or retreat of ice sheets.

lignin a chemical compound that makes up approximately one-quarter to one-third of the dry mass of wood.

magnetometer a geophysical instrument that measures changes in the magnetic field within soil or sediment, which might be caused by subsurface features or human activities such as burning.

material remains the buildings, tools and other artefacts surviving from former societies or individuals.

matrix the physical material within which artefacts are found.

Mesolithic an Old World chronological period beginning around 10,000 years ago, between the Palaeolithic and Neolithic, and associated with the use of small stone tools (microliths).

midden the accumulation of debris and domestic waste resulting from human use: a rubbish dump.

multibeam swath bathymetry a development of the echo-sounder, providing depth measurments in a thin strip below and to the side of the boat on which the equipment is based.

Neolithic an Old World chronological period characterized by the development of agriculture and the establishment of more permanent year-round settlements. In Europe this conventionally spans from *c.*5000 BC to the start of the Bronze Age.

Old World the parts of the world known to Europeans prior to the voyages of Christopher Columbus – Europe, Asia and Africa.

Palaeolithic the period before 10,000 BC, characterized by the earliest known stone-tool manufacture.

photomosaic A picture of a site or feature made by joining several different photographs together.

planning/drawing frame a frame, of a predetermined size, subdivided to produce a grid of equal squares, placed over an area to aid recording. The archaeological materials within each square are drawn to scale.

pre-disturbance survey non-intrusive survey work carried out before any excavation takes place, providing information on the condition of the site at that time.

provenance the source or place of origin of something.

radioactive decay the regular process by which radioactive isotopes break down into their constituent components, with a half-life specific to each isotope.

radiocarbon dating an absolute dating method that measures the decay of the radioactive isotope of carbon in organic material.

relative dating the determination of age without reference to a specific timescale, but instead through relationships with other examples/features (e.g. stratigraphy and typology).

remote sensing the imaging of phenomena from a distance (i.e. satellite imaging, airborne/seaborne surveys).

rescue/salvage archaeology the identification and recording of archaeological sites which are under threat.

research design the systematic planning of an archaeological research programme, which usually includes undertaking site studies and the subsequent publication of results.

resolution denotes the level of detail in an image/film.

rigid inflatable boat (RIB) a composite design incorporating a rigid hull, normally fibreglass or aluminium, with inflatable tubing sides, providing stability and buoyancy.

ROV remotely operated vehicle. These are used extensively for underwater search, surveys and working, especially in deeper waters (*see also* AUV).

salvage the act of recovering vessels or their cargo from loss at sea.

sediment individual grains or aggregate grains of mineral and organic material, which have been or are in the process of being eroded and transported from site of origin and deposited by water, wind, ice or people.

sedimentology the investigation of the structure and texture of sediments.

sidescan sonar a geophysical instrument similar in operation to an echo-sounder but having two fan-shaped beams that are projected either side of the towfish for greater coverage. The return signals are interpreted and shown graphically.

site-formation processes those processes affecting the way in which archaeological materials became buried and any subsequent changes that occur thereafter.

Site Surveyor/Recorder a three-dimensional survey programme used extensively in maritime archaeology (*see also* Direct Survey Method).

spoil soil/sediments that have been removed during excavation.

stratification the laying down of layers (strata) one above the other. A succession of layers should provide a relative chronological sequence.

stratigraphy the study of the formation, composition, sequence and correlation of stratified sediment.

sub-bottom profiler a geophysical instrument that works on the same principle as an echo-sounder but with a lower operating frequency, allowing the penetration of sand and silt to detect buried archaeological material.

theodolite a surveying instrument comprising a focusing telescope (which is able to pivot on both horizontal and vertical planes), scales for measuring vertical and horizontal angles, and some type of levelling device.

total station a land survey instrument that can compute the position of targets relative to itself by measuring the distance, azimuth and elevation to each target.

towfish an instrument (e.g. side-scan sonar) that is towed behind a vessel.

treasure-hunting the search for and excavation of sites with the aim of monetary reward.

treenail cylindrical or chamfered hardwood pegs used to secure the planks of a wooden ship's sides and bottom to her frames (pronounced trunnel or trennel).

tree-ring dating *see* dendrochronology.

triangulation method of measuring three points relative to each other based on angles – creating a triangle between the points.

trilateration method of measuring three points relative to each other based on measured lengths and often used as a simple measuring technique using a baseline. NB, in the US this can be called triangulation. (*See also* datum offsets.)

typology the process of using descriptions to define the characteristics of an object and the subsequent association of similar objects based on these characteristics.

underwater acoustic positioning system used to provide positions under water in the same way that GPS is used on land.

WGS84 World Geodetic System 1984. An internationally agreed global reference framework of coordinates, within which a point can be defined anywhere in the world.

References and Further Reading

Ackroyd, N., and Lorimer, R., 1990, *Global Navigation: A GPS user's guide*. London.

Adams, J., 1985, *Sea Venture*, a second interim report, part 1, *International Journal of Nautical Archaeology* **14**.4, 275–99.

Adams, J., 2002a, Excavation methods underwater, in C. Orser (ed.), *Encyclopaedia of Historical Archaeology*, 192–6. London.

Adams, J., 2002b, Maritime archaeology, in C. Orser (ed.), *Encyclopaedia of Historical Archaeology*, 328–30. London.

Adkins, L. and Adkins, R. A., 1990, *Talking Archaeology: A Handbook for Lecturers and Organizers. Practical Handbooks in Archaeology, No. 9*. Council for British Archaeology, London.

Ahlstrom, C., 2002, *Looking for Leads: Shipwrecks of the Past Revealed by Contemporary Documents and the Archaeological Record*. Finnish Academy of Science and Letters, Helsinki.

Altes, A. K., 1976, Submarine Antiquities: A Legal Labyrinth, *Syracuse Journal of International Law and Commerce*, 77–97.

Anon (ed.), 2007, *Dive Straight In! A Dip-In Resource for Engaging the Public in Marine Issues*. CoastNet, UK.

Archaeological State Museum of Mecklenburg-Vorpommern, 2004, *Management plan of shipwreck site Darss Cog, MOSS Project*.

Atkinson, K., Duncan, A. and Green, J., 1988, The application of a least squares adjustment program to underwater survey, *International Journal of Nautical Archaeology* **17**.2, 119–31.

Aw, M. and Meur, M., 2006 (2nd edn), *An Essential Guide to Digital Underwater Photography*. OceanNEnvironment, Carlingford, NSW, Australia.

Bannister, A., Raymond, S. and Baker, R., 1992 (6th edn), *Surveying*. New Jersey.

Barker, P., 1993 (3rd edn), *Techniques of Archaeological Excavation*. London.

Bass, G. F., 1990, After the diving is over, in T. L. Carrell (ed.), *Underwater Archaeology: Proceedings of the Society for Historical Archaeology Conference*. Tucson, Arizona.

Battarbee, R. W., 1988, The use of diatom analysis in archaeology: a review, *Journal of Archaeological Science* **15**, 621–44.

Bettes, F., 1984, *Surveying for Archaeologists*. Durham.

Bevan, J., 2005, *The Professional Diver's Handbook*. London.

Birch, S. and McElvogue, D. M., 1999, *La Lavia*, *La Juliana* and the *Santa Maria de Vison*: three Spanish Armada transports lost off Streedagh Strand, Co. Sligo: an interim report, *International Journal of Nautical Archaeology* **28**.3, 265–76.

Blackmore, H. L., 1976, *The Armouries of the Tower of London: Vol. 1, The Ordnance*. HMSO, London.

Bolton, P. (ed.), 1991 (3rd rev. edn), *Signposts for Archaeological Publication: A Guide to Good Practice In the Presentation and Printing of Archaeological Periodicals and Monographs*. Council for British Archaeology.

Boyce, J. I., Reinhardt, E. G., Raban, A. and Pozza, M. R., 2004, Marine Magnetic Survey of a Submerged Roman Harbour, Caesarea Maritima, Israel, *International Journal of Nautical Archaeology* **33**, 122–36.

Boyle, J., 2003, *A Step-by-Step Guide to Underwater Video* (www.FourthElement.com).

Bradley, R., 1990, *The Passage of Arms: An Archaeological Analysis of Prehistoric Hoards and Votive Deposits*. Cambridge.

Brice, G. and Reeder, J., 2002 (4th rev. edn), *Brice on Maritime Law of Salvage*. London.

Brown, D. H., 2007, *Archaeological Archives: A Guide to Best Practice in Creation, Compilation, Transfer and Curation*. Archaeological Archives Forum (www.archaeologists.net/modules/icontent/inPages/docs/pubs/Archives_Best_Practice.pdf).

Brown, A. and Perrin, K., 2000, *A Model for the Description of Archaeological Archives*. English Heritage (www.eng-h.gov.uk/archives/archdesc.pdf).

Buglass, J. and Rackham, J., 1991, Environmental sampling on wet sites, in J. M. Coles and D. M. Goodburn (eds), *Wet Site Excavation and Survey. WARP Occasional Paper No. 5*. Exeter.

Camidge, K., 2005, *HMS Colossus: Stabilisation Trial Final Report*. Unpublished report for English Heritage (www.cismas.org.uk/docs/colossus_stab_trial_final.pdf).

Caminos, H., 2001, *Law of the Sea*. Aldershot.

Carpenter, A. C., 1993, *Cannon: The Conservation, Reconstruction and Presentation of Historic Artillery*. Tiverton.

Cederlund, C. O. (ed.), 2004, *MOSS Final Report*, web-published final report (www.nba.fi/internat/MoSS/download/final_report.pdf).

Coles, J. M., 1988, A Wetland Perspective, in B. A. Purdy (ed.) *Wet Site Archaeology*, 1–14. Boca Raton, FL.

Coles, J. M., 1990, *Waterlogged Wood*. English Heritage, London.

Conolly, J. and Lake, M., 2006, *Geographical Information Systems in Archaeology*. Cambridge.

Cooper, M., 1987, *Fundamentals of Survey Measurement and Analysis*. Oxford.

Cross, P. A., 1981, The computation of position at sea, *Hydrographic Journal* **20**, 7.

Dana, P. H., *The Geographer's Craft Project*. Dept of Geography, The University of Colorado at Boulder (www.colorado.edu/geography/gcraft/notes/datum/datum.html, updated 2003).

Dean, M., Ferrari, B., Oxley, I., Redknap, M. and Watson, K. (eds), 1992, *Archaeology Underwater: The NAS Guide to Principles and Practice*. London.

Dean, M., 2006, Echoes of the past: Geophysical surveys in Scottish waters and beyond, in R. E. Jones and L. Sharpe (eds), *Going over Old Ground – Perspectives on archaeological geophysical and geochemical survey in Scotland*, 80–7. BAR British Series 416, Oxford.

Delgado, J. P. (ed.), 2001 (new edn), *Encyclopaedia of Underwater and Maritime Archaeology*. London.

Det Norske Veritas Industry A/S, 2006, *Recommended Practice RP B401: Cathodic Protection Design*. Laksevag, Norway.

Dinacauze, D. F., 2000, *Environmental Archaeology, Principles and Practice*. Cambridge.

Dorrell, P., 1994, *Photography in Archaeology and Conservation*. Cambridge.

Drafahl, J., 2006, *Master Guide for Digital Underwater Photography*. Amherst.

Dromgoole, S., 1996, Military Remains on and around the Coast of the United Kingdom: Statutory Mechanisms of Protection, *International Journal of Marine and Coastal Law* **11**.2, 23–45.

Eckstein, D., 1984, *Dendrochronological Dating, Handbooks for Archaeologists No. 2*. European Science Foundation, Strasbourg.

Edge, M., 2006 (3rd edn), *The Underwater Photographer: Digital and Traditional Techniques*. Oxford.

Eiteljork II, H., Fernie, K., Huggett, J., and Robinson, D., 2003, *CAD: A Guide to Good Practice*. Oxford.

English Heritage, 1991, *Management of Archaeological Projects (MAP2)*. London (www.eng-h.gov.uk/guidance/map2/index.htm).

English Heritage, 1996, *Waterlogged Wood: Guidelines on the Recording, Sampling, Conservation and Curation of Waterlogged Wood*. London.

English Heritage, 2004a, *Geoarchaeology: Using Earth Sciences to Understand the Archaeological Record*. London (www.english-heritage.org.uk/upload/pdf/Geoarchaeology-2007.pdf).

English Heritage, 2004b, *Dendrochronology: Guidelines on Producing and Interpreting Dendrochronological Dates*. London.

English Heritage, 2006a, *Guidelines on the X-radiography of Archaeological Metalwork*. Swindon.

English Heritage, 2006b, *Management of Research Project in the Historic Environment (MoRPHE)*. London (www.english-heritage.org.uk).

English Heritage, 2006c, *MoRPHE Technical Guide: Digital Archiving and Digital Dissemination*. London.

Erwin, D. and Picton, B., 1987, *Guide to Inshore Marine Life*. London.

Ffoulkes, C., 1937, *The Gunfounders of England*. Cambridge.

Fish, J. P. and Carr, H. A., 1990, *Sound Underwater Images: A Guide to the Generation and Interpretation of Side Scan Sonar Data*. Boston, MA.

Fletcher, M. and Locke, G., 2005 (2nd edn), *Digging Numbers: Elementary Statistics for Archaeologists*. Oxford.

Fletcher-Tomenius, P. and Forrest, C., 2000, The Protection of the Underwater Cultural Heritage and the Challenge of UNCLOS, *Art Antiquity and Law* **5**, 125.

Fletcher-Tomenius, P. and Williams, M., 2000, When is a Salvor Not a Salvor? Regulating Recovery of Historic Wreck in UK Waters, *Lloyd's Maritime and Commercial Law Quarterly* **2**, 208–21.

Forrest, C. J., 2000, Salvage Law and the Wreck of the R.M.S. *Titanic*, *Lloyd's Maritime and Commercial Law Quarterly* **1**, 1–12.

Frost, H., 1963, *Under the Mediterranean*. London.

Gamble, C., 2006 (new edn), *Archaeology: The Basics*. Oxford.

Gawronski, J. H. G., 1986, *Amsterdam Project: Annual Report of the VOC-Ship Amsterdam Foundation 1985*. Amsterdam.

Gibbins, D., 1990, Analytical approaches in maritime archaeology: a Mediterranean perspective, *Antiquity* **64**, 376–89.

Green, J., 2004 (2nd edn), *Maritime Archaeology: A Technical Handbook*. London.

Green, J., and Gainsford, M., 2003, Evaluation of underwater surveying techniques, *International Journal of Nautical Archaeology* **32**.2, 252–61.

Greene, K., 2002 (4th rev. edn), *Archaeology: An Introduction*. Oxford (www.staff.ncl.ac.uk/kevin.greene/wintro3/).

Gregory, D. J., 1999, Monitoring the effect of sacrificial anodes on the large iron artefacts on the Duart Point wreck, 1997, *International Journal of Nautical Archaeology* **28**.2, 164–73.

Gregory, D. J., 2000, *In situ* corrosion on the submarine *Resurgam*: A preliminary assessment of her state of preservation, *Conservation and Management of Archaeological Sites*, **4**, 93–100.

Grenier, R., Nutley, D. and Cochran, I. (eds), 2006, *Underwater Cultural Heritage at Risk: Managing Natural and Human Impacts*. ICOMOS.

Griffiths, N., Jenner, A. and Wilson, C., 1990, *Drawing Archaeological Finds: A Handbook*. London.

Hampshire and Wight Trust for Maritime Archaeology, 2007, *Dive into History: Maritime Archaeology Activity Book*. Southampton.

Harland, J., 1984, *Seamanship in the Age of Sail: An Account of the Shiphandling of the Sailing Man-of-War 1600–1860, based on Contemporary Sources*. London.

Harris, E. C., 1989 (2nd edn), *Principles of Archaeological Stratigraphy*. London.

Health and Safety Executive (HSE), 2002, *The Control of Substances Hazardous to Health Regulations* (Statutory Instrument 2002 No. 689). London.

Health and Safety Executive (HSE), 2004, *Guidelines for safe working in estuaries and tidal areas when harvesting produce such as cockles, mussels and shrimps* (www.hse.gov.uk/pubns/estuary.htm).

Henson, D. (ed.), 2000, *Guide to Archaeology in Higher Education*. Council for British Archaeology.

Historic American Buildings Survey/Historic American Engineering Record, 2004 (3rd edn), *Guidelines for Recording Historic Ships*. National Parks Service, Washington.

Hodgson, J. M. (ed.), 1997 (3rd edn), *Soil Survey Field Handbook. Soil Survey Technical Monograph No. 5*. Harpenden, UK.

Hogg, D. F. G., 1970, *Artillery: Its Origin, Heyday and Decline*. London.

Holland, S. E., 2005 and 2006, Following the yellow brick road. *Nautical Archaeology Society Newsletter* **2005**.4 and **2006**.1: 7–9.

Holt, P., 2003, An assessment of quality in underwater archaeological surveys using tape measurements, *International Journal of Nautical Archaeology* **32**.2, 246–51.

Holt, P., 2004, *The application of the Fusion positioning system to marine archaeology* (www.3hconsulting.com/downloads.htm#Papers).

Holt, P., 2007, *Development of an object oriented GIS for maritime archaeology* (www.3hconsulting.com/downloads.htm#Papers).

Howard, P., 2007, *Archaeological Surveying and Mapping: Recording and Depicting the Landscape*. London.

Institute of Field Archaeologists, 2001a (rev. edn), *Standards and Guidance for Archaeological Excavation*. Reading.

Institute of Field Archaeologists, 2001b, *Standards and Guidance for the Collection, Documentation, Conservation and Research of Archaeological Materials*. Reading.

Institute of Field Archaeologists, 2001c (rev. edn), *Standards and Guidance for Archaeological Desk Based Assessment*. Reading.

Institute of Field Archaeologists, 2001d (rev. edn), *Standards and Guidance for Archaeological Field Evaluation*. Reading.

Institute of Field Archaeologists, 2007, *Draft Standard and Guidance for Nautical Archaeological Recording and Reconstruction*. Reading (www.archaeologists.net).

Institute of Field Archaeologists, forthcoming, *Standards and Guidance for the Creation, Compilation, Transfer and Deposition of Archaeological Archives*.

Jameson, J. H. Jr and Scott-Ireton, D. A. (eds), 2007, *Out of the Blue: Public Interpretation of Maritime Cultural Resources*. New York.

Joiner, J. T., 2001 (4th edn), *National Oceanic and Atmospheric Administration Diving Manual: Diving for Science and Technology*. Silver Spring, Maryland.

Judd, P. and Brown, S., 2006, *Getting to Grips with GPS: Mastering the skills of GPS Navigation and Digital Mapping*. Leicester.

Kaestlin, J. P., 1963, *Catalogue of the Museum of Artillery in the Rotunda at Woolwich. Part 1: Ordnance*. London.

Kennard, A. N., 1986, *Gunfounding and Gunfounders*. London.

Kennedy, W. and Rose, R., 2002, *The Law of Salvage*. London.

Kist, J. B., 1991, Integrating Archaeological and Historical Records in Dutch East India Company Research, *International Journal of Nautical Archaeology* **19**.1, 49–51.

Larn, R. and Whistler, R., 1993 (3rd edn), *The Commercial Diving Manual*. Melksham.

Lonsdale, M. V., 2005, *United States Navy Diver*. Flagstaff, Arizona.

McElvogue, D. M., 1999, A breech loading swivel gun from the *Curacao*, 1729, wreck-site, *International Journal of Nautical Archaeology* **28**.3, 289–91.

McGrail, S. (ed.), 1984, *Aspects of Maritime Archaeology and Ethnology*. London.

MacLeod, I. D., 1995, *In situ* corrosion studies on the Duart Point wreck, 1994, *International Journal of Nautical Archaeology* **24**.1, 53–9.

Martin, C. J. M., 1995a, The Cromwellian shipwreck off Duart Point, Mull: an interim report, *International Journal of Nautical Archaeology* **24**.1, 15–32.

Martin, C. J. M., 1995b, Assessment, stabilisation, and management of an environmentally threatened seventeenth-century shipwreck off Duart Point, Mull, in A. Berry and I. Brown (eds.), *Managing Ancient Monuments: An Integrated Approach*, 181–9. Clywd Archaeology Service, Mold.

Martin, C. J. M., 1997, Ships as Integrated Artefacts: the Archaeological Potential, in M. Redknap (ed.), *Artefacts from Wrecks: Dated Assemblages from the Late Middle Ages to the Industrial Revolution*, 1–13. Oxford.

Martin, C. J. M. and Martin, E. A., 2002, An underwater photomosaic technique using Adobe Photoshop, *International Journal of Nautical Archaeology* **31**.1, 137–47.

Milne, G., McKewan, C., and Goodburn, D., 1998, *Nautical Archaeology on the Foreshore: Hulk Recording on the Medway*. RCHM, Swindon.

Mook, W. G. and Waterbolk, H. T., 1985, *Radiocarbon Dating, Handbooks for Archaeologists No. 3*. European Science Foundation, Strasbourg.

Morrison, I., 1985, *Landscape with Lake Dwellings: The Crannogs of Scotland*. Edinburgh.

Muckelroy, K., 1978, *Maritime Archaeology*. Cambridge.

Munday, J., 1987, *Naval Cannon*. Shire Album 186. Princes Risborough, Bucks.

Murphy, L. E., 1990, *8SL17: Natural Site Formation Processes of a Multiple-Component Underwater Site in Florida*, Submerged Resources Center Professional Report No. 12, National Park Service, Santa Fe, New Mexico.

Naish, J., 1985, *Seamarks: Their History and Development*. London.

Nayling, N., 1991, Tree-ring dating: sampling, analysis and results, in J. M. Coles and D. M. Goodburn (eds), *Wet Site Excavation and Survey*. WARP Occasional Paper No. 5. Exeter.

Neal, V. and Watkinson, D., 1998 (3rd edn), *First Aid for Finds*. London.

Nelson Curryer, B., 1999, *Anchors: An Illustrated History*. London.

O'Keefe, P., 2002, *Shipwrecked Heritage: A Commentary on the UNESCO Convention on Cultural Heritage*. Leicester.

Oldfield, R. (illustrator), 1993, in BSAC *Seamanship for Divers*, London.

Oxley, I., 1991, Environmental sampling underwater, in J. M. Coles and D. M. Goodburn (eds), *Wet Site Excavation and Survey*. WARP Occasional Paper No. 5. Exeter.

Palma, P., 2005, Monitoring of Shipwreck Sites, *International Journal of Nautical Archaeology* **34**.2, 323–31.

Papatheodorou, G., Geraga, M., and Ferentinos, G., 2005, The Navarino Naval Battle Site, Greece: an integrated remote-sensing survey and a rational management approach, *International Journal of Nautical Archaeology* **34**, 95–109.

Parker, A. J., 1981, Stratification and contamination in ancient Mediterranean shipwrecks, *International Journal of Nautical Archaeology* **10**.4, 309–35.

Pearson, V. (ed.), 2001, *Teaching the Past: A Practical Guide for Archaeologists*. Council for British Archaeology, York.

Perrin, K., 2002, *Archaeological Archives: Documentation, Access and Deposition: A Way Forward*. English Heritage, London (www.english-heritage.org.uk/upload/pdf/archives.PDF).

Peterson, H. L., 1969, *Roundshot and Rammers: An Introduction to Muzzle-Loading Land Artillery in the United States*. New York.

Petrie, W. M. F., 1904, *Methods and Aims in Archaeology*, London.

Quinn, R., Breen, C., Forsythe, W., Barton, K., Rooney, S. and O'Hara, D., 2002a, Integrated Geophysical Surveys of the French Frigate *La Surveillante* (1797), Bantry Bay, County Cork, Ireland, *Journal of Archaeological Science* **29**, 413–22.

Quinn, R., Forsythe, W., Breen, C., Dean, M., Lawrence, M. and Liscoe, S., 2002b, Comparison of the Maritime Sites and Monuments Record with side-scan sonar and diver surveys: A case study from Rathlin Island, Ireland, *Geoarchaeology* **17**.5, 441–51.

Quinn, R., Dean, M., Lawrence, M., Liscoe, S. and Boland, D., 2005, Backscatter responses and resolution considerations in archaeological side-scan sonar surveys: a control experiment, *Journal of Archaeological Science* **32**, 1252–64.

Renfrew, C. and Bahn, P., 2004 (4th edn), *Archaeology: Theories, Methods and Practice*. London.

Richards, J. and Robinson, D. (eds), 2000 (2nd edn), *Digital Archives from Excavation and Fieldwork: A Guide to Good Practice*. Archaeological Data Service, Oxford (http://ads.ahds.ac.uk/project/goodguides/excavation/).

Robinson, W. S., 1998 (3rd edn), *First Aid for Underwater Finds*. London and Portsmouth.

Roskins, S., 2001, *Excavation*. Cambridge.

Rule, N., 1989, The direct survey method (DSM) of underwater survey, and its application underwater, *International Journal of Nautical Archaeology* **18**.2, 157–62.

Schofield, W., 2001 (5th edn), *Engineering Surveying*. Oxford.

Scientific Diving Supervisory Committee (SDSC), 1997, Advice notes for the Approved Code of Practice (www.uk-sdsc.com).

Smith, R. and Brown, R. R. (eds), 1988, Guns from the sea. *International Journal of Nautical Archaeology* **17**.1, special issue.

Smith, S. O., 2006, *The Low-Tech Archaeological Survey Manual*. PAST Foundation, Ohio.

Spence, C. (ed.), 1994 (3rd edn), *Archaeological Site Manual*. London.

Steiner, M., 2005, *Approaches to Archaeological Illustration: A Handbook*. CBA Practical Handbook 18. Council for British Archaeology, York.

Sutherland, A., 2002, Perceptions of marine artefact conservation and their relationship to destruction and theft, in N. Brodie and K. Walker Tubb (eds), *Illicit Antiquities: The Theft of Culture and Extinction of Archaeology*. London.

Throckmorton, P., 1990, The world's worst investment: the economics of treasure hunting with real life comparisons, in T. L. Carrell, (ed.), *Underwater Archaeology: Proceedings of the Society for Historical Archaeology Conference 1990*. Tucson, Arizona.

Tomalin, D. J., Simpson, P. and Bingeman, J. M., 2000, Excavation versus sustainability *in situ*: a conclusion on 25 years of archaeological investigations at Goose Rock, a designated historic wreck-site at the Needles, Isle of Wight, England, *International Journal of Nautical Archaeology* **29**.1, 3–42.

Tyers, I., 1989, Dating by tree-ring analysis, in P. Marsden (ed.), A late Saxon logboat from Clapton, London Borough of Hackney, *International Journal of Nautical Archaeology* **18**.2, 89–111.

Upham, N. E., 2001 (2nd rev. edn), *Anchors*. Princes Risborough.

Uren, J. and Price, W., 2005 (4th rev. edn), *Surveying for Engineers*. London.

Watts, G. P. Jr, 1976, Hydraulic Probing: One solution to overburden and environment, *International Journal of Nautical Archaeology* **5**.4, 76–81.

Wentworth, C. K., 1922, A scale of grade and class terms for clastic sediments, *Journal of Geology* **30**, 377–92.

Wheatley, D. and Gillings, M., 2002, *Spatial Technology and Archaeology: the Archaeological Application of GIS*. London.

Williams, M., 2001, Protecting Maritime Military Remains: A New Regime for the United Kingdom, *International Maritime Law* **8**.9, 288–98.

Zhao, H., 1992, Recent Developments in the Legal Protection of Historic Shipwrecks in China, *Ocean Development and International Law* **23**, 305–33.

Index